CRYOBIOLOGY

CRYOBIOLOGY

John Roggema
Ralph Barristow

Kruger Brentt
Publishers
2023

Kruger Brentt Publishers UK. LTD.
Company Number 9728962

Regd. Office: 68 St Margarets Road, Edgware, Middlesex HA8 9UU

ISBN: 9781787150423

For information on all our publications visit our website at http://krugerbrentt.com/

PREFACE

Cryobiology is a multidisciplinary science, studying the physical and biological behaviors of living materials at low temperatures. Cryobiology contains many disciplines such as, cellular biology, theriogenology and molecular biology, engineering and mathematics, veterinary and human medicine, intensive and extensive farming on land and in watery environments. Optimization of the cryopreservation procedure of spermatozoa needs all the above-mentioned disciplines because of the complex cellular structure, activation and capacitation mechanisms of spermatozoon. Cryopreservation is a technique of preserving viable and fertile gamete for a long period under freezing temperature. This gamete can be used again for producing off springs and different genetic strains, to create seed bank etc. Since most identified metabolic processes stop at temperatures below the glass transition phase (the transformation from liquid to a glassy state), cryopreservation reduces the risk of microbial contamination or cross contamination with other tissue or cell samples.

The pioneering studies in cryobiological science starting from the middle of the last century is based on the idea that whether life span and longevity could be extended by storing various frozen tissues and organs or even human body for many years. Possible cellular and tissue destruction during the freeze-thaw processes became the subject of curiosity. Cryoinjury is the most important area of research for checking the response of cell changes according to inner and outer environment (Mazur, 1984). It also considered the properties of freezing and defrosting. Important parameters which involve in these research areas are diffusion, osmosis, Cryoprotectants, cooling and thawing process. Cryopreservation method comprises conversion of cell maintenance media to culture media which have cryopreservation agent, like dimethyl sulfoxide (DMSO). Then Cells are cooled at temperature of -80°C in specific cooling container. After cooling cells are transferred to very low temperature storage of below -135°C. Liquid nitrogen is commonly used for this extreme low temperature.

The present book contains nineteen chapters covering all related disciplines. These chapters include Cryobiolgy: An Introductio, Cryopreservation: Methods, Equipment and Concerns, Cryopreservation and its Applications, Basics of Cellular Cryopreservation: Research & Clinical Use, Cryobiology and Medical Sciences, Cryogenic Preservation Technology, Future of Cryopreservation, Mammalian Oocyte Cryopreservation, Cryobiology and Cryopreservation of Sperm, Immunocytochemical Staining of the Metaphase Ii Spindle in Mammalian Oocytes, Fundamental of Gamete Cryobiology, Cryopreservation An Emerging Paradigm Change, Dolly and Polly, Plant Cryopreservation: Current Issues, Impact of Cryopreservation Associated Damage on Different Parameters of Human Spermatozoa, Cryopreservation and in Vitro Storage, Cryopreservation for Preservation of Potato Genetic Resources, Oocyte Cryopreservation and Potato Shoot Tip Cryopreservation. This book elucidates the concepts and innovative models around prospective developments with respect to cryobiology and cryopreservation. Related terminology is given at the end for ready reference.

We are grateful to all those persons as well as various books, manuals, periodicals, magazines, journals etc. that helped in the preparation of this book. In spite of the best efforts, it is possible that some errors may have occurred into the compilation and editing of the book. Further queries, constructive suggestions and criticisms for the improvement of the book are always welcome and shall be thankfully acknowledged.

John Roggema

Ralph Barristow

CONTENTS

1

CRYOBIOLGY: AN INTRODUCTION

1.1 CRYOBIOLOGY: AN OVERVIEW

Cryobiology is the scientific study of the effects of freezing and sub-freezing temperatures on biological fluids, cells, and tissues. It is an extension of cryogenics, which is the study of the properties of matter at very low temperatures. Cryobiological techniques have application in genetic research, livestock breeding, infertility treatment, and organ transplantation.

Cryopreservation is a technology for preservation of biological materials at extreme low temperatures, sometimes in solid carbon dioxide (dry ice) at -80°C (-112°F) or typically in liquid nitrogen at -196°C (-321°F). Theoretically, at cryogenic temperatures (lower than -100°C) biological materials cease their activities and metabolism (Pictet 1893, cited from Davenport 1897) and are capable of being revived with minimal changes after thawing to normal physiological temperatures. In fact, the reduction of body temperatures below 0°C ("cryo-hibernation") occurs in some amphibian species in nature. These species have developed protective mechanisms to survive extreme winters.

- A related field, cryogenics, is devoted to the study of the effects of low temperature. Cryogenic preservation of humans has been proposed; it is envisioned that maintaining someone in what could be considered a state of suspended animation could allow those with a currently non-curable disease to be revived at some point in the future when a cure has been found. Currently, however, despite claims to the contrary, this process is impossible because of the irreparable cell damage that occurs during the freezing process.
- The terms cryobiology and cryogenics are derived from the Greek kryos, meaning icy cold. Temperatures used in cryogenics range from–148°F (–100°C) to near absolute zero–459.67°F (–273.15°C). Such ultra-low temperatures can be achieved

by the use of supercooled gases. The study of these gases dates back to 1877, when Swiss physicist Raoul Pictet (1846–1929) and French physicist/chemist Louis Cailletet (1832– 1913) first learned how to liquefy oxygen. Although they worked independently and used different methods, both men discovered that oxygen could be liquefied at 297.67°F (–183.15°C). Soon after, other researchers liquefied nitrogen at–321.07°F(–196.15°C).

- Other breakthroughs in cryogenics included James Dewar's invention of the vacuum flask in 1898. Dewar's double-walled vacuum storage vessel allowed liquefied gases to be more readily studied. In the last 100 years, a variety of other methods for insulating super-cooled fluids have been developed.
- In the twentieth century, scientists began applying cryogenic techniques to biological systems. They explored methods for treating blood, semen, tissue, and organs with ultra-low temperatures. This research has resulted in advances in genetic research, livestock breeding, infertility treatment, and organ transplantation.
- Cryobiology has provided an inexpensive way to freeze and store the embryos of different strains of research laboratory animals, such as mice. Maintaining a breeding colony of research animals can be expensive, and cryogenic storage of embryos can reduce cost considerably. When the animals are needed, they can be thawed and implanted. It is this ability that has inspired some to consider cryobiology as a means of preserving humans.
- In agriculture, cryopreservation allows livestock breeders to mass produce the embryos of genetically desirable cattle. For example, hundreds of eggs can be harvested from a single prize dairy cow and frozen for later implantation in other mothers. Using similar techniques, pigs that are too fat to reproduce on their own can be artificially implanted with embryos developed from frozen eggs and sperm. In addition, cryobiologists are examining the possibility of increasing buffalo herds by freezing bison embryos and later implanting them into cows to give birth.
- Cryobiology has met with great success in the treatment of human infertility. The use of frozen sperm, eggs, and embryos increases the success rate of fertility treatments because it allows doctors to obtain a large number of samples that can be stored for future fertilization. Techniques for freezing sperm were relatively easy to develop, and in 1953, the first baby fertilized with previously frozen sperm was born. The process for freezing embryos is much more complicated, however. It involves removing water from the cells and replacing it with an organic antifreeze that prevents the formation of ice crystals that can cause cells to burst. Advances over the last few decades have made this technique highly successful, and in 1984, the first baby was born from a previously frozen embryo. Freezing

eggs is an even more difficult challenge because the fragile membrane structure that surrounds the eggs makes them difficult to freeze without causing severe damage. However, eggs have been successfully frozen using a chemical solution similar to the ovaries' natural fluids. The technique has been used to freeze eggs, which were later thawed and used to impregnate a woman. In 1997, the first birth resulting from frozen eggs was recorded.

- Organ storage is another important area of cryo-biological research. Using conventional methods, organs can only be stored for very short periods of time. For example, a kidney can be kept for only three days, a liver for no more than 36 hours, and hearts and lungs for no more than six hours. If these organs could be frozen without subcellular damage, storage times would be lengthened almost indefinitely. Although researchers have made great advances, they have not yet perfected the process of freezing and reviving organs. The problem they face is that the formation of ice crystals can damage fragile tissue. However, researchers have devised a cryopreservant liquid that protects the organs during the freezing process. In another development, a so-called antifreeze protein has been successfully used to freeze and revive rat livers. The protein is derived from fish living in the Arctic that has evolved to survive in very cold water. An antifreeze protein alters the structure of ice crystal in order to be less damaging to cells. A technique called vitrification cools organs so quickly that their molecules do not have time to form damaging ice crystals. Continued success in these areas may one day lead to reliable methods of freezing and storing organs for future transplant.
- **Cryobiology**: Study of the effects of low temperatures on organisms (often for the purpose of achieving cryopreservation).
- **Cryogenics:** A discipline within physics. The study of the production and behavior of materials at cryogenic temperatures.
- **Cryonics:** This addresses the frozen storage of human corpses or heads with the speculative intention for future reanimation. This is usually considered to be outside of the realm of serious scientific research.
- **Cryosurgery:** A form of surgical technique applying cryogenic temperatures to destroy malignant tissues or cells.
- **Cryotherapy:** The local or general use of low temperatures in medical therapy. The scientific basis concerning efficacy of these approaches, especially whole-body cryotherapy in cold chambers, has not received long-term or significant scrutiny.
- **Cryoelectronics:** The study of electronic phenomena at cryogenic temperatures, such as superconductivity (useful for aerospace and telecommunication technologies). The practical application of superconductivity is called Cryotronics.

1.2 PRINCIPLE OF CRYOPRESERVATION

Cryopreservation is *the process of preservation or conservation of biological materials at cryogenic temperatures (usually lower than –180°C). Cryopreservation of genetic materials of germ cells, such as seed, sperm, or eggs, is called* **Germplasm Cryopreservation.** *The cryopreservation of biological materials at an extremely high cooling speed without ice crystal formation is also called* **vitrification.**

Cryopreservation is a process of preserving or storing cells, tissues, organs or any other biological materials from any potential damage by maintaining the materials at very low temperature (typically -80 °C using solid CO2 or –196 °C using liquid Nitrogen. In cryopreservation, a very low temperature is used to preserve living cells and tissues and maintain their viability. Unprotected freezing is normally lethal.Cryopreservation is based on the conversion of water present in the cells from a liquid to a solid state. When cooling below 0°C, the biological effects are dominated by the freezing of water, which typically constitutes at least 80% of the tissue mass. The cell water requires much lower temperature to freeze (even up to -68°C) due to the presence of salts and organic molecules in the cells, in comparison to the freezing point of pure water (around 0°C). The metabolic processes and biological divisions in the cells/tissues are almost stopped when stored at low temperature.

1.3 PROCESS OF CRYOPRESERVATION

The cryopreservation of plant cell culture followed by the regeneration of plants involves the following steps:

- Development of sterile tissue cultures
- Addition of cryoprotectants and pre-treatment
- Freezing
- Storage
- Thawing
- Re-culture
- Measurement of viability
- Plant regeneration

The features of the above steps are described as follows:

- **Development of sterile tissue culture:** One of the important steps is the selection of plant species with reference to morphological and physiological characters .It directly influences the ability of explant to survive cryopreservation. Any tissue from a plant can be employed for cryopreservation e.g. meristems, endosperms, embryos, ovules, seeds, cultured plant cells, calluses, protoplasts. Out of these, meristematic cells and suspension cell cultures which are in the late lag phase or log phase are most appropriate.

- **Addition of cryoprotectants and pre-treatment:** The compounds that can prevent the damage caused to cells by freezing or thawing are called as cryoprotectants. Cryoprotectants reduce the freezing point and super-cooling point of water. As a result, the ice crystal formation is delayed during the process of cryopreservation. Cryoprotectants used are dimethyl sulfoxide (DMSO), glycerol, ethylene, propylene, sucrose, mannose, glucose, proline and acetamide. Among them, DMSO, sucrose and glycerol are most commonly used. Generally, a mixture of cryoprotectants instead of a single one is preferred for more effective cryopreservation without damage to cells/tissues.
- **Freezing:** The sensitivity of the cells to low temperature is variable and largely relies on the plant species.

  The different types of freezing methods used are as follows:

  i. **Slow-freezing method:** The tissue or the essential plant material is allowed to slowly freeze at a slow cooling rates of 0.5-5°C/min from 0°C to -100°C. Then it is transferred to liquid nitrogen. Slow-freezing method facilitates the flow of water from the cells to the outside. This avoids intracellular freezing and promotes extracellular ice formation. Because of this, the plant cells are partially dehydrated and can survive better. The slow-freezing technique is successfully employed for the cryopreservation of suspension cultures.
  ii. **Rapid freezing method:** This process is quite simple. In this technique, the vial containing plant material is plunged into liquid nitrogen. During rapid freezing, reduction in temperature from -300° to -1000°C/min occurs. The freezing process occurs so quickly that small ice crystals are formed within the cells. In addition to it, the growth of intracellular ice crystals is also minimum. Rapid freezing technique is applied for the cryopreservation of shoot tips and somatic embryos.
  iii. **Stepwise freezing method:** This technique is a combination of slow and rapid freezing procedures having the advantages of both, and occurs in a stepwise manner. Firstly, the plant material is cooled to an intermediate temperature. Then it is kept there for about 30 minutes. Finally, it is rapidly cooled by plunging it into liquid nitrogen. Stepwise freezing method has been successfully applied for cryopreservation of suspension cultures, shoot apices and buds.
  iv. **Dry freezing method:** It has been reported that the non-germinated dry seeds can survive freezing at very low temperature in comparison to water-imbibing seeds which are sensitive to cryogenic injuries. In a similar way, dehydrated cells are observed to have a better survival rate after cryopreservation.
- Storage: The frozen cultures should be maintained at the specific temperature. Generally, the frozen cells/tissues are maintained at temperatures in the range of -70 to -196°C for storage. Although, with temperatures above -130°C, ice crystal growth may take place inside the cells which decreases viability of cells. The ideal storage is done in liquid N2 refrigerator at 150°C in the vapour phase, or

at -196°C in the liquid phase. The final aim of storage is to halt all the cellular metabolic activities and preserve their viability. The temperature at -196°C in liquid nitrogen is regarded as ideal for long term storage. A regular and constant supply of liquid nitrogen to the liquid nitrogen refrigerator is necessary. It is essential to check the viability of the germplasm time and again in some samples. Proper documentation of the germplasm storage should be done.

- **Thawing:** Thawing is usually performed by plunging the frozen samples in ampoules into a warm water (temperature 37-45°C) bath with robust swirling. By this process, rapid thawing (at the rate of 500- 750°C min^{-1}) takes place, and this preserves the cells from the damaging effects from ice crystal formation. As soon as the thawing occurs (ice completely melts), the ampoules are transferred to a water bath at temperature 20-25°C at the same instant. The cells get damaged if left in warm (37-45°C) water bath for long time. For the cryopreserved material (cells/tissues) where the water content has been decreased to an optimal level before freezing, the process of thawing becomes less vital.
- **Re-culture:**To remove cryoprotectants, the thawed germplasm is washed various times.Following standard procedures, this material is then re-cultured in a fresh medium.In some cases, the direct culture of the thawed material is preferred without washing. It is so because certain vital substances, released from the cells during freezing, are assumed to enhance in vitro cultures.
- **Measurement of viability:** The measurement of survival or viability of the frozen materials can be performed at any stage of cryopreservation or after thawing or re-culture.
- The techniques used to determine viability of cryopreserved cells are the same as applied for cell cultures. The commonly used techniques are staining techniques using triphenyl tetrazolium chloride (TTC), Evan's blue and fluorescein diacetate (FDA).The entry of cryopreserved cells into cell division and regrowth in culture is the best indicator to measure the viability of them. This can be evaluated by the using following expression.
- **Plant regeneration:** The regeneration of the desired plant is the ultimate purpose of cryopreservation of germplasm.The cryopreserved cells/tissues have to be carefully nursed, and grown for appropriate plant growth and regeneration .Along with maintenance of proper environmental conditions, addition of certain growth promoting substances is often essential for successful plant regeneration.

1.4 ADVANTAGES OF CRYPRESERVATION

- Helps in the preservation of biological materials.
- By this the biological materials can be preserved for long time.
- Sperm, gametes, embryos, tissues, bone marrow, organ can be preserved.

- Helps to study the adapting nature of plants and animals under the low temperature.

1.5 LIMITATIONS FOR CRYOPRESERVATION

- An individual with good technical and theoretical knowledge of living plant cells as well as cryopreservation method is required.
- Migration of water, causing extracellular ice formation, can also cause cellular dehydration. The associated stresses on the cell can cause damage directly.
- The physiological status of the plant material is also important. Cryopreservation protocols for many plant species are not available.

1.6 PRECAUTIONS FOR CRYOPRESERVATION

- The formation of ice crystals inside the cells should be prevented as they are responsible for causing injury to the organelles and the cell.
- Cells might be damaged if the intracellular concentration of solutes is high.
- Leakage of certain solutes from the cell during freezing should be checked.
- The physiological status of the plant material is also essential.

2

CRYOPRESERVATION: METHODS, EQUIPMENT AND CONCERNS

2.1 INTRODUCTION

This chapter reviewed the cryobiology principles, the cryoinjury mechanisms, the role of cryoprotective agents, the state of the art cryopreservation methods and equipment, and concerns regarding the use of LN_2 as long- term storage systems. Cryopreservation of cells and tissues is a slow developing field, due the complexity of biological samples. The development of more efficient freezing protocols and cryogenic storage systems, which will ensure the proper preservation of biological materials, is crucial. The goal of this review paper is to provide researchers with a better understanding of the physical events occurring during cryopreservation, in respect with the performance of the cryogenic storage equipment.

The technique of cryopreservation aims to maintain the biological samples in a metabolic suspension state, by which all chemical, physical and biological changes are diminished. The biology principles behind cryopreservation are widely investigated in the literature. However, little scientific interest has been focused to the current equipment used as cryogenic storages and how they affect the viability and functionality of the cells during their preservation. This review paper attempts to bridge the gap between the biology and engineering aspects of cryopreservation. The basic principles of cryobiology, the cryoinjury mechanisms, the role of cryoprotective agents, the state of the art cryopreservation methods and equipment, and concerns regarding the use of liquid nitrogen as cryogenic refrigerant are covered. Further to our knowledge there is no complete review in the literature describing the connection between the current cryogenic equipment and the physical events taking place during cryopreservation. This chapter aims to provide researchers with the key knowledge of cryopreservation, not only from a biological point of view, but from a technical aspect, too, in order to contribute to the development of the cryobiology field.

There has been a steady and continuous growth of the cryobiology field. Scientists try to gain better understanding of the effects of the effects of low temperatures and of freezing of living cells. Significant development of robust protocols and devices aim to overcome the freezing injury, in order to achieve viable storage conditions and to ensure the maximum rate of cells' post-recovery. Long term preservation of cells for scientific, industrial and medical applications is achieved when cells are stored at cryogenic temperatures.At this temperature range all biological, physical and chemical processes of any aqueous system (e.g. cells) are diminished, as there is insufficient thermal energy for chemical reactions and water does not exist in liquid phase below -130°C. Consequently, several medical and biological applications have benefited from storing biological materials at cryogenic temperatures. Red blood cells and platelets can be transfused after years of storage, artificial insemination, banking of cells and tissues, bone marrow transplantation, in-vitro fertilization, facilitated transport of cells and tissues, storage of seeds for reproduction purposes and many more applications are nowadays possible with cryopreservation. However, one of the challenges in the field of cryopreservation is the development of effective cryogenic storage systems, which guarantee the cells' viability and functionality after their recovery.

2.1 PRINCIPLES OF CRYOBIOLOGY

There are four major steps in cryopreservation and the recovery process of cells. First, cryoprotective agents (CPAs) are added to the cells, in order to protect them from the freezing process. Then, the cells are cooled,in a controlled manner, until they reach a temperature below -130°C, at which all the metabolic processes of the cell stop. The recovery of the cells follows the reverse procedure; firstly the cells are thawed following a defrost protocol and then the CPAs are removed. The viability of the cells can be impacted by each of these steps. Many experimental findings indicate that cryopreserved living cells have been damaged or even destroyed by the freezing process; a type of damage called cryoinjury. Although, the cryoinjury of the cells is not caused by the cell's inability to endure low-cryogenic storage temperatures, cells are damaged due to the effects of an intermediate zone of temperatures in the range of -15 to -60°C, which they must traverse twice; once during cooling and once during thawing. The cause of damage in these temperatures is the phase change of water in the extracellular and intracellular solution. Water is the main substance of all living cells. When all water is converted into ice, the cellular metabolism stops. Thus, it is important for the refrigeration system to provide slow cooling, in order to freeze the external of the cell without developing intracellular ice. As ice forms at the extracellular environment of the cell, an osmotic imbalance occurs between the intracellular and extracellular mediums, which cause the discard of water out of the cell. As a result, an increase of solute concentrations outside and inside the cell occurs, which often means the death of the cell. On the other hand, when the cell maintains too much water on its intracellular medium, ice crystals begin to form, which will recrystallize during thawing and damage the cell.

2.2 BASIC MECHANISMS OF CRYOINJURY

As the cell is cooled down to approximately -5°C, both its intracellular and extracellular mediums remain unfrozen and supercooled. Supercooling is a phase during which the temperature of the sample is below its freezing point, but its cellular infrastructure does not form ice crystals. Between -5°C and approximately - 10°C, the chemical potential of the supercooled water inside the cytoplasm is increasing and causes the waterto flow out the cell. With further reduction of the temperature, ice begins to form in the extracellular solution, but the intracellular solution remains supercooled and unfrozen, as the plasma membrane restrains the ice crystals growth inside the cytoplasm. The physical events that occur below -10°C, depend on the cooling rate to which the cell is being exposed. These mechanisms have been extensively described by Peter Mazur in his 'two-factor hypothesis'. Mazur's theory suggests that if an optimal cooling rate is not applied, the cells are damaged due to the effects of water's transport rates.

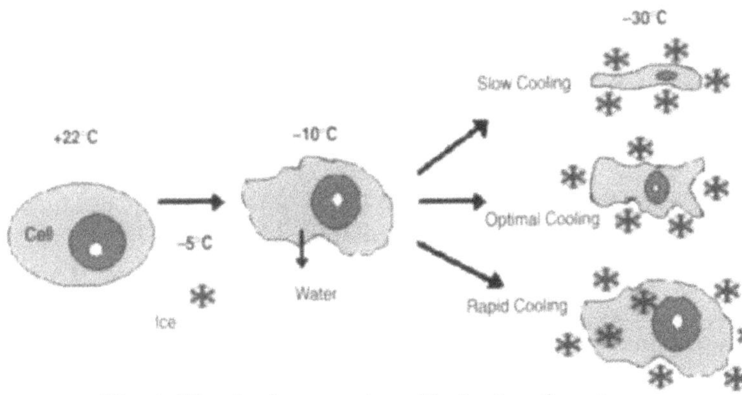

Fig. 1: Physical events in cells during freezing

In the case of rapid cooling, the water inside the cell does not have enough time to be transferred in the extracellular medium, as the chemical potential of the extracellular water reduces faster than the rate by which the intracellular water can flow out of the cell. As a result, the cytoplasm becomes increasingly supercooled and the only way to reach an osmotic equilibrium with the extracellular solution is to freeze internally. This mechanism of the cell is known as *Intracellular Ice Formation* (IIF) and has lethal effects on cell viability .

Figure 2 shows photomicrographs of IIF in mouse oocytes cooled at 100°C/min. On the left of the figure 2 (A) the intact oocytes prior freezing are shown. A well-defined boundary between the extracellular and intracellular mediums is clearly shown. The middle photo (B) shows the frozen oocytes, as ice has propagate from the external solution to the intracellular solution; the blackening of the oocytes indicates the IIF and the darken background is due to the ice of the CPA solution. Finally, the right photo (C) shows the recovery of oocytes, after they have been thawed. It is apparent that the cell membrane which separated the extracellular and intracellular medium does no longer exist, indicating the damage of cell's morphology.

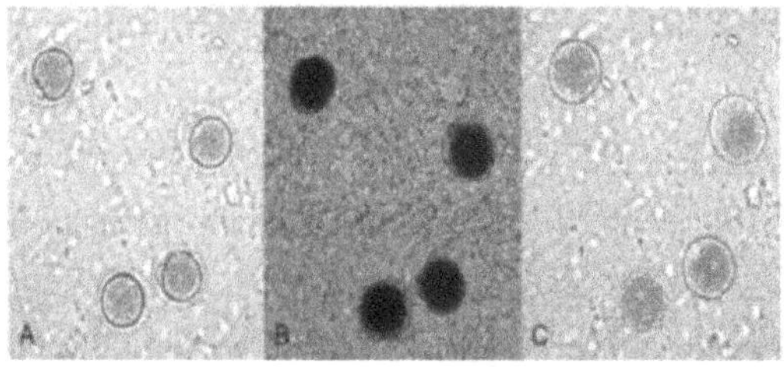

Fig. 2: Rapid cooling injury of mouse oocytes

In the case of slow cooling, the rate of water transport from the cytoplasm towards the extracellular medium is fast enough to eliminate excessive levels of supercooling, to dehydrate the cell and to eventually preclude IIF. However, several other events take place, causing the damage of the cell, which are known as *solution effects*. This general damage classification includes extreme dehydration of lipid vesicles and cell's membrane, high solute concentrations mainly electrolytes, which cause the denaturation of lipoproteins of cell's membrane, mechanical damage due to formation of 'cell channels' between adjacent ice crystals and finally, severe volume shrinkage, which applies physical compression forces in the cell's membrane and causes loss of cell's membrane area. Mazur and Cole studied the survival rate of red blood cells,after they have been slowly frozen, and they found that the cause of damage changes according to the unfrozen fraction of water in the cell; when the unfrozen water fraction of the cell is greater than 20% then, cell damage is caused by osmotic dehydration and the cell's survival depends on the electrolyte concentration. On the other hand, if the unfrozen water fraction of the cell is between 5 to 15% then, cell damage is caused by mechanical effects of cell-to-cell interaction, as they aggregate into small channels as they are excluded from the ice . Figure 3 shows photomicrographs of frog red blood cells, for different value of unfrozen fraction of water, during slow freezing. One can notice that cells are confined to channels of unfrozen solution between the formed ice crystals and, as the unfrozen fraction of water reduces, the channels' diameter and cells' volume decrease, as well .

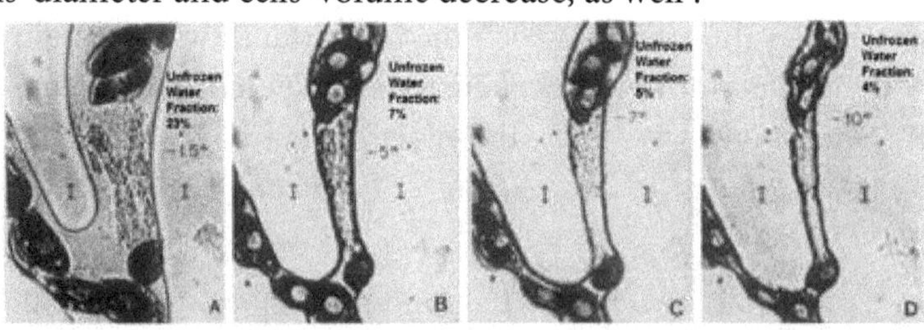

Fig. 3: Slow cooling injury – Cells' aggregation into channels

However, if the cooling rate complies with the optimal cooling rate of the cell, they can be recovered withoutshowing any significant changes to their morphology, function and viability. At the intermediate zone of temperatures (-15 to -60°C) and while ice starts to form in the extracellular medium and the intracellular medium remains unfrozen and supercooled, an increase of the osmolality of the extracellular medium occurs,as very little solute can be concentrated into the ice [3]. As long as the intracellular and extracellular mediums maintain their osmotic equilibrium as the temperature continues to drop, more ice is progressively formed and the solution concentrations of both mediums continue to increase, until they reach either a eutectic temperature, in which the remaining solution solidifies or a glass transition temperature, in which the remaining solution vitrifies. At this point, the cells can be stored at cryogenic temperatures for long-term periods, without being cryoinjured.

The 'two-factor hypothesis' and the cell's survival rate can be graphically presented as a function of the cooling rate. Figure 4 illustrates the trend of cells' survival according to optimal, rapid or slow cooling rate. The survival curve follows a characteristic inverted U-shape, which its peak corresponds to the optimal cooling rate. This ideal cooling rate allows the optimal survival of the cell, as it is slow enough that precludesIIF and rapid enough to protect the cell form the solution effects.

The optimal cooling rate differs for each cell type, as it depends on the water permeability of cell's membrane, which has unique value for every cell type. The following graph shows the relationship between percentage of survival and cooling rate for three types of cells; mouse bone marrow stem cells, hamster oocytes and human red blood cells. All of the curves have an inverted U-shape, only the values of theoptimal cooling rate changes, as it can be seen in Figure 5.

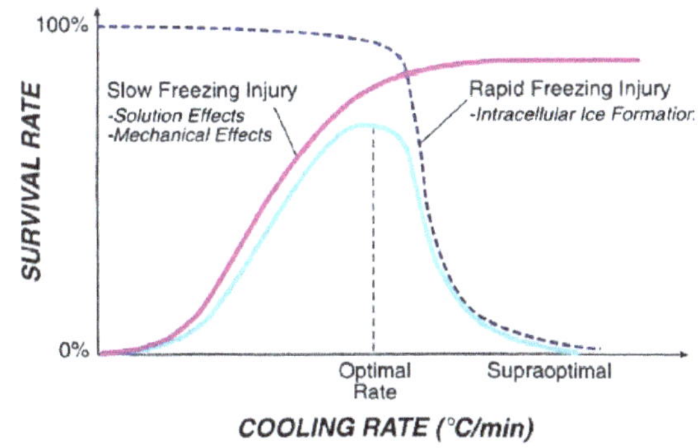

Fig. 4: Relationship between the cooling and survival rates ofhypothetical cells

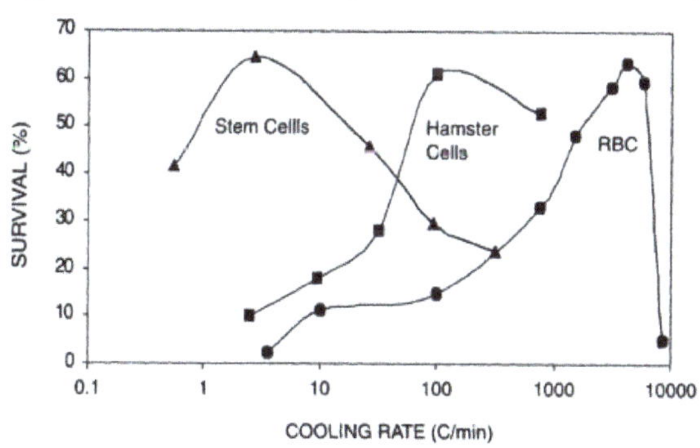

Fig. 5: Relationship between percentage of survival and coolingrate for three types of cells

2.3 CRYOPROTECTIVE AGENTS

The role of cryoprotective agents (CPAs) is to diminish the effects of cryoinjury during freezing or vitrification. CPAs are chemicals added to the cellular solution and alter the physical behaviour of the cell. There are numerous CPAs of different types, which can create hundreds combinations of CPA solutions. Themost widespread CPAs are dimethyl sulfoxide (Me_2SO), glycerol, sucrose and trehalose. CPAs are generally categorised into two types, according to their ability to penetrate through the cellular membrane; permeating CPAs, such as Me_2SO, glycerol, 1,2-propanediol, and non-permeating CPAs, mainly polymers (polyvinyl pyrrolidone, hydroxyethyl starch) and sugars. Low concentrations of CPAs are capable of protecting the cellsagainst the solution effects, rather than the IIF, however their mechanism of protection depends on their type.

Non-permeating CPAs act by increasing the osmolality of the extracellular solution, which forces the cell to shrink before freeze. In this case, the initial freezing of the cell may require more time, but the intracellular composition, solute concentration and size will not be altered, and they will remain a function of the extracellular osmolality, as if the CPA were never added. Permeating CPAs have the ability to decrease the concentration levels of electrolytes and the amount of extracellular ice formation in the cell. However, they cause the increase of cells' final size, as they penetrate the intracellular membrane, adding more components in cell's volume.

High concentrations ofCPAs increase the viscosity of the extracellular and intracellular medium, which reduces dramatically the ice formation process and the crystal growth. As a result, the rate of ice formation isdecreasing along with the temperature. Usually, in glass-transition temperatures the viscosity is high enough to stop all phase transitions inside the cell. Ice may forms at any temperature below the melting point of the cell, but when the temperature is above the glass-transition temperature, the ice formation rate can be controlled, by controlling the freezing time between the cooling down from the melting point temperature to the glass-transition temperature. However, high CPA concentrations correspond to high toxicity levels, which damage the cells instead of protecting them]. The toxicity of CPAs can be reduced either by using lower concentrations of CPAs or by reducing the period of time of cell's exposure to the CPA solution. Moreover, studies have shown that CPAs increase the retention of intracellular water and thus, enhance the probability of IIF. Finally, cells can be damaged by excessive osmotic forces developed during the addition and removal of the CPAs, before and after their cryopreservation.

2.4 METHODS OF CRYOPRESERVATION

The viability and extend of cryoinjury in a cell depend on the amount of water in it and the ability of that water to crystallize during the freezing process. The variables that influence the survival rate of a cell are the concentration of CPAs, the cooling rate, the storage temperature and the rewarming rate used. The effects of cooling rates, the use of cryoprotective agents and their mechanisms of action have been extensively described in the previous section. In this part the significance of the storage temperature and the current equipment system technologies are outlined.

2.4.1 Initial Freezing

All cryopreservation processes, despite the long-term storage method used afterwards, begin with the initial freezing of the sample, during which the sample is transferred from its physiological environment to a cryoprotective solution, to minimise the cryoinjury effects. This phase is also known as *hypothermic phase* and the way it works is by diminishing the cell's metabolism and the chemical toxicity of the CPAs. Biological materials can be introduced into the hypothermic phase either by being directly packaged into the precooled (0 to 4°C) cryoprotective solution and then transferred to a cryopreservation chamber of the same temperature range, or by being placed into a control rate freezer. However, the later method has been proven to result in better post-cryopreservation survival rates, as it provides robust temperature compensation for therelease of the latent heat produced by the cells. This heat release is continuous and caused by the freezing process, as ice formation is an exothermic process.

2.4.2 Equilibrium Freezing

Two are the main methodologies currently used for the initial freezing of cells; equilibrium freezing and verification. During equilibrium freezing, cells are exposed to relatively low CPA concentrations and then slowly cooled at a controlled cooling rate, in order to gradually freeze the extracellular solution and in this way to reduce the probability of IIF. During equilibrium freezing, the cells are exposed to low concentrations of CPAs, equilibrated at room temperature for a short period (10 to 25 min), placed into containers (straws or vials), sealed and stored into acontrolled-rate freezer. Thus, the cells are exposed to gradual increase of CPAs concentrations, as more extracellular ice is formed.

To avoid the lethal consequences of IIF and solute concentrations, the controlled freezing systems must facilitate the equilibration of the sample and its surrounding medium, at a temperature marginally below the sample's freezing point. To do that, the sample is forced to freeze in specific regions, by introducing several nucleation points in the cell. This process is called *seeding* and is induced extracellularly either by contacting the cell with a cold instrument or by a short-lived introduction of liquid nitrogen (LN_2) into the preservation chamber. As a result a solute gradient is created between the intracellular and extracellular medium, which forces the freezing water to flow out of the cell, minimizing that way the chances of IIF.

Equilibrium freezing has three major disadvantages. First of all, the ice crystals formed in the extracellular solution may cause direct mechanical damage to the cell membrane or other fine structures (e.g. sperm tails) and lead to the loss of cell's biophysical function and viability. Secondly, in order to achieve the highest survival rates of the preserved cells, a very tightly controlled optimal cooling rate is required, which many existing cryogenic systems find difficulties to obtain. Finally, the cooling requirements differ from one cell type to another, as the optimal cooling rate depends on individual cell characteristics. As a result, different cell types require different cooling devices, a fact that limits the application of the equilibrium freezing method.

2.4.3 Vitrification

An alternative approach is vitrification, by which any signs of ice formation are prevented. This method maintains the cell as a whole in a vitreous state and precludes both intracellular and extracellular ice formation. In order to achieve that, high concentration of CPAs and a relatively fast cooling rate in excess of the critical cooling rates (the minimum cooling rate to vitrify a solution) are used. During the process, the liquid phase of the cell eventually becomes so viscous that the water molecules do not have enough energy to form crystalline structures, and as a result they start to become amorphous solid or glass. The temperature at which this phenomenon takes place is known as glass transition temperature and below this temperature the cell considers to be biologically inert.

Vitrification with no or low concentrations of CPAs would be suitable for almost all cell types, but it would require the freezing of the samples at ultra-fast cooling rates, on the order of tens of thousands of degrees Celsius per minute. The only method that approaches ultra-fast freezing is the drop of a small cell suspensionvolume directly into LN_2, but the cooling rate cannot be controlled. A thin vapour coat film forms around the surface of the sample when plunged in LN_2, insulates the sample and prevents a more rapid temperature loss (Leidenfrost effect). Vitrification protocols that require the direct contact of the sample to the LN_2 raise concerns of potential viral contamination, while the whole process of vitrification is very challenging, due to the limited amount of time of cell handling and loading. New techniques of vitrification, though, show very promising results, which overcome the current shortcomings of rapid freezing. For examples, S^3 vitrification exposes the sample to a series of CPA solutions, loaded into sterile straws (0.25 ml), heat sealed them at bothends and vitrifies them. In this way, there is sufficient time for exposure in the CPA solution, loading and vitrification, while the relatively large sized container and the significant lower cooling rates, simplifies the vitrification process and diminishes the amount of LN_2 used.

2.4.4 Laser Pulse Vitrification

In 1998 *Fowler and Toner* developed a protocol for successful recovery of rapidly frozen red blood cells by vitrification, without the use of CPAs. There technique included the use of a laser pulse prior thawing, which selectively targeted the intracellular ice, so that it was melted and resolidified into a glass phase. The erythrocytes were loaded into straws (1mm x 1μm), plunged into methanol slurry (-96°C) and transferred into LN_2. The average cooling rate of the samples was approximately 10,000 °C/min. Then, the sampleswere removed from the LN_2, placed on a cryostage (-170°C) and a laser pulse (7 ns width, 532 nm wavelength) irradiated the haemoglobin, which exists in the intracellular solution. The short laser pulse was able to be absorbed only by the intracellular medium, which to the increase of its temperature above 0°C, while the extracellular solution remained at -170°C. Up to 80% of cells treated that way remained viable after thawing.

2.4.5 Liquidus Tracking (LT)

The principle of LT is the dynamic control of CPA concentrations, throughout the cooling process, in order to maintain the cell just above its freezing point at all times, without the formation of ice. As a result, vitrification can occur without the extremely high CPA concentrations and without the need of ultra-rapid cooling rates. LT systems adjust the CPA concentrations, temperature and time program to the characteristics of the stored cells at all stages of the process. LT systems consist of a conventional controlled rate freezer, with a cooling rate of 0.1 to 0.3 °C/min, while the solution composition is controlled by standard peristaltic pumps.

2.4.6 Isochoric and Hyperbaric Cryopreservation

All the above cryopreservation methods are performed under constant pressure of 1atm of the chamber (isobaric process). A new approach is the freezing under pressure, i.e. the freezing under constant volume of the chamber. Two techniques are currently used, the isochoric and hyperbaric cryopreservation, and both of them use none or few amounts of CPA solutions. The isochoric method has proved to minimise significantly the increase of solute concentration during the freezing process, by almost an order of magnitude, compared to the isobaric cryopreservation. Isochoric and hyperbaric methods follow the same theoretical principles (changes in pressure), but they are different processes. The isochoric cryopreservation is a two-phaseequilibrium process, in which ice and liquid exist simultaneously at equilibrium under constant temperature and volume, while hyperbaric cryopreservation the solution is maintained in a single phase, as liquid. In hyperbaric cryopreservation, the elevating pressure changes and rapid cooling rates prohibit the ice formation in the solution, thus the biological sample maintains its original composition. However, cells and organs cryopreserved by the hyperbaric method do not show very high survival rates. Isochoric cryopreservation offers a more robust method of freezing, by which the system adjusts itself to the minimal pressure for a particular temperature, instead of continuously increasing the pressure of the chamber [35]. Erythrocytes in 5% v/v Me_2SO or 8% v/v glycerol solutions, under 120 MPa pressure and cooling rate of 35°C/min or 160°C/min showed survival rates of 92% or more .

2.5 CRYOPRESERVATION EQUIPMENT

2.5.1 LN2 - Direct Temperature Feedback Method

These systems monitor and control the temperature of the chamber. More advanced systems are additionally monitoring the samples' temperatures and they adjust the chamber's operational temperature accordingly. T- type thermocouples are continuously monitoring and controlling the chamber's and samples' temperatures, compare them to the programmed optimal temperatures and the appropriate amount of LN_2 is determined and injected into the chamber. The delivery of the LN_2 in the chamber is provided usually by dual injection valves for more precise temperature control, faster cooling rates and as a back-up system in case of failure of one of the valves. Solenoids can also be used, but they tend to wear over time.

2.5.2 LN_2 - Timed Pulse Method

Similarly to direct temperature feedback systems, timed pulse systems control the chamber temperature. They usually consist of a microprocessor control system, a pump, a heater and one or more solenoid valves with openings, which meter a timed pulse of LN_2 into the chamber. The appropriate amount of LN_2 is determined and injected into the chamber by several variables, such as the valves' sizes, the tank's pressure and the number of solenoid openings. The heater heats the LN_2 to its boiling point, in order to provide constant tank pressure and to ensure that the appropriate amount of LN_2 is provided to the chamber. The disadvantage of timed pulse systems is their low reliability, as the same cooling program can produce different cooling curves of the cells over time.

2.5.3 LN_2 - Plunge Freezing Method

This is the simplest method of controlled rate freezing with the use of LN_2 and it is usually used for small numbers and low volume straws and vials. A heat block is loaded with the biological samples and then is submerged into a tank of LN_2. The heater increases or decreases its power in order to achieve a controlled freezing rate of the chamber.

2.5.4 Step Down Method - Mechanical Freezers

This is a non-automated freezing method, but is still in practise. The samples are initially cooled down by being placed in a refrigerator for several hours, then transferred to a mechanical freezer and finally stored either into a LN_2 freezer or continue their storage into an Ultra-Low Temperature (ULT) freezer. ULT freezers follow the principles of any mechanical refrigeration system and consist of the same basic components; a compressor, a preservation area, evaporator, condenser and refrigerant. The refrigeration of the system is provided by the basic vapour-compression cycle, in which the refrigerant circulates through the system and passes through a number of processes. The temperature range of an ULT freezer is typically between -56 to -86°C, but it is important to be able to operate within the range of -70 to -80°C. There is no conventional refrigerant that can cover the temperature gap between ambient room and operational temperatures; for that reason ULT freezers are equipped with two refrigeration circuits, each one with its individual compressor and refrigerants, using a cascade technique. Based on the cascade technique, the first refrigeration cycle is used to cool down the system from ambient to an intermediate temperature, while the second refrigeration circuit is able to further reduce the temperature of the preservation area to the required operational conditions.

The disadvantages of step down method are that is a long time process, is difficult to repeat and cannot provide controlled cooling rates. Moreover, the post-thaw recovery rates of cells maintained into ULT are very low, even for robust cells, due to power outages, mechanical failures or human caused mishaps, such as frequent door openings or improper closing of the freezer's door. As far as energy consumption is concerned,ULT systems account for more than 50% of the total energy used in a laboratory space.

2.5.5 Alcohol Bath Freezers

These units are commonly used in laboratories, for preservation of cells at temperatures above -80°C. Alcohol bath freezers are basically refrigerated circulators, in which an amount of alcohol is used as cooling medium to pass through a cooling system and exchange temperature, in order to reduce it. These systems require a relatively large volume of cooling medium, in order to provide stable temperatures and accurate controlled cooling rates, due to medium's high heat capacity. However, alcohol bath freezers are not capable of achieving rapid cooling rates.

2.5.6 Vitrification Systems

These systems usually include the rapid transition of the sample to a glassy state, by direct submersion into LN_2. The cooling rate of the equipment can be increased either by the use of negative pressure systems, which depress the freezing point of LN_2 to -205°C, or by minimising the sample's volume. Other available systems freeze the cell by getting them into contact with an aluminium block cooled in LN_2, in order to avoidthe direct contact of the sample and the LN_2, for contamination reasons and to eliminate the Leiden frost effects.

2.5.7 Stirling Engines Freezers

The original idea of Stirling engines is to convert thermal energy to mechanical work. However, their operation principle can be used in reverse and thus cryopreservation systems use Stirling engines to convert mechanical energy to thermal energy; in other words, Stirling engines operate as heat pumps. Most refrigeration systems operate on the Rankine cycle and their cooling capacity depends on the properties of their refrigerant. Triple stage Rankine refrigerators are currently the state of the art of mechanical freezers, but they can achieve temperatures of only -140°C. Sterling cycle is by definition less efficient than Rankine cycle, but is capable of providing lower temperatures than the Rankine cycle. Moreover, they can freeze small amounts of samples and their cooling capacity is much lower than the LN_2 systems. However, due to personal safety and cross-contamination concerns of LN_2 systems, Sterling engines freezers find great application in clean rooms and laboratories where the use of LN_2 is not possible. Liquidus Tracking

2.5.8 Freezers

LT method is relatively a novel technique, thus there are not many types of equipment available. The typical configuration is a conventional slow cooling freezer coupled with a LT controller and two dynamic peristaltic pumps. The cooling is provided by the same way as in the conventional slow rate freezing process,but the LT controller monitors the temperature inside the chamber and adjusts the speed of the peristaltic pumps, in order to alter the concentration of CPA solution of the cells. The decrease of the temperature in thechamber, increases the CPA concentration of the cell, but as the temperature of the cell decreases, the toxicity of the CPA decreases, allowing the sample to tolerate the high CPA concentrations. The only requirements of the LT freezers are the large volumes of CPAs and their proper mixing around the sample.

2.5.9 Isochoric Systems

Isochoric systems are quite simple and inexpensive; however they are not commercially available. One of thesystems described in the literature was used by *Rubinsky et al.* and it consists of a constant volume chamber that is hermeticaWlly sealed and in which the pressure is monitored with a pressure gauge. The chamber is filled with fluid and is cooled by immersion in a constant temperature bath. The major advantage of isochoric cryopreservation is that the solution concentrations during the freezing are lower at each temperature by almost an order of magnitude, compared to the isobaric cryopreservation [35]; a fact that mayincrease the post recovery success rates of the cells.

2.6 LONG-TERM STORAGE IN LN_2 - CONTAMINATION AND SAFETY ISSUES

The temperatures at which biological materials are stored have great influences on their shelf life. Generally, the lower the storage temperature, the longer the length of time after which the biological material can be recovered. Ultimate cell viability and stability is achieved, when samples are preserved below -130°C the glass transition temperature of water. It is believed that at this temperature all metabolic processes are retarded and biological stability is obtained, which makes possible the intact storage of samples even for 2,000 years. The only consideration is the slow accumulation of background ionizing radiation, but the effects on the cells will appear after centuries of storage.

Storage at temperatures of -150°C is typically recommended, because it provides a 20°C safety margin between the storage and glass transition temperature of cells. Storage in LN_2, with a boiling point of -196°C, is a convenient method to maintain low operating temperatures and to provide even greater safety margin for better sample security. The reasons of why LN_2 finds such great application in cryopreservation systems are due to the extremely low cryogenic temperatures that offers, the fact that it is chemically inert, relatively inexpensive, readily available, non-flammable and does not require electrical power to provide freezing; power supply is only used for the electronic features of the systems, such as auto-filling, monitoring and alarms.

However, besides the advantages of LN_2 systems, it is important to outline that the exposure to an environment with high levels of LN_2 in the atmosphere can cause dizziness, unconsciousness and even death, as the molecules of nitrogen displaces the molecules of oxygen. Thus the handling of LN_2 requires extra care, as it is colorless, odorless, and tasteless and it is breathed as if it were air, making its detection by the human senses impossible.

LN_2 storage systems can provide cooling in either liquid or vapour phase. Storage in liquid LN_2 offers uniform temperature profile across the chamber, while the vapour phase storage shows a temperaturegradient and relatively higher temperatures than the liquid phase. However, storage in liquid phase LN_2 is contains the risk of contamination and explosion of the samples, as the vials used for packing the biological materials may leak and allow the interaction between the LN_2 and material. Thus, the rest of the samples stored in the chamber can become cross contaminated with microbes and LN_2 can enter the leaking vial. As a result, when the leaking sample is retrieved, the trapped LN_2 will expand, transform into gas, change its volume and cause a small explosion, and the destruction of the sample.

LN_2 usually has a very low microbial count but, during storage and distribution may become contaminated with microorganisms and potentially infect the stored material by transmitting viruses, bacteria, fungi and animal cells in the holding tank. By opening the liquid nitrogen container, the water vapour in the air above the container becomes cooled and forms small ice crystals with high electrostatic charge, which capture airborne microorganisms and fall into the container. However, the contamination can be controlledby the appropriate emptying and cleaning of the vessel. Other potential contaminations can occur from contaminated surfaces or leaking samples.

It is suggested that LN_2 should be treated as biohazardous substance, as the screw cap of the plastic vials allowed contact between the contaminated LN_2 and the sample. Another example of contamination is the Hepatitis B Virus (HBV) outbreak caused by LN_2 storage. It also reported reported that how the bone marrow and stem cells harvested from one patient with HBV led to infect other five patients, who received transplants stored in the same tank . Also, a survey of microbial contamination of hematopoietic stem cells stored are LN_2 freezers was conducted by *Fountain et al.* and showed that of the 583 cultures tested, 1.2% were found to be contaminated by microorganisms . Four out of five freezers examined contained low level of microbial contamination, however the fifth freezer was heavily contaminated with Aspergillus, the same microbes found in the contaminated samples.

A more recent assay showed that small particles, can be transmitted in the vapour used to cool programmable vapour LN_2 freezers, and as the immediate storage environment for bulk storage and transport of cryopreserved samples and remain suspended in the vapour phase of LN_2 for at least 24 hours. The longer a sample has been held in a storage tank then the greater the risk of contamination from theaccumulating microbial load of the bulk cryogen.

Attempts to eliminate the danger of cross contamination include cooling in LN_2 passed through a 0.2µm pore-size filter and placing the carrier (vial, straw, etc.) into a container that partially or completely isolates itfrom the LN_2 during storage.

Storage in vapour LN_2 can address the contamination and the explosion problems, however cannot provide the uniform temperature distribution in the chamber, developing that way different temperature gradients along the storage area. However, new designs of vacuum insulated chambers, with low evaporation rates of LN_2 are showing promising results and vapour LN_2 temperatures of -190°C. These vapour LN_2 systems restrict the amount of energy entering the preservation area, ensuring that the temperature of the vapour is maintained close to the liquid LN_2 temperature, and the heat exchange surface extends from the bottom to thetop of the freezer.

3

CRYOPRESERVATION AND ITS APPLICATIONS

3.1 INTRODUCTION

Cryopreservation is a process that preserves organelles, cells, tissues, or any other biological constructs by cooling the samples to very low temperatures. The responses of living cells to ice formation are of theoretical interest and practical relevance. Stem cells and other viable tissues, which have great potential for use in basic research as well as for many medical applications, cannot be stored with simple cooling or freezing for a long time because ice crystal formation, osmotic shock, and membrane damage during freezing and thawing will cause cell death. The successful cryopreservation of cells and tissues has been gradually increasing in recent years, with the use of cryoprotective agents and temperature control equipment. Continuous understanding of the physical and chemical properties that occur in the freezing and thawing cycle will be necessary for the successful cryopreservation of cells or tissues and their clinical applications. In this review, we briefly address representative cryopreservation processes, such as slow freezing and vitrification, and the available cryoprotective agents. In addition, some adverse effects of cryopreservation are mentioned.

Biological and chemical reactions in living cells are dramatically reduced at low temperature, a phenomenon that can lead to the possible long-term preservation of cells and tissues. However, freezing is fatal to most living organisms, since both intra- and extracellular ice crystals are formed and results in changes to the chemical setting of cells that lead to cellular mechanical constraints and injury. The major hurdle for cells to overcome at low temperatures is the water-to-ice phase transition. Cell injury at fast cooling rates is attributed to intracellular ice formation, whereas slow cooling causes osmotic changes due to the effects of exposure to highly concentrated intra- and extracellular solutions or to mechanical interactions between cells and the extracellular ice (Fig. 1).

3.1.1 Cryopreservation

Cryopreservation is a process that maintains biological samples in a state of suspended animation at cryogenic temperature for any considerable period and is used to preserve the fine structure of cells. The freezing behavior of the cells can be altered in the presence of a cryoprotective agent (CPA; also called cryoprotectant), which affects the rates of water transport, nucleation, and ice crystal growth. Numerous research papers on cryopreservation have studied the underlying physical and biological factors affecting the survival of cells at low temperatures during the cooling and warming processes. Unlike in single cell suspensions, bulk tissues have different heat and mass transfer effects that occur during the cryopreservation of bulk tissues and thus these factors make it more difficult to achieve rapid cooling and warming rates as well as an equal distribution of CPAs. Cryopreserved cells or tissues possess some advantages for basic research and current and future clinical applications. With the constant availability of cryopreserved cells and tissues, extensive quality testing can be performed to determine the suitability of the cells or tissue for transplantation without the need to obtain fresh samples. The successful cryopreservation of cells and tissues has been gradually increasing with the use of CPAs and temperature control equipment (Table 1).

Briefly summarize the principles of cryopreservation and its application in clinical areas.

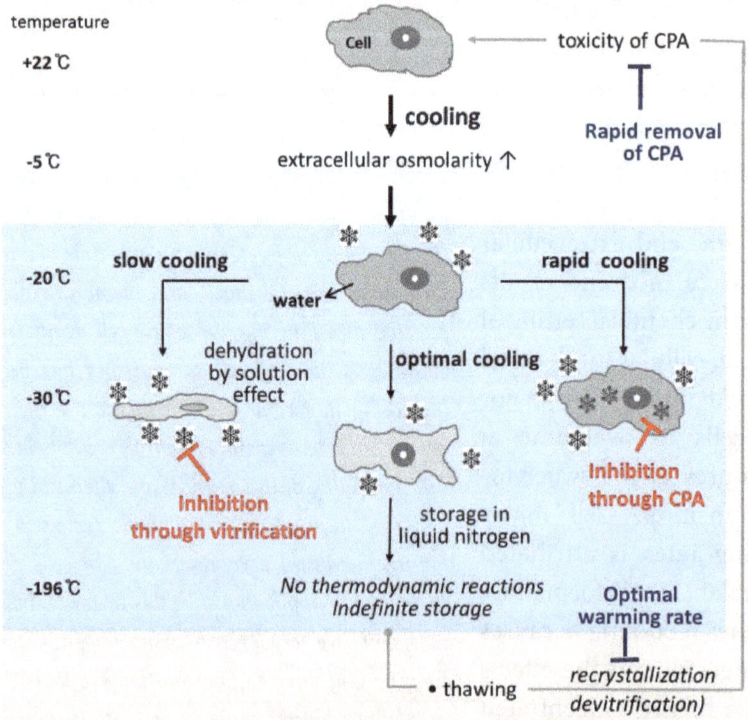

Fig. 1: Physical events and cryoinjury of cells during freezing and thawing. Cryoinjuries are caused, at least in part, by the solution effect (leading to osmotic shock) and intracellular ice formation (leading to breakdown of intracellular structures).

Table 1 Comparison between the slow-freezing and vitrification methods

Characteristic	Procedure	
	Slow freezing	**Vitrification**
Working time	More than 3 h	Fast, less than 10 min
Cost	Expensive, freezing machine needed	Inexpensive, no special machine needed
Sample volume (μL)	100–250	1–2
Concentration of CPA	Low	High
Risk of freeze injury, including ice crystal formation	High	Low
Post-thaw viability	High	High
Risk of toxicity of CPA	Low	High
Status of system	Closed system only	Opened or closed system
Potential contamination with pathogenic agents	Low	High
Manipulation skill	Easy	Difficult

3.2 CRYOPRESERVATION PROCEDURE

Cryopreservation is the use of very low temperatures to preserve structurally intact living cells and tissues for a long period of time. Depending on the cell types or given cells among different mammalian species, there is great diversity in cryobiological response and cryosurvival during the freezing and thawing cycle (Fig. 1 and Table 1).

Cryopreservation processes can generally be grouped into the following types:

- Slow freezing;
- Vitrification, which involves the solidification of the aqueous milieu of the cell or tissue into a noncrystalline glassy phase;
- Subzero nonfreezing storage; and
- Preservation in the dry state.

Generally, the storage of mammalian cells in the dry state is not readily possible because of difficulties in introducing the disaccharide trehalose (disaccharide of glucose, 342 Da) and amino acids (used as preservatives in plants) into the intracellular region.

The major steps in cryopreservation are:

- The mixing of cpas with cells or tissues before cooling;
- Cooling of the cells or tissues to a low temperature and its storage;
- Warming of the cells or tissues; and
- Removal of CPAs from the cells or tissues after thawing. The appropriate use of CPAs is therefore important to improve the viability of the sample to be cryopreserved.

3.3 MECHANISM OF CRYOINJURY

The exact mechanism of cryoinjury, which is the damage of cells associated with the phase changes of water in both extra- and intracellular environments at low temperatures, has not been clearly established. The cooling and thawing velocities can largely affect the physicochemical and biophysical reactions, altering the survival rate.

Cryoinjury mechanisms involving osmotic rupture caused by extra- or intracellularly concentrated solutes and intracellular ice formation have been highly suggested, both processes of which are dependent on the cooling rate (Fig. 1). In addition, cell viability limits are defined primarily in terms of an intact plasma membrane that retains normal semipermeable properties. Indeed, conditions that allow the plasma membrane to survive may not allow the survival of critical organelles within cells.

3.3 CPAs

The CPA, which is usually a fluid, reduces the freezing injury from the cryopreservation process (Fig. 1). CPAs should be biologically acceptable, be able to penetrate the cells, and have low toxicity.2 Various CPAs have been developed (Table 2) and are used to reduce the amount of ice formed at any given temperature, depending on the cell type, cooling rate, and warming rate. In order to achieve the best survival rate of cells and tissues, the sample volume, cooling rate, warming rate, and CPA concentrations should be optimized depending on the different cell types and context of tissues. It should be mentioned that the macroscopic physical dimension of the tissue is a major point to be defined in a cryopreservation protocol because of heat and mass transfer limitations in these bulk systems.

CPAs can be divided into two categories:

(1) cell membrane-permeating cryoprotectants, such as dimethyl sulfoxide (DMSO), glycerol, and 1,2-propanediol; and

(2) nonmembrane-permeating cryoprotectants, such as 2-methyl-2,4-pentanediol and polymers such as polyvinyl pyrrolidone, hydroxyethyl starch, and various sugars.

Unlike synthetic chemicals, biomaterials such as alginates, polyvinyl alcohol, and chitosan can be used to impede ice crystal growth, along with traditional small molecules. The direct inhibition of ice crystal formation and application of antioxidants and other compounds have been used to attempt to reduce cell death from processes such as apoptosis during the freezing and thawing cycle. Common CPAs are briefly addressed in the following subsections and Table 2.

Table 2 commonly used cryoprotective agents and their uses

Cryoprotective agents	Membrane permeability	Possible toxicity
Cell Banker series	Yes	Unknown but less than that of DMSO
Dimethylsulfoxide (DMSO)	Yes	Reduction in heart rate Toxicity to cell membrane
Ethylene glycol (EG)	Yes	Gastrointestinal irritation Pulmonary edema Lung inflammation
Glycerol	Yes	Renal failure
Trehalose	No	Relatively less toxic
Propylene glycol (1,2-propanediol)	Yes	Impairment in the developmental potential of mouse zygotes

3.3.1. Glycerol

This polyol compound remained the most effective of additives until the protective effect of DMSO was demonstrated . Glycerol is a nonelectrolyte and thus may act by reducing the electrolyte concentration in the residual unfrozen solution in and around a cell at any given temperature. It is widely used in the storage of bacteria and animal sperm.

3.3.2. Dimethylsulfoxide (DMSO)

DMSO has been commonly used for the cryopreservation of cultured mammalian cells because of its low cost and relatively low level of cytotoxicity. Like glycerol, DMSO acts by reducing the electrolyte concentration in the residual unfrozen solution in and around a cell at any given temperature. However, a decline in the survival rate and the induction of cell differentiation caused by DNA methylation and histone alteration have been reported. These negative effects of DMSO in cryopreservation create some difficulties for its use in routine clinical applications.

3.3.3. Polymers

The entrapment of CPAs within a capsule during cell resuspension in an encapsulating material is another strategy for the modulation of cell location.4 Among encapsulating materials, synthetic nonpenetrating polymers can provide cryoprotection of cells within the scaffold, thereby bypassing the limitations of diffusion in higher-dimensional cryopreservation. Vinyl-derived polymers, such as polyethylene glycol ($C_{2n}H_{4n+2}O_{n+1}$, molecular weight: 200–9500 Da), polyvinyl alcohol [$(C_2H_4O)_n$, molecular weight: 30–70 kDa), and hydroxyethyl starch (130–200 kDa), have a capacity to decrease the size of formed ice crystals.

3.3.4. Proteins

Sericin is a water-soluble sticky protein (~30 kDa) isolated from the silkworm cocoon and has been developed as a fetal bovine serum- or DMSO-replacing CPA for human adipose tissue-derived stem or progenitor cells, or hepatocytes. Small antifreeze proteins derived from marine teleosts or fishes have also attracted attention as CPAs.

3.3.5. Cell Banker Series

A newly developed Cell Banker series allows for rapid cell cryopreservation at −80 °C, and has been shown to achieve better survival rates following freezing and thawing. The Cell Banker series of cryopreservation media contain 10% DMSO, glucose, a prescribed high polymer, and pH adjustors. Serum-containing Cell Bankers 1 and 1+ can be used for the cryopreservation of almost all mammalian cells. Indeed, conventional cryopreservation media include fetal bovine serum, which contains a mixture of growth factors, cytokines, and undefined substances such as bovine exosomes, rendering its use forbidden in the establishment of a standardized cryopreservation protocol for clinical use in humans. In this aspect, the nonserum-type Cell Banker 2 is optimal for the cryopreservation of cells in serum-free culture conditions. Cell Banker 3 (or Stem cell Banker) is composed of 10% DMSO and other inorganic compounds (US20130198876) and satisfies the criterion of a chemically defined known ingredient that is xeno free and is thus suitable for the preservation of somatic stem cells and induced pluripotent stem cells.

3.4 FREEZING METHOD: CONVENTIONAL SLOW FREEZING AND VITRIFICATION

Cryopreservation can be accomplished by slow freezing and vitrification methods (Table 1). The major differences between the two are the concentrations of CPAs and the cooling rates used. Theoretically, if cooling is sufficiently slow, cells could efflux intracellular water rapidly enough to eliminate supercooling and thus prevent intracellular ice formation. As a result of differences in the capacity of different cells to move water across the plasma membrane, optimal cooling rates will be dependent on cell types. Slow freezing first substitutes the water within the cytoplasm with CPAs which reduces cell damage and adjusts the cooling rate in accordance with the permeability of the cell membrane. Slow-cooling protocols involve a typical cooling rate of about 1 °C/min in the presence of less than 1.0M of CPA, with use of a high-cost controlled-rate freezer or a benchtop portable freezing container. The advantages of slow freezing are that it has a low risk of contamination during the procedures and does not demand high manipulation skills. However, slow freezing has a high risk of freeze injury due to the formation of extracellular ice (Table 1).

As an alternative to the slow-freezing technique, vitrification is a process by which cell suspensions are transformed directly from the aqueous phase to a glass state after direct exposure to liquid nitrogen.35 The process requires cooling of the cells or tissues to deep cryogenic temperatures (i.e., with liquid nitrogen) after their exposure to high concentrations of CPA (in the ratio of 40–60%, weight/volume), with subsequent rapid cooling to avoid ice nucleation.

Vitrification is largely dependent on three factors: (1) viscosity of the sample; (2) cooling and warming rates; and (3) sample volume. Thus, a delicate balance must be maintained among all the relevant factors to ensure successful vitrification.

There are two methods of vitrification: equilibrium and nonequilibrium. Equilibrium vitrification requires formulation of multimolar CPA mixtures and their injection into the cell suspensions. Nonequilibrium vitrification, which is further divided into carrier-based (including the former plastic straws, quartz microcapillaries, and cryoloops for obtaining a minimum drop volume18) and carrier-free systems, uses an extremely high freezing rate along with lower concentrations of the CPA mixture. A major advantage of vitrification is the low risk of freeze injury, thereby ensuring a sufficiently high cell survival rate. However, the high potential of contamination with pathogenic agents is present, and the technique requires good manipulation skills.

3. 4. APPLICATIONS OF CRYOPRESERVATION

The applications of cryopreservation can be categorized into the following areas:

- Cryopreservation of cells or organs
- Cryosurgery;
- Biochemistry and molecular biology;
- Food sciences;
- Ecology and plant physiology; and
- Many medical applications, such as blood transfusion, bone marrow transplantation, artificial insemination, and *in vitro* fertilization (IVF).

Some suggested advantages of cryopreservation include the possible banking of cells for human leukocyte antigen typing for organ transplantation, the allowance of sufficient time for transport of cells and tissues among different medical centers, and the provision of research sources for identifying unknown transmissible diseases or pathogens. Furthermore, the long-term storage of stem cells is still the initial step toward tissue engineering, which holds promise for the regeneration of soft tissue esthetic function and for the treatment of known diseases that have currently no therapy option.

3.4.1. Oocytes and Embryos

The first case of embryo cryopreservation for fertility preservation took place in 1996, with the application of a natural IVF cycle prior to chemotherapy in a woman diagnosed with breast cancer. Cryopreservation of mature oocytes is a proven technique for preserving the reproductive capacity. Results from a retrospective study of 11,768 cryopreserved human embryos that underwent at least one thaw cycle from 1986 to 2007 showed that there was no significant impact of the duration of storage on clinical pregnancy, miscarriage, implantation, or live birth rate, whether from IVF or oocyte donation cycles. Since oocytes are highly prone to chilling injury cryopreservation of immature oocytes and ovarian tissue is a promising approach-with reports of live births-but the need for investigational improvements remain.

3.4.2. Sperm, Semen and Testicular Tissue

Germ cell depletion caused by chemical or physical toxicity, disease, or genetic predisposition can occur at any age. Fertility preservation is of great importance to guarantee the quality of life of patients facing chemo and radiotherapy. Sperm and semen can be used almost indefinitely after proper cryopreservation. There are new trials for cryopreserving testicular tissues in the form of cell suspensions, tubular pieces, and entire gonads, but this technique is still premature. Overall, cryopreservation can be used as a first-line means of preserving fertility for men undergoing vasectomy or treatments that may compromise their fertility, such as chemotherapy, radiotherapy, or surgery.

3.4.3. Stem Cells

Adult stem cells are capable of differentiating into multiple types of specific cells and can be obtained from various locations other than bone marrow, including fat tissue, the periosteum, amniotic fluid, and umbilical cord blood. Stem cells can be subdivided into embryonic stem cells, mesenchymal stromal cells, and hematopoietic stem cells, all of which are considered as goldmines for potential application in regenerative medicine. Clearly, the fields of tissue engineering, gene therapy, regenerative medicine, and cell transplantation are largely dependent on the ability to preserve, store, and transport these stem cells without modification of their genetic and/or cellular contents.

3.4.4. Hepatocytes

Primarily isolated hepatocytes have found important applications in science and medicine over the past 40 years in a wide range of areas, including physiological studies, investigations on liver metabolism, organ preservation and drug detoxification, and experimental and clinical transplantation.7, 11 In addition, there is currently increasing interest in the applications of liver progenitor cells across a range of scientific areas, including both regenerative medicine and biotechnology, which raises the need for cryobanking.

3.4.5. Others

Although primary neuronal cells and cardiomyocytes are routinely used for neuroscience and cardiology research, a gold standard protocol for the preservation of these cells has not yet been developed. With the discovery of glucocorticoid-free immunosuppressive regimens,51 pancreatic islet transplantation may be considered as an alternative for the treatment of type 1 diabetes. For this reason, the development of islet cryopreservation methods has been ongoing, but results are still suboptimal, with a survival rate of less than 50%.

3.5. LIMITATIONS OF CRYOPRESERVATION

Although numerous usages of the cryopreservation technique exist, both in basic and clinical research, some limitations still exist. Cells metabolize almost nothing at low temperatures such as −196 °C (i.e., in liquid nitrogen), which has inevitable side effects, including a genetic drift toward biological variations of cell-associated changes in lipids and proteins that could result in the impairment of cellular activity and structure. If there were no limit to the amount of CPA that could be used, cells would be preserved perfectly. In conventional settings, however, CPAs themselves can be damaging to cells, especially when used in high concentrations. For example, there is a possibility that DMSO may alter chromosome stability, which can lead to a risk of tumor formation. Apart from endogenous changes in cells, the possible infection or contamination with cells such as tumorous ones should be prevented.

3.6. PERSPECTIVES

Improvements in the freezing and thawing rates, osmotic conditions, choice and concentration of CPAs, and equilibration times in the CPAs might result in better survival and functionality of human tissue and cell samples, permitting their successful future clinical application. The freezing methods (slow freezing vs. vitrification) used in cryopreservation and several CPAs were briefly addressed. New CPAs are constantly being investigated owing to the inherent toxicity of many known agents. A better understanding of the chemistry and biology behind freezing and thawing will be necessary for future development of this process and for finding the safest and most effective cryopreservation method.

The successful cryopreservation of biological samples may play a pivotal role in research connected with clinical utility for all kinds of human trials. Collectively, the most prominent future goals of cryopreservation should focus on the development of procedures that minimally affect the intactness of cryopreserved cells or tissues, followed by the standardization and optimization of the technique for routine use.

4

BASICS OF CELLULAR CRYOPRESERVATION: RESEARCH & CLINICAL USE

4.1 INTRODUCTION

The primary purpose of cryopreservation is to preserve biological specimens. Individual cells and biological tissues may be cryopreserved in a living state of suspended cellular metabolism at the temperature of liquid nitrogen (-196°C). This practice is crucial for biomedical research, clinical medicine, zoology, botany, and biotechnology. When frozen and kept properly, specimens may remain in a state of suspended cellular metabolism indefinitely and can be thawed as needed. Before the development of freezing media, cells were frozen in their own growth media or high serum media.

Cryoprotective agent development has improved cell viability post-thaw as well as stem cell pluripotency. Increased knowledge of the physiochemical processes involved during cell cryopreservation has resulted in reproducible protocols. Freezing techniques exist for tissues, single cells, embryos and microorganisms. Challenges with cryopreservation include post-thaw cell growth, viability, maintenance of pluripotency, changes in gene expression, and the ability to differentiate post-thaw.

WHY SHOULD YOU FREEZE YOUR CELLS?

- *Store your cells for future study.*
- *Cell storage acts as an insurance policy in case of contamination, failure, or cell shortage.*
- *Actively growing your cells over a long period of time may alter gene expression and differentiate cells. Continued growth and passage of cells may cause them to lose their original features. This endangers reliable research results!*
- *Freezing and storage of cells is a crucial step to ensure long-term cell use and reproducible results.*
- *Save time and money. Obtaining new cells is costly and time-consuming!*

By overcoming the challenges of cell freezing, researchers and clinicians will continue to bring valuable treatments to patients.

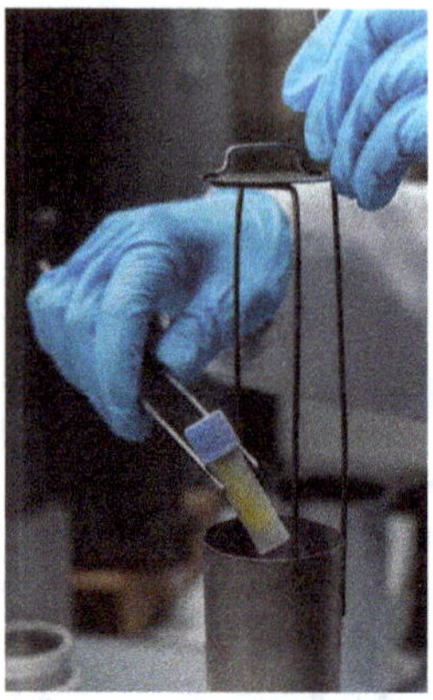

4.2 BASICS OF CRYOPRESERVATION

The purpose of cryopreservation is to store cells indefinitely by halting the cell's metabolism with ultralow temperatures. The freeze-thaw process is stressful to all cells and tissues. Therefore, effective techniques were developed to prevent cell death and damage. One common cryopreservation technique involves changing cell maintenance media to culture media containing a cryopreservation agent, such as dimethyl sulfoxide (DMSO). Cells are then cooled at a rate of -1°C/min (for mammalian cells) by transferring to -80°C in a specialized cooling container. After cells have cooled to -80°C, they are transferred to ultralow temperature storage of below -135°C. The most common ultralow temperature storage is liquid nitrogen in liquid or vapor form. After freezing, cells are stored as original seed and working stock assets. Ultralow storage temperatures work by maintaining cells below the glass transition temperature (Tg) of pure water (Miller, 1969). This suspends all molecular processes and prevents free radical generation that negatively effects cry preserved cultures. Cells survive by cooling the cultures to a "glassy" state slowly, so that less ice forms within the cell. Too much intracellular ice can cause mechanical damage to the cell when thawing. In addition, cells must not freeze too slowly because cells will shrink and dehydrate. Physical stress of shrinkage and dehydration results in high solute concentrations that increase toxicity.

Damage is also prompted by pH shifts and protein degradation that compromises the cell membrane or metabolic pathways. Technologies for the cryopreservation of cells and tissues are constantly improving. The best cryopreservation results are obtained through research-based approaches that are adjusted for the specific purpose of cryopreservation. There are six applications of cryopreservation that include preservation of cells or organs, cryosurgery, biochemistry or molecular biology, food sciences, ecology or plant physiology, and medical applications (transfusion, bone marrow and cells transplantation, artificial insemination, and in vitro fertilization). This guide will cover the basics of cellular cryopreservation for research and clinical use.

4.3 MAJOR ADVANTAGES OF CRYOPRESERVATION

There are many advantages to cell freezing. Proper cryopreservation of cells is the best way to save money, improve reproducibility of experiments, and protect important stocks of cells that may be difficult to replace. Cost savings occur because regular cell stock orders will not be necessary since cells may be split and multiplied. Expanding cells for ryopreservation will accommodate long experiments or industrial high throughput screening efforts. Keeping cell passage numbers low increases the likelihood that populations will remain homogenous between experiments because changes occur when populations are expanding. Therefore, keeping a cryopreservation master cell bank and working stocks will improve reproducibility between similar experiments. Gene expression or other cellular changes may be documented if passages are frozen down, tested, and compared. In case of contamination, errors, or unintended variables occur then the original stocks can be revisited.

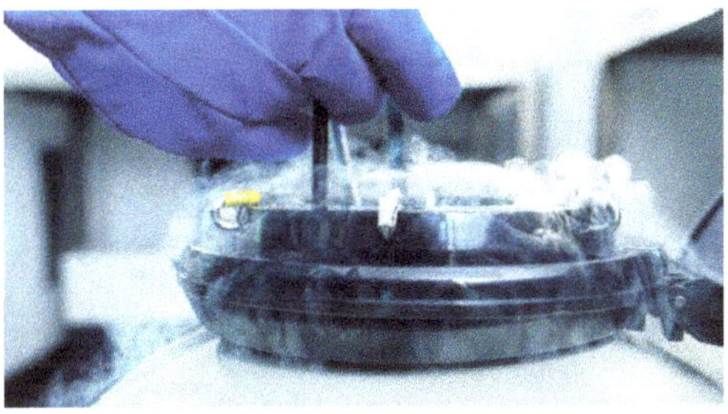

4.4 CHALLENGES OF CRYOPRESERVATION

Government regulations of cells for medical use have increased the need for documented cell handling processes. To account for these regulations, protocols for cryopreservation must be established and adhered to by researchers and medical personnel. While optimized cryopreservation protocols exist for most areas of research and medicine, technical issues persist. These issues include low survival, altered gene expression and morphology, loss of cellular function, epigenetic changes, and differences in protein composition. If optimal conditions are not provided during cell freezing then damage and cellular apoptosis or necrosis may occur post-thaw. Freezing cells includes intracellular and extracellular complex events that are not fully understood. During the freezing process ice forms in and out the cell and this is very much dependent on the cooling rate. If ice forms externally to the cell, water migrates out of the cell causing dehydration, shrinkage and finally cell death. If too much water remains inside the cell during the freezing process, intracellular ice crystals form that damage cellular organelles and pierce the cell membranes during the thawing process. Rapid cooling minimizes the solute imbalance between the internal and external of the cell, but more intracellular ice is formed. Slow cooling intracellular water migrates out of the cell resulting dehydration and shrinkage (Elliot, Wang, & Fuller, 2017). Cells may be successfully frozen when the cooling rate is slow enough to prevent intracellular ice formation but fast enough to prevent dehydration and damage.

Osmotic pressure imbalance is the main concern during cryopreservation as it may cause dehydration, deformation, and injury by effecting the water movement in or out the cell. When freezing cells at too low cooling rate, water may leave the cell through the cell membrane to join the ice on the exterior. Since solutes are excluded from ice crystals, the solute concentration increases and water moves from low solute concentration to high solute concentration to achieve osmotic balance. To mitigate osmotic imbalance, freezing solutions with specific cryoprotectants are used to prevent dehydration by replacing water and stabilizing the cell. These cryoprotectants are added to cryopreservation media that may be individually formulated or bought commercially. Rapid cooling may be used for cryopreservation by the vitrification technique. Vitrification is a fast cryopreservation method that avoid ice crystals formation within the cell. Vitrification method is usually successful when cryopreservation solutions contain high concentration of cryoprotective agents and it is used routinely for the cryopreservation of gametes and embryos.

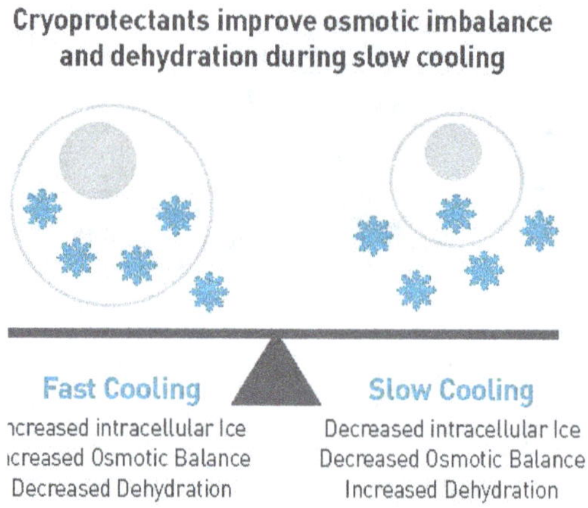

Fig. 1: A delicate balance must be maintained while freezing cells.

4.5 FREEZING METHODS AND STORAGE

There are two main methods to cryopreserve cells for long-term: controlled rate freezers and step-down freezing. The more traditional approach of step-down freezing uses a specialized freezing container that is designed to cool cells at -1°C/min in -80°C freezers (e.g. Mr. Frosty). For this method cryovials are placed in specialized freezing containers in a -80°C freezer, and the next day cells are transferred to liquid nitrogen storage. Controlled rate freezers may be programed for precise and active cooling protocols so that cells are frozen typically at

Should you Store your Cryovials in the Vaporous or Liquid Nitrogen Layer?

- *The current vogue is that the vaporous layer is better. The difference in temperature between layers is not great and the liquid phase is at greater risk for contamination.*

Recommended:

- *After 24 hours you should check one of the vials for viability and recovery.*

-1°C/min. The temperature is actively monitored while the freezer is cooled with liquid nitrogen. Some controlled rate freezers do not require any consumable cryogens. More recently, controlled rate freezers for the bench top have become available. Once cells are frozen, they may be kept in liquid nitrogen or mechanical storage as long as it is below -135°C.

Several types of storage containers for step-down cooling exist. Previous types of containers used alcohol to insulate and control cooling. Modern containers use insulation with a thermo-conductive alloy core to control the rate of freezing and they are proven to freeze at -1°C/minute. Cryovials are vessels designed for cell freezing and storage in cryopreservation solution. Most plastics and other materials might shatter or crack at low temperatures. Aliquots of cells are often made in 1ml per tube. Polypropylene plastic, screw-capped, 2ml, externally threaded vials, with labels are recommended. These vials are designed to fit easily in freezing/storage containers and they withstand ultralow temperatures. Before committing to a vessel type, make sure the specifications meet your requirements.

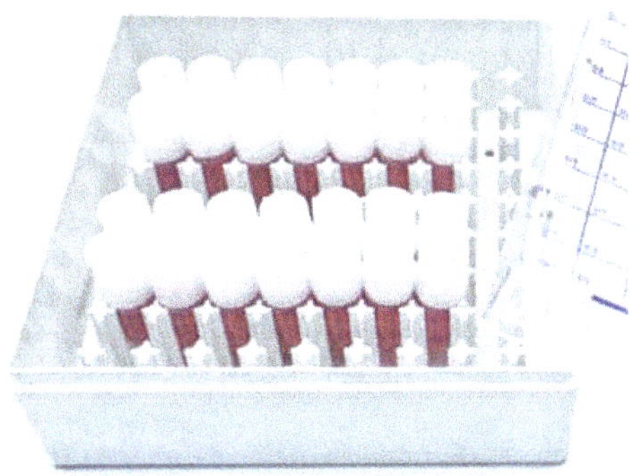

Fig. 2: Freezing mammalian cells in cryovials and freezing/storage containers

4.6 CREATING A MASTER CELL BANK AND WORKING CELL BANK

Biobanking is an important asset to both industrial and academic institutions. Successful cell banking will preserve primary cells and cell lines for research, cells for medical use, keep experiments reproducible, improve effectiveness of treatments, and keep operating costs low. Before generating a master or working stock, it is important to test incoming cells for malicious agents such as mycoplasma, bacteria, or fungi. Once cells have passed all testing during a quarantine period then they may be split for banking and further use.

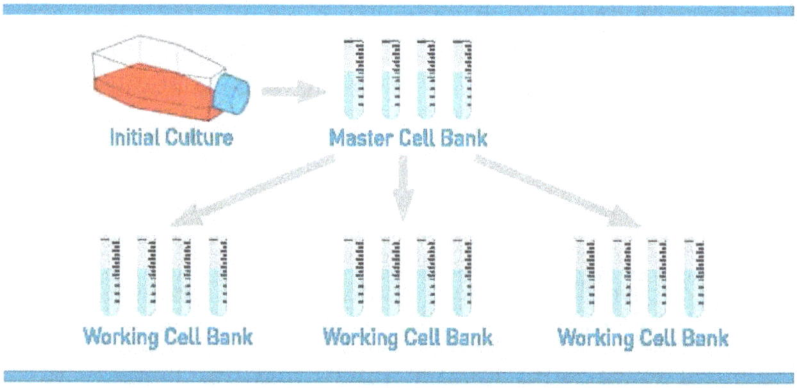

Fig. 3: Preparing a master and working cell bank

Cryopreservation of a large master stock is important in case the working stocks are lost or contaminated. The number of cells needed for experiments or treatments should be kept in mind to determine the number of master stock tubes. Typically, master tubes can be any number from 10-100. Quality tests should be repeated at this level and if the master stocks pass then they may be used for the preparation of working stocks, which are typically 20-200 tubes. Quality control tests with these stocks are recommended to determine cell viability, quality, and authentication. Working stocks should be kept at the same low passage number and experiments should be completed in a similar passage range. This way, if stocks run out then the smallest passage number may be split for more. When master stock tubes dwindle, more should be split and stored.

TIPS FOR PREPARING YOUR MASTERAND WORKING CELL BANK

- *One cryovial may contain 1-20 million cells in 1 ml. Start by freezing 1-5 million cells in 1 ml, depending on your cell type.*
- *When a healthy set of cells is ready for sub-culturing prepare your master cell bank (seed stock), consisting of 10 to 100 vials of cells. Use this seed stock to prepare two or more working cell banks (working stock), consisting of 20 to 200 vials of cells each.*

4.6.1 Choosing a Cryopreservation Media

In order to lower the solute concentration during the freezing process, cryoprotectants must be used. Cryoprotectants are compounds that protect cells from intracellular ice formation. DMSO, glycerol, ethylene glycol, and propylene glycol are all permeating cryoprotectants. Their mechanism of action involves entering the cell freely and replacing water, lowering the amount of ice formed, and acting as a secondary solvent for salts. Other cryoprotectants include high molecular weight compounds. Hydroxyethylstarch and polyvinyl pyrollidone are non-permeating ryoprotectants that are thought to stabilize the cell membrane. The type of cryoprotectant desired depends on the cell type, tissue, or organism being frozen. DMSO and glycerol are the most commonly used cryoprotectants, with DMSO being more common for mammalian cells. DMSO may be used at a concentration of up to 10% w/v (1.28M) and can be filtered for cell culture use. If single addition of DMSO media causes toxicity then multi-step addition may be used to gradually increase DMSO concentration and decrease osmotic damage. Another common cryoprotectant is glycerol. Glycerol may act as a nonelectrolyte to decrease electrolyte concentration in cell freezing solutions and is most commonly used for preservation of microorganisms, red blood cells and spermatozoa. While these agents protect cells during the slow freezing process they can also cause cell toxicity. The use of serum may help to protect the cells but this is not recommended for clinical or commercial uses that may require serumfree culture conditions. As an alternative to homebrew of cryopreservation media, commercially available media provide safe and reliable cryopreservation results. Defined freezing media, such as CryoStem™ (Biological Industries, cat. no. 05-710-1), are proven effective for human embryonic stem cell and human induced pluripotent stem cell cryopreservation. Cells frozen with this medium maintain pluripotency, attachment ability, and viability. However, even defined and safe media should be removed promptly after thawing cells to prevent toxicity.

4.6.2 Protocols of Cryopreservation

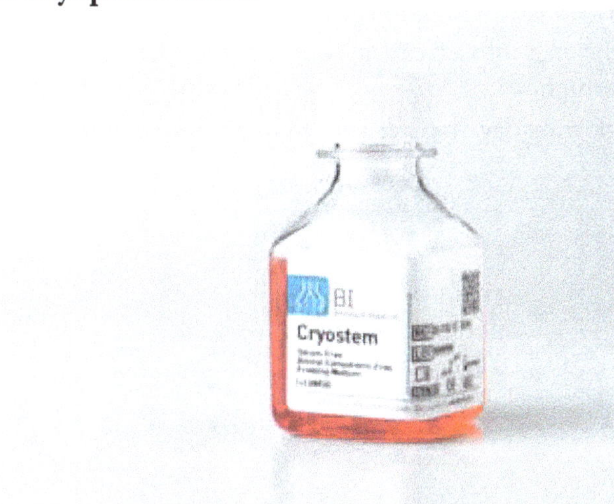

Freezing human pluripotent stem cells (hPSCs) using CryoStemTM Freezing Medium

Notes:

- hPSC may be freezed as clumps or single cells with high viability and minimal differentiation post thaw. The single cells hPSC can be thawed onto recombinant Laminin coated culture ware without the addition of Rock inhibitors. In case of using other matrices (e.g. Matrigel™) rock inhibitor is required.
- Keep CryoStemTM freezing medium on ice at all times.

  i. Aspirate hPSC culture medium.

ii. Rinse wells with Dulbecco's PBS w/o Ca & Mg (Cat# 02-023-1), or DMEM:F-12 (1:1) (Cat# 01-170-1).
iii. Add dissociation solution as desired. Cells may be detached using the enzyme and method that the culture has been routinely passaged with. In case of using collagenase, dispase or EDTA, incubate at 37°C or at room-temperature until the edges of the colonies begin to loosen from the plate. Incubation time will vary between cell lines, colony size and detachment solution used. Begin checking the culture after 3 minutes.
iv. Cells cultured on laminin may be detached using Recombinant Trypsin-EDTA Solution (Cat# 03-079-1) to yield a single cell suspension.
v. Transfer the clumps or cell suspension to a conical tissue culture centrifuge tube.
vi. Centrifuge at 200 x g for 5 minutes at room temperature. Remove and discard supernatant.
vii. Gently suspend the pellet in ice cold CryoStem™. The final volume is the number of vials desired multiplied by one ml In case of aggregates, do not break up cell clumps any more than necessary, two or three gentle pipeting motions are usually sufficient.
viii. Quickly transfer 1 ml into each cryogenic vial.
ix. Place the vials into a freezing container (e.g. Mr. Frosty™) and transfer to -80°C for overnight.
x. The following day, transfer vials to liquid nitrogen storage (vapor phase).

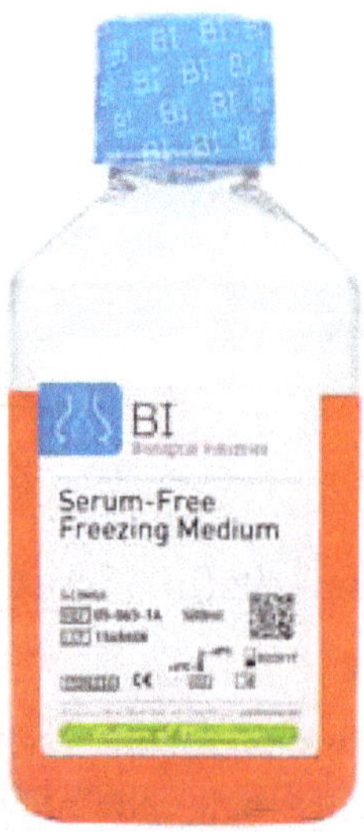

4.6.3 Cryopreservation of serum-free cultures using Serum-Free Freezing Media

On multiple cell lines, including 3T3, BGM, vero cells, MRC-5, HEK- 293, HEp-2, BSC-1, BF16-F10, BHK-21, CHO, HELA, MA-10, and mesenchymal cells.

- For freezing adherent cells, detach cells using dissociation solution according to the manual instructions. For freezing of cells in suspension skip to step 2.
- Centrifuge to pellet the cells (200-300g, 3-5 minutes).
- Suspend the pellet in ice-cold Serum-Free Freezing Medium at a concentration of 3-5 million cells per ml.
- Freeze the cells gradually (1-2°C per minute) in a freezing container (e.g. Mr. Frosty) and store them in liquid nitrogen.

4.6.4 Cryopreservation Protocols

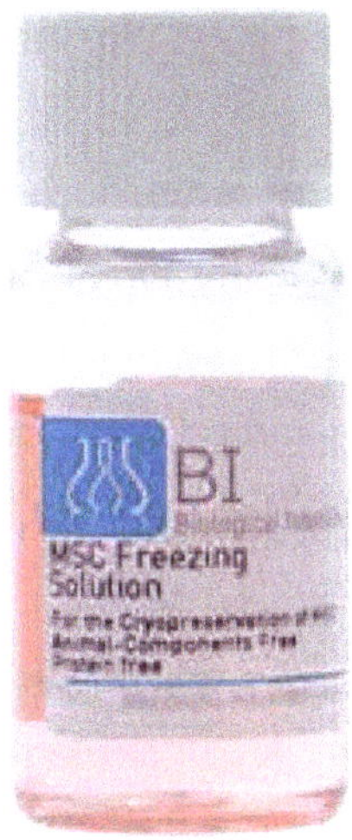

Cryopreservation of human Mesenchymal Stem Cell (hMSC) Cultures using MSC

4.6.5 Freezing Solution

- Detach adherent hMSC using MSC Dissociation Solution (Cat. No. 03-075-1) or Non-Enzymatic MSC Dissociation Solution (03-077-1) according to manual instruction.
- Pellet hMSC by centrifugation at 300-400g for 4-5 minutes. Carefully discard supernatant.
- Re-suspend the pellet in ice-cold MSC Freezing Solution at the desired concentration (recommended: 0.5-1.0x106cells/ml; 1ml/vial).
- Immediately place the cryovials in appropriate freezing container (e.g., "Mr. Frosty") and store at -80°C overnight.
- Transfer the cryovials into liquid nitrogen.

> - Always use a reliable freezing media that is serum-free and animal-component free to insure that animal products will not influence the human cells.
> - Serum is not necessary when freezing hPSCs.
> - Cryoprotectants, such as DMSO or glycerol, in freezing media help to protect cells from damage caused by ice crystal formation.

4.7 TIPS FOR BEST RESULTS WHILE FREEZING HUMAN PLURIPOTENT STEM CELLS

- Follow institutional regulation and guidelines for all hPSC work.
- Cells may be frozen as aggregates or single cells.
- It is best to cryopreserve when cells are at their maximum growth rate (log phase). Cryopreservation at ~80% confluence should work best.
- Mycoplasma testing is recommended before freezing.
- Label your vials (cell, passage, lot #, date, your name) with a marker that will withstand alcohol and liquid N2. Printed labels also work well. Keep the records online as well as a hardcopy.
- Follow your cryopreservation protocol carefully. Work quickly and step by step.
- Add the freezing media gradually. Do not allow freezing media to return to room temperature.
- DO NOT sacrifice sterility and personal safety for speed!

4.7.1 Safety Tips and Considerations :

Close and prolonged contact with liquid nitrogen during cell storage may cause serious burns or asphyxiation if inhaled. Personal protective equipment (PPE) must be worn when storing cells. PPE includes a face shield, thermal gloves, a lab coat, and closed toed shoes. Nitrogen levels should be checked regularly to make sure that complete evaporation doesn't occur. Vessels should be clear of frost buildup and any leaks. Alarm systems to monitor the performance of the liquid nitrogen vessels and the oxygen levels in the storage rooms may be installed as a precaution. Always use the correct storage vessel designed to withstand liquid nitrogen storage. Threaded cryovials should always be used to prevent exploding tubes.

DMSO is a combustible and toxic material that is easily absorbed through the skin (Cayman Chemical Company, 2017). It must be kept in a safety cabinet and appropriate PPE that includes gloves, eyewear, and a lab coat must be worn when handling. If DMSO gets on your glove, promptly remove the glove. If there is skin or eye contact, the skin must be washed with soap and the eyes must flushed with water for 15 minutes. DMSO, and soiled materials, must also be treated as chemical waste and should be disposed properly.

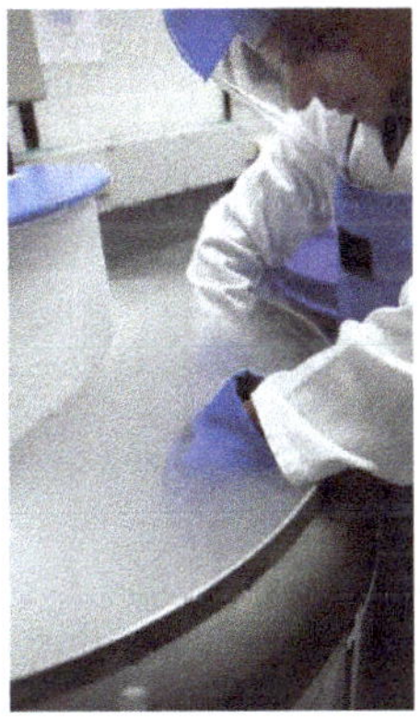

4.7.2 Troubleshooting :

The practice of cryopreservation has become routine in research. Many troubleshooting methods are employed when issues arise.The most common concern with cell freeze/ thaw is poor cell viability post-thaw. Viability issues are caused by cryoprotectant, improper cooling rate, post-freeze temperature flux, improper thaw, or incorrect cell density (Table 1). If contamination or other issues are detected in working stock cultures then return to master stock. If master stock also exhibits the same issues then the company or source of the cultures should be notified.

Cells that do not pass validation should be rejected once the source cells responsible for issues are identified. Proper maintenance of stored cells and tissues will result in reproducible experiments and treatments.

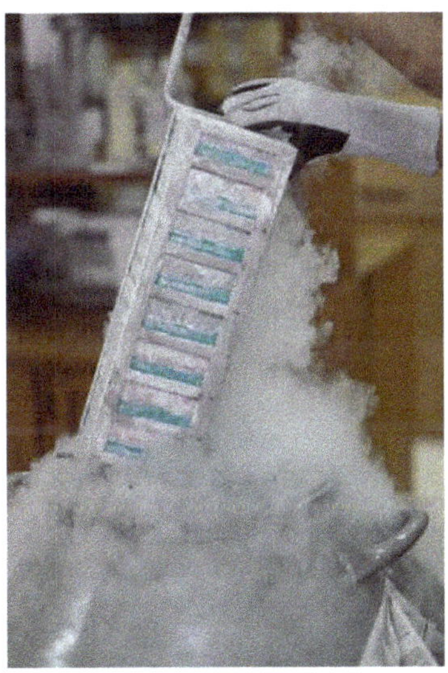

The practice of cryopreservation has become routine in research. Many troubleshooting methods are employed when issues arise. The most common concern with cell freeze/ thaw is poor cell viability post-thaw. Viability issues are caused by cryoprotectant, improper cooling rate, post-freeze temperature flux, improper thaw, or incorrect cell density (Table 1). If contamination or other issues are detected in working stock cultures then return to master stock. If master stock also exhibits the same issues then the company or source of the cultures should be notified. Cells that do not pass validation should be rejected once the source cells responsible for issues are identified. Proper maintenance of stored cells and tissues will result in reproducible experiments and treatments.

Table 1. Troubleshooting: viability issues after cryopreservation

Issue	Suggestions
Toxic Cryoprotectant	Use commercially available and defined media according to manufacturers' instructions. Remove cryoprotectant promptly after thaw. Do not allow cells to remain at room temperature in cryoprotectant media.
Improper Cooling Rate	Use a gradual cooling rate of -1°C/min. To attain this rate, use a thermally insulated freezing container or a controlled rate freezer.
Post-Freeze Temperature Flux	Maintain the cryogenic temperature of cell vials after they reach <-130°C. Keep cells on dry ice when transporting and make sure liquid nitrogen tanks are filled properly.
Improper Thawing Rate	Cells must be thawed quickly. Use a 37°C water bath or dry thawer to thaw vials.
Incorrect Cell Density	Freeze and plate cells at the appropriate density for the cell type. Cell lines should have published density requirements. Typical density per frozen vial is 1x10^6 - 10x10^6 cells/ml. Testing for optimal freeze-thaw density may be necessary.

5

CRYOBIOLOGY AND MEDICAL SCIENCES

5.1 INTRODUCTION

It is "the study of the effects of freezing and low temperatures on living organisms". While the main focus of cryobiology is on living organisms (as suggested above), cryobiological techniques have been expanded to include treatment of nonliving things as well. This addition of non-living objects has expanded the focus to not only include plants, animals, and humans, but also to include machinery and other metal objects.

Cryobiology can be many different things, including treatment, preservation, as well as, controversy. The different cryobiological techniques each have their own equipment, procedures, and ethical issues associated with them, yet are so similar. While there is much unrest about whether cryobiology is justifiable, a personal choice is still the governing force in the matter. The theory of Cryotransport and embryo cryopreservation have very high expectations for the technology of the future, and that is too much of a risk for me to take.

- A common synonym of cryobiology is "cryogenics". According to the cryogenics is defined as "the branch of physics dealing with the behavior of matter at very low temperatures, and with the production of those temperatures." As can be seen, cryogenics is a slightly more focused study of the cryobiological field dealing primarily with production methods of those freezing temperatures. Another sub-study field, if you will, is that of cryonics. Cryonics is defined as the practice of freezing humans who are not curable by current medical technology, in the hope that ways may be found to bring them back to life at some future time when ways of repairing the damage caused by the freezing process have been developed, as well as cures of the diseases or other causes of death which necessitated their cryonics suspension.
- As suggested above, cryobiology is primarily centered on the effects on those living organisms or non-living objects, although the methods and procedures

are also very important. All around the world new and improved techniques are being tested to make this a better medical and preservational procedure. Not only are the different techniques and procedures getting increased attention, but also many accompanying controversial issues are as well. Low temperature biology, or cryobiology, has the possibility to very easily and very drastically affect everyoneís lives in the future. Through rapidly approaching cryobiological medical procedures and techniques, the presence of intense ethical issues may play an important role in many everyday decisions that do not now exist.

5.2 CRYOBIOLOGY TECHNIQUES

There are many different cryobiological techniques that exist and are being used today. Such techniques include cryosurgery, embryo and gamete preservation, tissue preservation and transplantation, blood and blood product preservation, and cryotransport. Some of these have been around for so long that they are now considered commonplace medical procedures, while others are still considered experimental in the medical world.

5.2.1 Cryosurgery

Cryosurgery, a technique that has been in operation for years, still has some aspects that are being extensively tested. Much of the questions and continual testing revolves around the injury or infection that needs to be treated and hopefully cured through cryosurgery. Many of these uses of cryosurgery are widely accepted from a medical standpoint while others are not widely accepted to the point that in some clinics and medical facilities they are not provided. According to "commonly accepted uses of cryosurgery are for the treatment of many types of skin lesions (including cancer), for benign and dysplastic mucosal lesions, for uterine cervicitis and intraepithelial neioplasia, and for cardiac surgery in the treatment of tacchyarrhythmias".

Cryosurgery involves different techniques in which selective necrosis of tissues is achieved, or hoped to be achieved, through freezing at extremely low temperatures. The achievement of these low temperatures is through a very precise cooling sequence, using a specific solution called cryogen. Each of the separate techniques involves its own specific procedure in order for the surgery to be successful. Also the goal of treatment is a determining factor on which cryosurgical technique is used. According to the three major goals of treatment to choose from are improvement, cure, or palliation. Once these initial factors are chosen, the technique to appropriately achieve the goals is used.

One of these cryosurgical techniques uses a dipstick apparatus. The dipstick, is simply a cotton-tipped applicator that is dipped into liquid nitrogen and applied to a lesion until adequate necrosis has occurred. Another technique involves an open-spray apparatus by which a fine spray of cryogen is applied to the lesion. Solidified carbon dioxide can also be applied directly to the skin with a mix of acetone to treat certain types of acne. This type of treatment is called slush therapy. Perhaps the most popular technique used in cryosurgery is the cryoprobe. This technique involves a precooled metal accessory that is directly applied to the lesion. The cryoprobe is also frequently accompanied by liquid nitrogen.

In most medical facilities, these procedures are routinely done and have shown both positive and negative results. Both the positive and the negative results have been reported on numerous accounts in scientific research papers and journals. An example of negative research results comes from a study involves cryosurgical procedures done on the liver as well as analysis of damage done. They found that the formation of ice in the hepatic sinusoids expands the sinusoids, dehydrates the hepatocytes, and then leads to possible structural damage. This particular study shows that not all cryosurgical procedures are effective, at least, without causing irreversible damage to the operation subject in the process.

An example of positive research results comes from a study of other surgical techniques in comparison to cryosurgery. In this study the other treatment techniques were laser, ultrasonic aspiration, microwave surgery, scalpel, and electric scalpel. The major finding was that while all techniques have advantages and disadvantages, cryosurgery was the best method for cosmetic effect and economy. It is concluded that cryosurgery has many added benefits besides just cosmetic and economic advantages. Examples include, suitability for office, nursing home, or outpatient facility; the lack of need for general anesthesia; optional need for local anesthesia; the lack of need for operative suits; the simplicity and safety of the procedure; the lack of physical restrictions afterward; availability for pregnant patients; and suitability for those patients who are fearful of surgical procedures.

Not only has there been research-based proof that cryosurgery is only positive or only negative, but also there can easily be a mixture of both. A study shows that cryosurgery for prostate cancer can have both positive and negative results. "Hypothermic injury of the rectal wall and bladder base was demonstrated" after prostate cancer treatment with cryosurgery. However in addition to this, "no residual prostatic tissue or tumor was found on follow-up biopsy".

5.2.2 Embryo and Gamete Cryopreservation

Although cryosurgery may easily be the most frequent cryobiological technique, embryo and gamete preservation comes in a close second. This particular technique has been used for many species of animals, starting with the mouse in 1972, to the goat in 1976, and the human in 1983. Not only has this technique been used for these types of animal species, in each it has been a successful operation; successful meaning that live young have been produced from frozen embryos. In fact, this type of procedure has been so successful throughout many species of the animal kingdom, the only species that has still been unable to undergo this procedure with success is swine. So many animals from all different species have been born through this process of embryo cryopreservation that it can be estimated that tens of thousands individual organisms have survived the storage in liquid nitrogen at 196 degrees Celsius.

Along the same lines as for cryosurgery, embryo and gamete cryopreservation involves a very precise and straightforward method of freezing, unique for each organism. Also along the same lines, many of the methods involve general steps in the freezing process. The first step is to expose the embryo to some sort of cryoprotective solution, for it has been shown that this is an essential step in successful cryopreservation. Also, this step is where a lot of further testing has occurred, and disagreement has been found, on which cryoprotective solution is the best. The next step is the first step in the actual cooling process. In this step, the embryo is cooled to slightly lower than zero degrees Celsius, followed by a seeding procedure. Seeding is used for inducing ice formation under controlled conditions. There is much disagreement on whether seeding is actually necessary, although most procedures require it. The next step, is to control the cooling of the embryo to an intermediate subzero temperature. Many studies have shown successful cryopreservation at temperatures as low as 80 to 100 degrees Celsius and temperatures as high as 20 to40 degrees Celsius. As mentioned before, what is being preserved and who was performing the procedure would determine this temperature. The final step in the freezing process involves a very rapid cooling to ñ196 degrees Celsius for long-term storage. It has been estimated that if the proper freezing steps have been taken, the embryo can remain "alive" for thousands of years.

Now that you have successfully cryopreserved an embryo, you must be able to thaw it out in a proper way too. In general, the warming and thawing of the frozen embryo is directly related to the amount of freezing that had initially been done; this is specific for each organism. After warming and thawing of the embryo is complete, the cryoprotective solution must be removed, because in most cases, if handled improperly, this solution is toxic to the embryo.

Katrina Forest and her colleagues of the University of Wisconsin have been studying the effects of the cryoprotective solution on the successful preservation of embryos, and have come up with some striking possibilities and observations. They have observed that immersion in large quantities of cryoprotective solution has caused the unsuccessful cryopreservation of many of the test organisms, for example the hamster. In a study used a technique in which they dipped a thin nylon loop into the cryoprotective solution and suspended the embryo on this thin layer of liquid. This allowed them to simply dip the loop containing the embryo into liquid nitrogen for instantaneous freezing. Using this method, they have been the only group to successfully get a hamster birth from a frozen embryo. They hope that this method would help improve the successfulness of human embryo preservations in the future.

Traditionally, the cryoprotective solution that has been used was supposed simulate the body fluids of the womb, consisting of a sodium ion- rich saline solution. Until recently, this was a standard procedure. Another study, it is points out that maybe the success of any cryopreservation is directly related to the amount of sodium ions in the cryoprotective solution. In his study, tested a new solution containing choline ions (a less abrasive ingredient) on mouse eggs, and got a survival rate of 90%. This compares to the 50% survival rate of the use of conventional saline solution. Numerous other studies on improvements of the cryoprotective solution have been made and many alternate choices are now available.

5.2.3 Tissue and Blood Cryopreservation:

Cryobiology technology has recently been focused and manipulated to aid in the care and well being of ill patients, in a different way than cryosurgery. Tissues and blood of the body can be removed, frozen, stored, and then later reintroduced into the body when they are needed or desired. One particular use of this technology is done for bone marrow transplants. Until this technology, these transplants relied on time to be effective, but now cells can be frozen and stored for years and still be effective. The Cryobiology Laboratory located at the Fred Hutchinson Cancer Research Center deals directly with this type of cell storage. According to the laboratory with this technology, "donor and recipient need no longer be the same ABO blood type; the red blood cells can be removed from the marrow". In this laboratory, the cells, stem cells or bone marrow cells are stored for short-term in a controlled freezer that can cool at a rate of one degree per minute, or for long-term in a large, liquid nitrogen freezer at 180 to195 degrees Celsius. In this same article, the laboratory has reported storage of about 1200 patients cells, giving a total of 4500 cell packets that are labeled and ready to use when needed. Since this article was published, the numbers have most likely increased.

Another form of tissue preservation is that of blood. In the same manner, blood or blood products can be stored in freezers and then used when needed by a patient. Current technology has designed methods of recycling placentas and umbilical cords that would normally be discarded after birth, to be used as blood and hematopoietic stem cell sources. According to procedure of saving the placenta and umbilical cord at birth is a very simple procedure that only takes minutes, and in turn could possibly save your life or the life of your child. These cells are used for many possible reasons including stem cell matching for your child, treatment of life-threatening diseases, or even gene therapy. Umbilical cord blood transplantation also has been shown to be an acceptable alternative to bone marrow transplantation.

5.2.4 CryoTransport:

Cryotransport is the newest procedure done in cryobiology. As defined earlier cryotransport, also known as cryonics, is the freezing of the entire body. Cryotransport also can be the freezing of any organ, for example the brain. This procedure is also known as neuropreservation. A clear-cut method of how cryonics is done is not in existence, because every cryonics company or industry has their own specific procedure. One of these cryonics companies is the Alcor Life Extension Foundation. They have not only precise methods of preservation, but also precise methods of training and requirements for training. Alcor's main objective is to "give people a plausible, 'if-all-else-fails' pathway to what they have always wanted, something likely to be achieved in any case, over the next century: an indefinitely long, healthy life".

Alcor's cryotransport procedure follows these steps when a person who has joined Alcor's Life Plan, dies. First, a standby team immediately starts cooling the body and giving intravenous medications. The medications are circulated through the body by external chest compressions or by a specialized circulation apparatus. This standby team includes four technicians, one of whom is certified as an EMT, and another who is an Alcor physician. From the site of death, the cryotransport patient is then moved to the site of surgery. Here a solution cooled to around 5 degrees Celsius is circulated through the body. This procedure is known as "washout", and is performed by a team augmented by a surgeon. This washout procedure must begin within 30 minutes after cardiac arrest and end within 2 hours to ensure decreased probability of brain damage to the patient. Once the body has been stabilized at 5 degrees Celsius, the body is transported to Alcor's main medical facility where it undergoes the remaining steps. Here the body's water is replaced with a cryoprotective agent (glycerol). Over a period of a few hours the concentration of the cryoprotective agent is increased from 4% to 75%. Upon completion of this cryoprotection procedure, the body is then placed into a dewar. A dewar is essentially a large "thermos flask", which is filled with liquid nitrogen. Liquid nitrogen is used because it is relatively inexpensive, but it also requires no refrigeration, thus cutting down on energy needed. The body can then be kept, preserved in the dewar, for years.

The resuscitation of preserved people has not yet been successful, although much testing has been done. "The main problem is that when cells are frozen, water seeps out of them and collects between them. As this water turns to ice, it forms crystals which puncture cell membranes" which cause irreversible damage. The Alcor Foundation, as well as their patients, believe that there is a way in which to revive these preserved people and that it will be found in the near future. For this reason, Alcor's preservational procedures are so precise and specific.

Although, there are many other cryonics companies, Alcor is the largest in the world in both popularity and number of patients. This is why much of the general cryonics procedures are based on their research.

5.5 SUPPORT OF CRYOBIOLOGY

Cryobiology techniques have achieved much support, especially in recent years. This support has come from the public, the industries dealing with cryobiology, and also from the medical field in general. For the most part, cryobiology has achieved many things in each of these areas. Not only has cryobiology given new and effective treatments for patients who need them, it has also provided a substantial gain of knowledge in the medical field. And this, in turn, allows industries to make more money and to grow. The medical gain of what cryobiology has put into our grasp is sufficient enough for many people to support this with not only their money, but with their participation. New techniques would not be effective or even operational if it wasn't for the persons using the service, or those having faith in the possibilities presented by them.

The knowledge gained through these newly evolved procedures in the medical field has been phenomenal. For example, the different types of cryoprotective solutions has given the medical field a better handle on how to properly treat certain tissues and cells in other areas of medicine. New techniques have been developed, based on prior cryobiological work, which have also improved the quality of life, and aided in the treatment of certain illnesses. Many new transplantation techniques and processes have been gained. The applications of these processes are getting well known as cryobiology is being further explored. Many other medical procedures would not even be possible if it hadn't been for cryobiology.

For many people, the major "plus" about cryobiology is its ability to lengthen a person's life. They have six major purposes for supporting cryonic suspension of a person. The first purpose is to give a child who has "died" a chance to grow. The Society believes that science is on the road to major discoveries in child illnesses and the treatment and cure of them, and for this reason they believe that it is a gift to the family to put a child in cryonic suspension. The second purpose is a chance to continue parenthood. This purpose comes into play when a parent "dies", but is not ready to go or to be left go. The third purpose is a chance to see great grandchildren. They point out that the scientific process of increasing the maximum life span is a very time consuming process, and, at times, in order to meet great grandchildren, you would have to undergo cryonic suspension. The fourth purpose is a chance for an extended life span. Many people, just feel that they are not ready to leave this world, even though their bodies have given out. The Society says that by undergoing cryonic suspension, one can literally lengthen their life span. The fifth purpose is a chance to be born. By this they actually mean that every person should have the right to be born into a loving and caring family and at the right time. Cryobiology has allowed unborn fetuses be kept under cryopreservation for years, until they are desired by the parents. The Society also believes that this procedure can easily take the place of many abortion problems, both physical and emotional. The sixth, and last, purpose is a chance to be whole. This purpose is geared towards those who have some sort of major disability, such as, the loss of a limb, the loss of eyesight, or being a quadriplegic. Through this purpose, as with the others, the Society feels that science will eventually come up with solutions for every medical condition (including the ones mentioned above), and cryonic suspension will allow these new technologies to affect everyone's lives.

Another form of support for cryobiology is the dispersal of cryobiological information. The **Society for Cryobiology** is a group with more than 400 members, whose main goal is to do just that (1999). This organization achieves this goal in two primary ways. Yearly, a scientific meeting is held, which focuses specifically on cryobiology and related topics. Also, the Society for Cryobiology has its own journal entitled Cryobiology. These types of informative measures not only inform the public, but also those who are involved in cryobiological science.

5.6 OPPOSITION TO CRYOBIOLOGY

Not only is there much support of cryobiology, there is also much opposition. This opposition comes in all shapes and sizes. Many people oppose cryobiology in some forms, but not in others. An example of this would be, when a person undergoes cryosurgery to repair a skin lesion, but then opposes cryonics of a human being. Not only is this type of point of view common, it is understandable. The parameters of cryobiology and its different procedures are so far apart that they are hard to lump together as being similar.

One of the major oppositions to cryobiology is one of the same reasons why there is much support for it. This factor would be the ability to increase the length of a life. Many persons believe that the length of life that we are destined to should not be tampered with, and this is exactly the reason why they oppose cryobiology techniques. The ability to live "forever" is one of the major points of emphasis by cryonics and other cryo-manipulative techniques, and this brings about much of the controversy. The definition of "death" is the backbone of this controversy, because everyone has their own definition. According to the , "death" for the whole body is not what actually happens when a person is pronounced clinically dead, because the individual cells and whole organs are still alive. Using this definition of "death", they have justified cryonics as a life-lengthening process. Those with other definitions, feel that cryonics does not extend life, but rather may even extend death. Since a successful method has not yet been found to resuscitate cryonics patients, persons opposing this technique believe it is not actually a "life-lengthening" procedure.

Another very important reason for opposing cryobiology is the potential damage that may occur. Any type of cryobiological procedure affects the cells and the surrounding areas, and, thus, could cause damage. In cryosurgery, damage has been reported right along with the success of treatment for the problem, as mentioned earlier. In most cases, this damage can cause future problems that will need to be treated, producing a viscous cycle. Damage may also occur in embryo or gamete cryopreservation. Viable births from cryopreserved embryos are, by far, the minority of all births. This minority is because of poor technique, simply inactive embryos, or damage sustained by cryoprotective solutions or the freezer. Birth defects have been known to be caused by cryopreservation techniques, and the long-term implications are still unknown to us. The fact that no successful resuscitations have been performed on cryonics patients shows that damage has occurred somewhere along the way, or else that resuscitation of a frozen human being is, in fact, impossible.

The overall price of most cryobiological procedures is very expensive for the average person to afford. At the same time, cryobiological procedures are becoming more of the "norm" in many medical clinics or facilities. This has the potential for great financial problems when payment is due, because most people would not have the money. On the other hand, this great risk of financial burden on those using cryobiology has been addressed by some of the cryobiology industries. For example, has formulated payment plans suitable to those inquiring their services. Alcor has also provided the "patients" with a fund known as the Patient Care Fund, which is for the benefit of the "patient" and their family. It seems that not only has Alcor made payment plans suitable for the "patient" and their family, but they have also increased the cost of these procedures in recent years, by considerable amounts. Alcor, in specific, claims to be a non-profit organization, while at the same time is charging a fee for cryonic suspension between $50,000 and $120,000. On top of these prices, a whole-life (insurance) policy costs around $500 a year, $150 for a sign up fee, and $360 per month for annual dues. These prices, along with normal living expenses, add up to a lot of money to pay, not to mention that the annual fees extend through the length of cryonic suspension. These seemingly large prices, are, yet, another reason why not everyone is supportive of cryobiology.

Legal issues may also arise from some of these techniques and procedures. For example, what if a married couple decides to have embryos cryopreserved for later, and more convenient, use, but before that time comes, they get a divorce. This introduces the problem of who gets the embryos that used to be "theirs", but no longer can be? The custody of frozen embryos is a growing issue, and really is a legitimate question. Who is in control of stored, frozen embryos and gametes? And who gets legal rights to them?

There is also the possibility of people going to extremes with cryopreservation. By this, I mean that perhaps people will begin preserving inanimate objects such as their favorite stuffed animal or their lucky shirt, using up valuable and expensive time, energy, and space. Perhaps this is a very far-fetched idea, but it is, in fact, possible.

6

CRYOGENIC PRESERVATION TECHNOLOGY

6.1 INTRODUCTION

To ensure reproducible results and continuity in research and biomedical processes, today's scientists are faced with the task of genetically stabilizing replicable materials such as living cells and organisms, and ensuring sub-cellular components such as nucleic acids and proteins are preserved unchanged. Serial subculturing of replicable materials is time consuming and can often result in contamination or genetic drift as smaller and smaller portions of a population are selected. Improper storage and handling of non-replicable materials can lead to divergent and irreproducible research results. However, a population of living cells or a suspension of subcellular components can be stabilized by subjecting them to cryogenic temperatures which, for all practical purposes, stops time.

The process of stabilizing biological materials at cryogenic temperatures is called cryopreservation, a practical application of cryobiology, or the study of life at low temperatures. Advances in cryopreservation technology have led to methods that allow low-temperature maintenance of a variety of tissues, cell types and subcellular materials. Techniques are available for the preservation of microorganisms, tissues, primary cells, established cell lines, small multicellular organisms, complex cellular structures such as embryos, as well as nucleic acid and proteins.

The freezing process involves complex phenomena that, even after decades of research, are not fully understood. Cryobiological studies have led to speculation on what occurs during the freezing of living cells and how adverse phenomena can be overcome. Since water is the major component of all living cells and must be available for the chemical processes of life to occur, cellular metabolism stops when all water in the system is converted to ice. Ice forms at different rates during the cooling process. Slow cooling leads to freezing external to the cell before intracellular ice begins to form1. As ice forms external to the cell, water is removed from the extracellular environment and an osmotic imbalance occurs across the cell membrane leading to water migration out of the cell. The increase in solute concentration outside the cell, as well as intracellularly as water leaves the cell, can be detrimental to cell survival.1 If too much water remains inside the cell, damage due to ice crystal formation and recrystallization during warming can occur and is usually lethal.

The rate of cooling has a dramatic effect on these phenomena (Fig. 1). Rapid cooling minimizes the solute concentration effects as ice forms uniformly, but leads to formation of more intracellular ice since water has not migrated out of the cell. Slow cooling on the other hand, results in a greater loss of water from the cell and less internal ice being formed, but results in an increase in the solution effects. Cell permeability affects the rate of water loss; more permeable cells are able to tolerate rapid cooling better than less permeable cells.2 Mazur et al.3 postulated that ice crystal formation and solution effects both play a role in cell damage, and that an optimum cooling rate minimizes the effect of each. With few exceptions, a cooling rate of 1°C per minute is preferred.

6.1.1 Cryoprotective Additives

Using cryoprotective additives, or chemicals that protect the cells during freezing also minimizes the detrimental effects of increased solute concentration and ice crystal formation.

6.1.2 Commonly Used Cryoprotective Agents

The most commonly used cryoprotective agents are dimethylsulfoxide (DMSO) and glycerol, although many other additives have been used for specific purposes. Additionally, maintaining frozen cells at the proper storage temperature and using an appropriate warming rate also contribute to minimizing damage to frozen cells and tissues.

A key element of a good cryopreservation program is standardization of the processes employed. Because of the complexity of the preservation process small variations in processing and storage can lead to subtle changes in the biological materials. By standardizing the methodologies there is greater assurance that research results will be consistent and comparable. Therefore, once a successful cryopreservation regimen is established efforts should be made to carefully document the methodology.

6.1.3 Seed Lot System

To aid in maintaining the genetic stability of cultured cells, the frequency of subculturing beyond the original established culture must be minimized. When freezing cells, use a system that ensures that early passage material is always available for producing new working stock. One method of preserving early passage material is to use a seed lot system. When preparing the first frozen lot of a culture, a portion of the lot is set aside as seed material. The vials designated as seed material are maintained separately from the working stocks to ensure that they remain unused and are not handled during retrieval operations (Fig 2). When the first lot of working stock is depleted, a vial is retrieved from the seed lot and used to prepare a second working stock. This continues until all seed vials, except one have been depleted.

The last seed vial is then used to prepare a second seed lot. The second seed lot remains only one or two passages from the original material, but may be separated by many years if the lots are adequately sized.

The seed stock concept is similar to cell banking practices whereby a master cell bank that is fully qualified is developed and maintained for use in preparing working cell banks used in a manufacturing process. A similar system could also be used for non-replicable materials, such as DNA and proteins, except for a different purpose. In this case, since the material is not renewable, the seed material would be available solely for comparative purposes when changes in the working material are suspected. Sufficient material should be retained as seed stock to allow for testing when necessary. In addition to seed material, a small portion of the original lot, and even portions of working stocks, should be segregated and maintained in a location remote from all other material. This practice assures that the segregated materials are not handled during stocking and retrieval activities, and remain at a constant temperature. Backup materials should be stored in an off-site location if possible to ensure that preserved materials are not lost in the event of a physical disaster at the primary location. Using seed lots and maintaining off-site reserve material are of primary importance in ensuring continuity and longevity in any well-managed collection of biological materials.

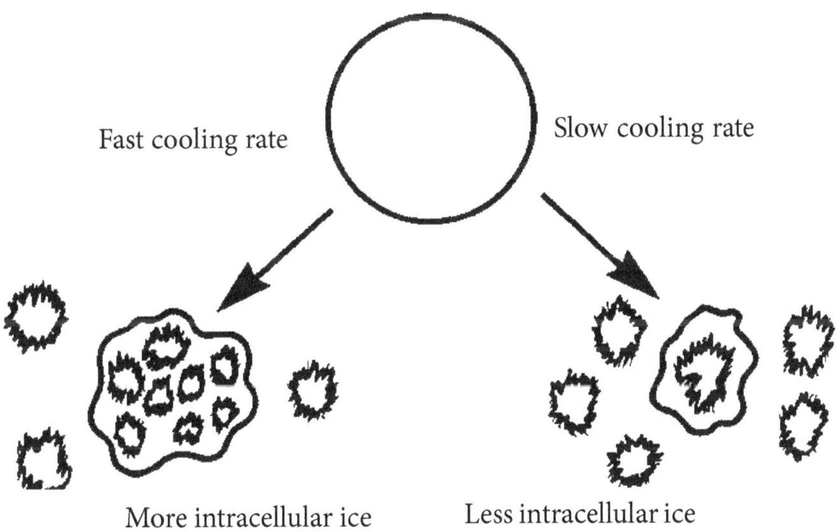

Fig. 1: Use a cooling rate of 1°C/minute and a cryoprotective agent to minimize damage due to osmotic imbalance and ice crystal formation.

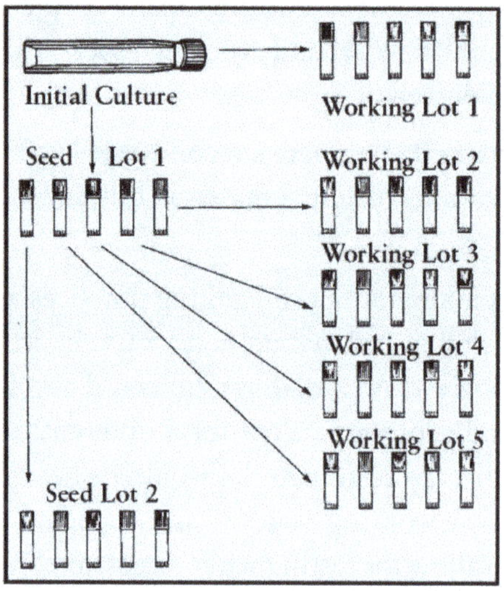

Fig. 2: Seed lot system.

6.2 CRYOPROTECTIVE AGENTS

Many compounds have been tried as cryoprotective agents, either alone or in combination, including sugars, solvents and even serum. Although there are no absolute rules in cryopreservation, glycerol and DMSO have been widely used and traditionally have been demonstrated to be the most effective agents for preserving living cells and organisms. Other cryoprotectants that have been used occasionally, either alone or in combination, include: polyethylene glycol, propylene glycol, glycerine, polyvinylpyrolidone, sorbitol, dextran and trehalose.

The need to preserve tissues and whole organs has led to the development of novel preservation methodologies that are also applicable to enhancing the recovery of frozen cells and organisms. These include variations on the concentration of cryoprotectants, and additives that protect cells against apoptosis, or programmed cell death. For many years the death of cells following freezing was assumed to be caused by events causing physical changes or damage to cells. More recently it has been discovered that more subtle events that contribute ultimately to cell death may be manageable with proper additives. Cryoprotective agents serve several functions during the freezing process. Freezing point depression is observed when DMSO is used which serves to encourage greater dehydration of the cells prior to intracellular freezing.

Cryoprotective agents also seem to be most effective when they can penetrate the cell, delay intracellular freezing and minimize the solution effects. The choice of a cryoprotective agent is dependent upon the type of cell to be preserved. For most cells, glycerol is the agent of choice because it is usually less toxic than DMSO. However, DMSO is more penetrating and is usually the agent of choice for larger, more complex cells such as protists. The cryoprotective agent should be diluted to the desired concentration in fresh growth medium prior to adding it to the cell suspension. This minimizes the potentially deleterious effects of chemical reactions, and assures a more uniform exposure to the cryoprotective agent when it is added to the cell suspension, reducing potential toxic effects. DMSO and glycerol are generally used in concentrations ranging from 5-10% (v/v), and are not usually used together in the same suspension with the exception of plant cells. The optimum concentration of the cryoprotectant varies with the cell type and for optimal results the highest concentration the cells can tolerate should be used. For some materials it may be advantageous to examine the sensitivity of the cells to increasing concentrations of the cryoprotective agent to determine the optimum concentration. The Quick-Reference Chart lists the recommended concentrations of cryoprotective agents for each group of cells and serves as a general guide to choosing the proper agent.

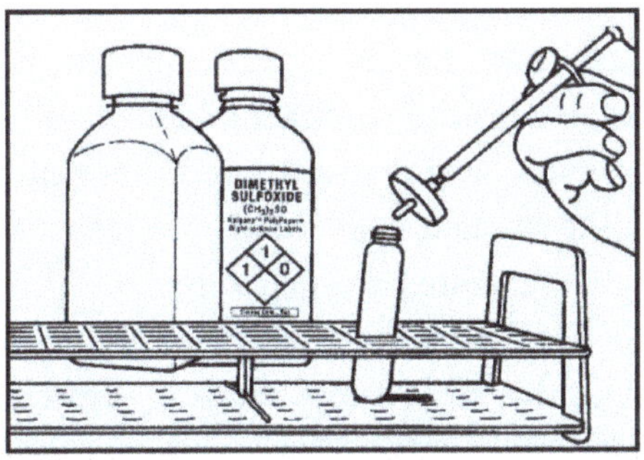

Fig. 3: Sterilization of DMSO by filtration.

Glycerol and DMSO should be of reagent grade or better, sterilized prior to use, and examined for undesirable properties. Each lot should be examined for toxic properties by exposing sensitive cells to concentrations previously used with success. Glycerol may be sterilized by autoclaving for 15 minutes at 121°C and 15 psig. Glycerol should be protected from light during storage. DMSO must be sterilized by filtration using a 0.2 micron nylon syringe filter or a Teflon® PTFE syringe filter which has been pre-washed with alcohol and rinsed with DMSO (Fig. 3).

Cryoprotective agents should be prepared in single-use volumes to minimize the risk of contamination and moisture introduction with repeated use from one container. Non-replicable materials generally do not require additives when frozen except when certain characteristics need to be protected. For example, tissue specimens that are frozen to preserve morphology may benefit from suspension in materials such as optimal cutting temperature (OCT) compound to optimize the results. Normally tissues are frozen in blocks, however combining freezing in OCT with storage in cryovials improves the handling of small tissue specimens.

6.3 PREPARATION OF BIOLOGICAL MATERIALS

How materials are processed in preparation for freezing can have an affect on the outcome of the preservation process. For non-replicable materials such as tissues, nucleic acids and proteins, the preparation process consists of ensuring that the materials are in the proper solution or freezing medium in order to maximize the intended use of the materials when recovered. However the stability and recoverability of living cells and organisms is affected by the growth conditions and pre-freezing processing. Several factors must be considered when preparing cells for cryopreservation. These include the type of cell, cell viability, growth conditions, physiological state of the cells, the number of cells, and how the cells are handled. When preparing the initial seed stock of a new isolate or cell line, the culture should be examined for identity and contaminating microorganisms at a minimum. This examination should be repeated after preservation and each time a new lot of the culture is prepared.

6.3.1 Microorganisms

Microbial cells, particularly bacteria and yeast, grown under aerated conditions demonstrate a greater resistance to the detrimental effects of cooling and freezing than non-aerated cells.1 T. Nei et.al.2 have demonstrated that cell permeability is greater in aerated cultures, and that the aerated cells dehydrate faster during cooling than non-aerated cells. Microbial cells harvested from late log or early stationary cultures also demonstrate greater resistance to the freezing process than younger or older cells. Generally, the greater the number of cells present initially, the greater the recovery. For most bacteria and yeast, approximately 107/mL cells are required to ensure adequate recovery. These can be conveniently harvested from agar slants or plates, or when greater quantities are required, grown in broth culture and harvested by centrifugation. In either case, cells are generally suspended in fresh growth medium containing the cryoprotective agent. Protists can also be concentrated by entrifugation, but are often suspended in the used medium and then diluted by adding an equal volume of fresh growth medium containing the cryoprotective agent.

Spore-forming fungi require harvesting of spores and suspension of the spores in fresh growth medium containing the cryoprotective agent. When freezing fungal spores, care must be taken not to delay the freezing process too long to ensure that germination does not occur prior to freezing.

For fungi that do not form spores, special procedures for harvesting mycelia prior to freezing must be utilized. For fungi with tough mycelia, the culture is harvested from agar growth by cutting and removing agar plugs containing the mycelia and placing the plugs into fresh growth medium containing the cryoprotective agent. Tough mycelia that do not adhere well to agar cultures are grown in broth culture and the mycelial mass is blended prior to freezing. The viability and an estimate of recovery should be determined both before and after freezing the culture. Viability is a measure of the culture's ability to grow and reproduce. For some material, such as protozoan cultures, this should include several passages to ensure stability. An estimate of the number of cells recovered can be made by several means including serial dilution, plate counts, or direct cell counting. A comparison of the counts prior to and after freezing gives an indication of the degree of recovery or the success of the preservation procedure.

6.3.2 Mammalian Cells

When preparing mammalian cells for cryopreservation, cell populations need to be adjusted to levels that ensure adequate recovery without unnecessarily growing large numbers numbers of cells. For most mammalian cells, a starting population between 106 to 107 cells/mL is optimum.7 The cell suspension should initially be prepared at a concentration twice that desired for preservation so that an equal volume of cryoprotectant (2 x cryoprotective agent + medium) can be added. Alternatively, the cell pellet can be resuspended in the cryoprotectant (1 x cryoprotective agent + medium) to the desired cell concentration. Gentle handling during cell harvesting and concentration procedures will ensure healthy cells prior to subjecting them to cold stress.

Vigorous pipeting and high-speed centrifugation should be avoided if possible. Where appropriate, the pH should be maintained by gassing with 5% or 10% C02. Factors which can affect the recovery of cryopreserved mammalian cells include: (a) type of cell, (b) the growth phase of the culture, (c) the stage of the cell in the cell cycle, and (d) the number and concentration of cells in the final suspension. Attempts to improve the viability of cryopreserved cells should consider these factors, as well as the nature of the cryoprotective agent and the freezing process.

Mammalian cell cultures are especially susceptible to contamination by other cells, such as HeLa8, and contaminating microorganisms. The species of origin of cell lines can be verified by isoenzyme analysis, karyotyping, immunological assays, or genomic analysis. These should be performed prior to and following preservation. Contamination by viruses and Mycoplasma sp. is of particular concern. A good characterization program for mammalian cell lines should include a check for contamination by bacteria, fungi, appropriate viruses, mycoplasma, and in some cases, protozoa.

6.3.4 Stem Cells

Stem cells are cryopreserved in a manner similar to other mammalian cells, with some exceptions to enhance recovery and clonogenic activity. Cryoprotection is normally afforded by using DMSO, sometimes in combination with serum, and freezing slowly is preferred. Trehalose can be used to reduce the potential toxicity associated with other cryoprotectants.10 Rapid warming is also preferred, and viability may vary depending on the cell type. Vitrification can also be used to preserve stem cells. The protocol involves suspending the cells in a concentrated mixture typically composed of more than one cryoprotectant. For vitrification, the freezing and re-warming processes are rapid to avoid ice crystal formation, and in some studies the vitrification process resulted in greater viability. DMSO, glycerol and propylene glycol have all been used successfully to cryopreserve stem cells.

6.3.5 Plant Cells

Plant cells respond to cryopreservation in a manner similar to other cells. The stage in the growth cycle from which they are harvested can affect their recovery, most optimum being late log phase. Also, cell density may play a role in recovery, the optimum cell density depending on the species being preserved. Combinations of cryoprotective agents are sometimes more effective than agents used singly. The cooling rate is important, and in many cases a two-step cooling process where the cells are held at -30°C to -40°C for a period of time before cooling to liquid nitrogen temperatures, is beneficial. This process enhances the dehydration of the cytoplasm prior to freezing. Rapid thawing is preferred, but there is evidence that slow warming is just as effective in some cases. Vitrification can also be used to preserve plant cells by using concentrated cell suspensios and rapid rates of cooling. Hardening of plants leads to greater tolerance of stressful conditions, such as experienced during the freezing process. Plants produce increased quantities of some compounds such as sugars and even glycerol which contribute to protecting the cells from osmotic stress during freezing. Undifferentiated callus tissue is often preserved in an effort to stabilize characteristics that can be affected by continued cultivation. Preservation of seeds is also an acceptable method of stabilizing plant germplasm, and the most common method is storage at low humidity and cool temperatures. However some seeds are tolerant of the increased desiccation associated with freezing and cryogenic storage, and can be stored at liquid nitrogen temperatures.

6.3.6 Viruses

Most viruses can be frozen as cell-free preparations without difficulty and do not require controlled cooling. The exceptions are those viruses cultured in viable infected cells which require controlled cooling. For cell-adapted viruses the preservation process should be applicable to survival of the host cell. When viruses are harvested from eggs, the high protein content of the allantoic fluid or yolk sac provides protection during the freezing process. Plant viruses can be preserved either in infected plant tissue or as purified virus preparations. The virus preparations are suspended in DMSO or another cryoprotectant prior to freezing. Recovery is generally best when the cooling rate is controlled, although most plant viruses will tolerate a rapid freezing procedure. Recovery of plant viruses simply involves thawing in a warm bath, followed by inoculation into the appropriate plant host.

6.3.7 Embryos

Embryos have been preserved both by controlled cooling and vitrification. Recovery depends on the stage of embryonic development, and is measured by successful implantation leading to fetal development.

6.4 GENETICALLY MODIFIED MATERIALS

Genetically modified cells and organisms can be cryopreserved in a manner similar to the unmodified host cell.

6.4.1 Non-Replicable Materials

Non-replicable materials such as whole blood, serum, tissues, nucleic acids and proteins do not usually have any special requirements for successful preservation. The materials are generally frozen without a cryoprotectant, and the freezing process can be rapid. However, the process used depends on the end-use of the material. Successful recovery of the properties of whole blood requires ryoprotection and controlled cooling, and the quality of frozen tissues can be improved by using a suspension such as optimal cutting temperature (OCT) compound.

Quick-Reference Chart (To be used as a general guide only			
Cell Type	No. of Cells	Cryoprotective	Temperature
Beteria	10^7/mL	Glycerol (10%)	-60°C*
Bacteriophage	10^2 pfu/mL	Glycerol (10%)	-80°C
Fungi Hyphae Spores	 † 10^6/mL	 Glycerol (10%) Glycerol (10%)	 -150°C -80°c
Yeast	10^7/mL	Glycerol (10%)	-150°C
Protonoa	10^5-10^7/mL	DMSO (5-10%)	-150°C
Algae	10^5-10^7/mL	Mehanol (5-10%) or DMSO (5-10%)	-150°C
Plant Cells	**	DMSO (5-10%) + Glycerol	-150°C
Animal Cells	10^6-10^7/mL	DMSO (5-10%) or Glycerol (5-10%)	-150°C
Hybridomas	10^7/ml	DMSO (5-10%) + Serum (20%)	-150°C
Stem Cells	10^5-10^6/mL	DMSO (5-10%) + Serum (20-90%)	-150°C
Sub/Non Cellular Materials Materials Plant Viruses	 ‡	 None	 80°C
Plasmids	10^6/mL	Glycerol (10%)	-150°C
Phage Libraries	‡	Glycerol (10%)	-150°C
DNA	◊	None	80°C
RNA	◊	None	80oC
Protein	◊	None	80oC
Serum	◊	None	80oC
Multicelluar			

Quick-Reference Chart (To be used as a general guide only			
Embryos	20	1,2-propandediol, glycerol or ethylene glycol	150oC
Tissues	◊	OCT	80oC
Blood	◊	Glycerol	150oC

6.4.2 Equilibration

The period of time between mixing the cryoprotectant with the cell suspension and beginning the cooling process is called the equilibration period. For most cells, equilibration should occur for at least 15 minutes, but no longer than 45-60 minutes. The cryoprotective agent may be toxic to the cells if the equilibration time is too long. For tissues frozen in OCT, not technically a cryoprotectant, the period of equilibration is generally not significant since OCT does not usually penetrate the tissue but simply provides support during freezing and subsequent sectioning of the tissue. Equilibration, which should take place at ambient temperature, allows time for the cryoprotective agent to penetrate the cells, with larger and less permeable cells as well as embryos requiring a longer equilibration period. During this period of equilibration the cell suspension may be dispensed into vials and otherwise manipulated in preparation for freezing. An optimal equilibration time should be determined empirically for the cells being cryopreserved to maximize later recovery.

6.4.3 Ampoules and Vials

A variety of small containers such as flame-sealed glass ampoules and screw-cap plastic vials can be used for storing cells at ultra-low temperatures. The most commonly used sizes are 1.2- to 2.0-mL cryovials, the size and configuration of which maximize storage capacity while retaining ease ofhandling during stocking and retrieval activities. Generally, 0.5-1.0 mL of the cell suspension is placed into each container. Several factors must be considered when selecting a container, including its cryotolerance, storage conditions, type of cells to be stored, and safety considerations.

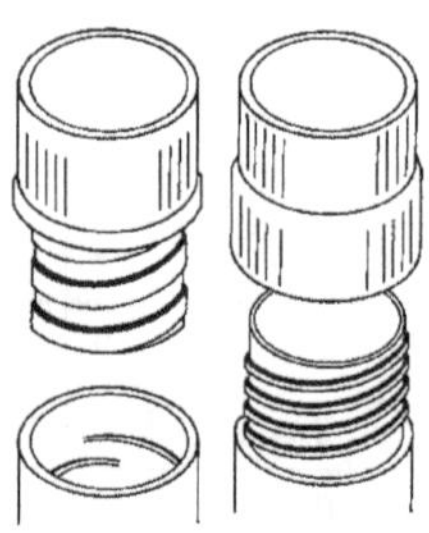

Fig. 4: Internal vs. External vial thread design

For temperatures above -100°C where low-temperature mechanical stresses are less severe, a variety of containers may be used. However, when storing material at liquid nitrogen temperatures, containers specifically designed to withstand cryogenic temperatures must be used. A variety of containers specifically designed for cryogenic use are available. Plastic vials have screw-on closures with external or internal threads (Fig. 4). The rate of warming may be affected by the type of container used since plastic vials usually require a longer warming period for complete thawing than glass ampoules. This difference in warming rate may be significant for some fastidious cells, but for most cells does not contribute to a loss of viability. Other containers can be used to store cryopreserved materials including straws traditionally used for embryos, and microtiter plates commonly used for freezing cell arrays or clones. The container of choice should be one that maximizes the ability to maintain viable material during storage, retrieval and handling. Glass ampoules may be flame sealed, however care must be taken that sealing is performed properly, since improperly sealed glass ampoules may have microchannels15 that lead to liquid nitrogen penetration over time. When these are retrieved from liquid nitrogen to ambient temperature, rapid conversion of the liquid nitrogen to vapor inside the ampoule can result in explosion of the ampoule. Plastic vials with screwtop closures are also susceptible to liquid nitrogen penetration,16 and while the explosion potential is minimized, liquid can spray from the cap/vial interface with potential dissemination of the vial contents during warming and handling.

Storage by immersion in liquid nitrogen is not advised.

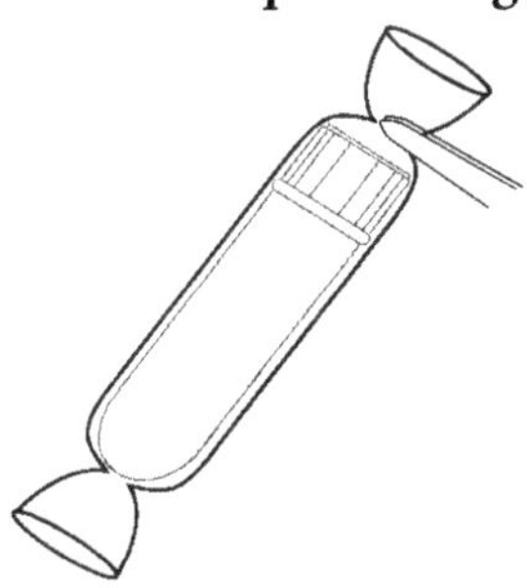

Fig. 5: If plastic vials are stored in the liquid phase of liquid nitrogen, the use of Thermo Scientific NuncCryoFlex is strongly recommended .

6.4.4 Rate of Cooling

Once the cells and the cryoprotectant have been combined and dispensed into vials, the next step is to cool the suspension. The rate of cooling is important since it affects the rate of formation and size of ice crystals, as well as the solution effects that occur during freezing. Different types of cells may require different cooling rates, however a uniform cooling rate of 1°C per minute from ambient temperature is effective for a wide variety of cells and organisms. Generally, the larger the cells, the more critical slow cooling becomes. Most bacteria and spore-forming fungi will tolerate less-than-ideal cooling rates and can be frozen by placing the material at -80°C for a period of time. More fastidious bacteria and non-sporulating fungi require more uniform rates of cooling. Protists, mammalian cells and plant cells often require even greater control of the cooling rate including special manipulation to minimize the detrimental effects of undercooling and the heat liberatedduring the phase change from water to ice.

Despite the control applied to the cooling of cells, most of the water present will freeze at approximately -2°C to -5°C. The change in state from liquid to crystalline form results in the release of energy in the form of heat; this is known as the latent heat of fusion. Warming of the sample occurs until the equilibrium freezing point is reached, at which temperature ice continues to form. To minimize the detrimental effects of this phenomenon, undercooling must be minimized by artificially inducing the formation of ice. This can be accomplished by seeding the suspension with ice or some other nucleating agent, or by rapidly dropping the temperature of the external environment to encourage ice crystal formation. To achieve uniform, controlled cooling rates, use a programmable-rate cell freezing apparatus. Simple units allow only the selection of a single cooling rate for the entire temperature range. More sophisticated units, however, allow a selection of variable rates for different portions of the cooling curve. Less costly and easier-to-use systems are available for simulating a controlled-rate cooling process by placing the vials in a mechanical freezer at -60°C to -80°C. In order to accomplish a uniform rate of cooling, the vials must be placed in specially designed containers.

Thermo Scientific Nalgene Mr. Frosty 1°C freezing container (Cat. No. 5100-0001) provides a simple-to-use system designed to achieve a rate of cooling very close to 1°C per minute. (Fig. 6) Typical cooling rates for homemade freezing systems lead to uncontrolled cooling that averages 1°C per minute but the cells actually experience more rapid rates of cooling during some parts of the cooling curve.16 Homemade freezing systems are also non-repeatable. Nalgene® Mr. Frosty® eliminates the need for direct immersion in an alcohol bath. This feature eliminates the potential for contamination due to wicking of the alcohol, as well as the presence of residual alcohol on the exterior of the vials. During handling at colder temperatures, the presence of alcohol on the vials makes the vials colder to the touch and extremely slippery.

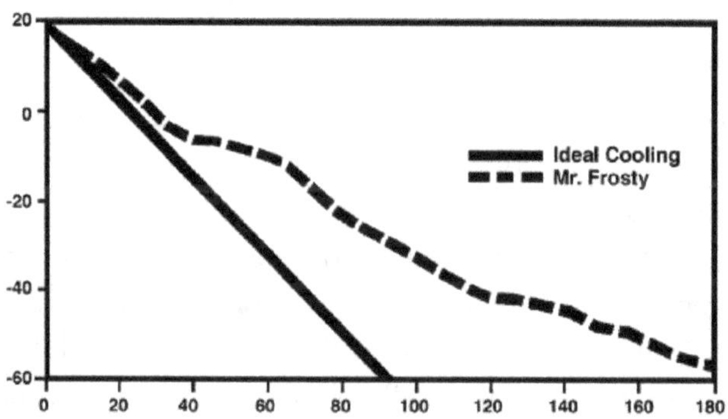

Fig. 6: The ideal cell cooling rate and the cooling rate provided by Nalgene Mr. Frosty 1°C freezing container.

Nalgene

Cryopreservation of embryos requires even greater control of the process because of their multicellular structure. In addition to the controlled rate freezing commonly used for single cells, a vitrification process is also used for preserving embryos.17 This requires suspending the embryos in a highly viscous solution, and rapidly cooling the suspension to eliminate the formation of ice crystals. The resulting frozen mass is a vitreous glass that requires storage at liquid nitrogen temperatures. If the storage temperature rises above -130°C ice crystals will form resulting in damage to the embryo.

6.5 STORAGE

When the sample has been frozen for 48 hours, a vial should be thawed to determine whether the cells are viable and able to establish a cell population, (i.e. if they survived the freezing procedure).

See Determination of Recovered Cells, below. The temperature at which frozen preparations are stored affects the length of time after which the material can be recovered. The lower the storage temperature, the longer the viable storage period. Ultimate stability of frozen cells cannot be assured unless the material is maintained below -130°C.18 Some bacteria and spore-forming fungi may tolerate storage temperatures of -60°C to -80°C for long periods of time. However, more fastidious cells, such as mammalian tissue cultures, hybridomas and stem cells must be maintained below -130°C to assure long-term stability. It has been demonstrated that some cells survive for less than one year when stored at -80°C.1 For ultimate security and maximum stability, living cells and embryos should be stored in liquid nitrogen freezers. However, there are risks associated with immersing vials directly into liquid nitrogen, as discussed previously. Liquid nitrogen units that provide all-vapor storage are ideal as long as the working temperature at the opening of the unit remains below -130°C.

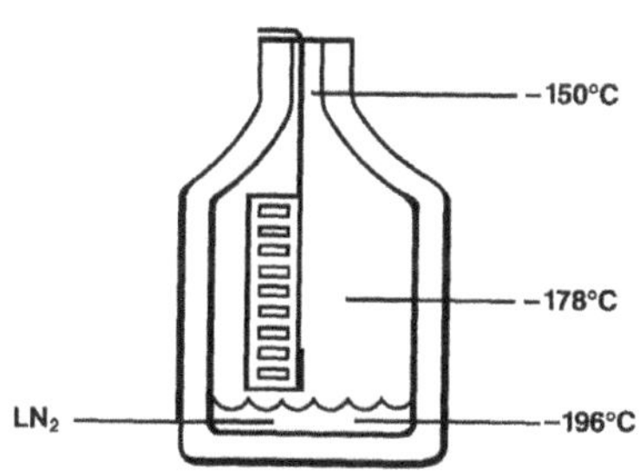

Fig. 7 Recommended temperatures in liquid-nitrogen storage system.

To assure that a liquid nitrogen freezer maintains the proper working temperature, the volume of liquid nitrogen in the unit should be adjusted to a level that results in a temperature of -150°C just above the stored material when the lid of the unit is removed19 (Fig. 7). An adequate working temperature can be attained in most liquid nitrogen freezers; however, the design of some models requires that the amount of liquid nitrogen necessary to attain the proper working temperature will reduce the amount of usable storage space. If vials are to be immersed in the liquid phase of liquid nitrogen, they must be correctly sealed in Nunc™ CryoFlex™ tubing to prevent penetration by liquid nitrogen. Improper use may cause entrapment of liquid nitrogen inside the vial and lead to pressure build up, resulting in possible explosion or biohazard release. Liquid phase LN penetration can also be a source of contamination for submerged samples not properly protected by Nunc Cryoflex tubing. In most cases, vapor phase storage at –130°C is adequate and avoids the hazards of liquid phase storage. Mechanical freezers that cool to -150°C are also available. Improper handling of material maintained at cryogenic temperatures can have a detrimental effect on the viability of frozen cells. Each time a frozen vial is exposed to a warmer environment, even briefly, it experiences a dramatic change in temperature. Storage systems should be designed to avoid exposure of stored material to warmer temperatures, as well as minimizing prolonged exposure of personnel during specimen retrieval. Box stacking systems (i.e. stainless steel racks) necessitate exposure of boxes at the top to warmer temperatures when retrieving boxes at lower temperatures. When box stacking systems are used, maintain a small number of vials of each preparation in the top box of the rack and store the remaining vials of each preparation in lower boxes. By doing this, a vial of one preparation can be retrieved without exposing all vials of any particular culture or lot.20 To maximize the available space in liquid nitrogen freezers and minimize exposure of material during retrieval, use small storage boxes or aluminum canes. Press the vials onto the canes, putting no more than one lot of one culture on each cane. Canes provide a flat surface for coding their position and easy identification during retrieval. Place the canes into cardboard or plastic sleeves to eliminate the potential for vials to fall from the canes. (Fig. 8) When retrieving vials from canes, the cane should be lifted only to a level that exposes the first available vial, without removing the remaining vials from the working temperature of the freezer.

Fig. 8: Clear plastic sleeve allows identification of vials in an aluminum cane.

6.6 RECONSTITUTION (THAWING)

For most cells, warming from the frozen state should occur as rapidly as possible until complete thawing is achieved. To achieve rapid warming, place the frozen vial into a 37°C water bath. Remember, material frozen in plastic vials will take longer to thaw than that in glass ampoules, and sometimes gentle agitation of the vial during warming will accelerate the thawing process. Care must be taken, however, not to vigorously agitate vials containing fragile cells such as protists and mammalian cells. As soon as the contents of the vial have been thawed, remove the vial from the water bath. To minimize the risk of contamination during reconstitution, disinfect the external surface of the vial by wiping with alcohol-soaked gauze prior to opening. Immediately transfer the contents of the vial to fresh growth medium following thawing to minimize exposure to the cryoprotective agent. For most cultures, the entire contents of the vial may be placed into fresh media, however further dilution may be necessary for cell lines. It is recommended that the cell suspension be centrifuged at 100 x g for 10 minutes after initial dilution, the supernatant removed, and the cells resuspended into fresh growth medium to remove residual cryoprotective chemicals. Some materials that are not sensitive to the cryopreservation process may tolerate thawing and re-freezing. Most replicable cells will not tolerate refreezing unless they are in a resistant form such as a spore. However, for non-replicable materials such as serum, nucleic acids, and proteins, thawing and re-freezing may be acceptable. Keep in mind that each time an aliquot is thawed and re-frozen subtle changes may occur in the character of the material that could impact future use. An alternative to thawing and refreezing is to store material in smaller aliquots for single use.

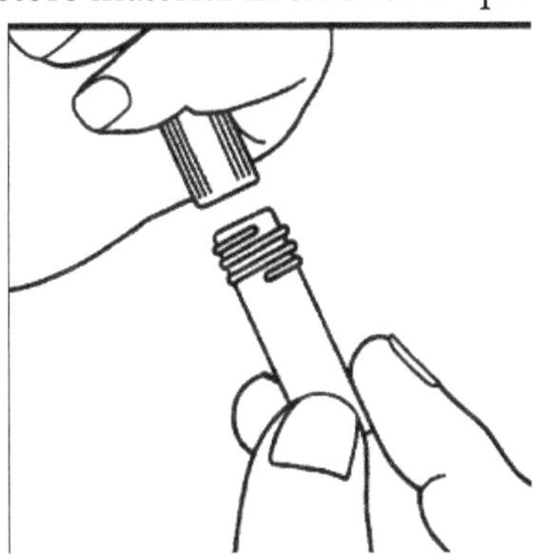

Aseptic technique minimizes possibility of contamination.

6.7 DETERMINATION OF RECOVERED CELLS

Methods used to estimate the number of viable cells recovered following freezing depend on the type of material preserved. Visual inspection alone can be deceptive, and although staining and dye exclusion are effective in determining the presence of viable cells for most mammalian cells, they do not indicate an ability to establish the cell population. For microbial cells, serial dilution and plate counts are effective in quantifying the population of cells recovered. Although there may be some vial-to-vial variation within a given lot, with constant storage conditions the number of recovered cells will generally be the same in all vials. Vial-to-vial variation may be an indication of problems occurring during storage and handling. For stem cells the recovery is determined in the same manner as for other mammalian cells by estimating the number of viable cells. In addition measuring the differentiation capacity and clone forming capability are also important in assuring complete stem cell recovery10. Embryo recovery can be determined via morphological examination, and verified by implantation and fetal development. Recovery of non-replicable materials is determined by acceptable results following intended use.

6.8 INVENTORY CONTROL

Appropriate record keeping is important in any laboratory and there are a number of methods available for keeping records on cry preserved materials.20 When establishing your own method, keep in mind that there is key information which will be important for future use:

- The preservation methodology used;
- The location and identification of the stored material;
- Preservation date; and
- Number of passages for replicable material.

The item number should be linked to associated data for that material, and for some purposes each container may require a unique identifying code linked to specific information for that particular aliquot.

Identification begins with proper labeling of the storage container. The label information should include a name or identification code for the frozen material, as well as a lot number. The information on the label should be kept with the inventory records that include the location code for each vial. These records can be maintained as paper documents, or preferably as electronic files. Duplicate inventory records should be maintained in a location separate from working records. Locator codes should be specific enough to allow rapid and easy retrieval for a specific lot and should include freezer unit number, a code for a freezer section or inventory rack, a box or canister number, and possibly even a grid spot within the box or a cane number when canes are used. Detailed locator codes minimize hunting for material which risks warming the freezer unit, exposure of other materials to warmer temperatures, and prolonged exposure of laboratory personnel to extremely cold temperatures.

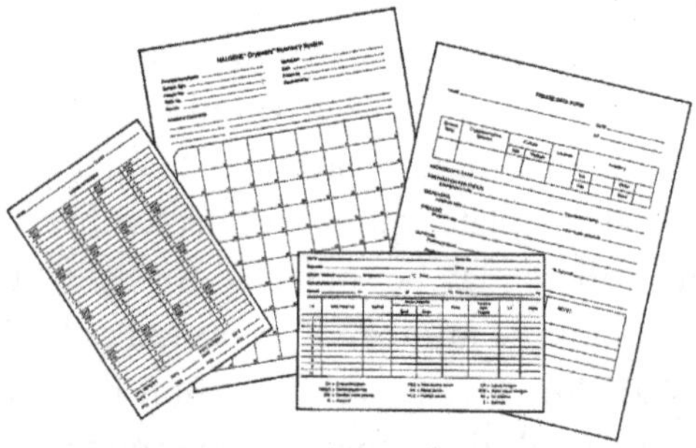

Accurate record keeping specifically for cryopreserved material. Full-size copies of these records are printed at the end of this manual and may be duplicated.

6.9 BIOLOGICAL MATERIALS MANAGEMENT

Despite its use in stabilizing living material, low temperature preservation can stress cells. Care must be taken to ensure that preserved material remains unchanged following preservation. A good preservation program should include effective characterization and cataloguing programs, both of which combined, provide optimum biological materials collection practices. There is little use in preserving material that is of little value, inadequately characterized, contaminated or misidentified. The first step in maintaining a collection of biological materials is to assess the material to be preserved to ensure that it is of use and is worth keeping. This practice should continue throughout the collection process to minimize the accumulation of preserved material to unmanageable levels. To avoid duplication of collection materials, a system of identification of each item should be established.

This can be done by devising unique numbering systems or number/letter combinations. Each identification number should be cross-referenced to other information about the biological material. Preservation of living cells ensures stability but does not correct any problems already present in the material.20 All material to be preserved should be examined thoroughly for contamination, proper identification, and other key characteristics unique to the cells, prior to preservation. Since freezing can stress cells and handling exposes cells to the risks of contamination, characterization must continue after successful preservation is accomplished. Each time a new lot of frozen material is prepared, complete characterization of the material should be carried out. Cataloguing and data record keeping are important aspects of all biological material collection programs. Cataloguing ensures that duplication of material does not occur and is especially useful when collection material is to be made available to others. Maintaining records on data generated during the characterization and preservation of collection materials ensures that any future problems can be adequately addressed. An important aspect of good biological materials management is constant assessment of the usefulness of the material, and removal of materials that are no longer needed.

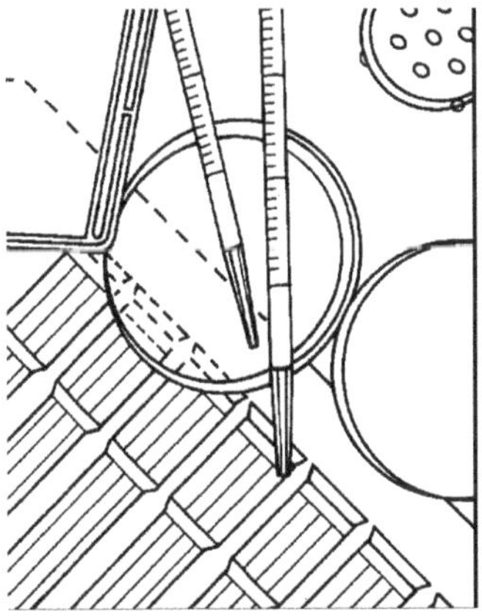

There are a number of tests you can perform to correctly identify cell lines. Correct identification 8 is important to prevent cross-contamination.

6.10 SAFETY CONSIDERATIONS

Safety precautions must be observed throughout the preservation and maintenance process. All work with hazardous cultures should be performed under proper containment, and U.S. Public Health service Biosafety Level guidelines should be adhered to at all times.

Human and other primate cells may contain adventitious viral agents that require special handling, and all primate cells that have not been thoroughly characterized should be handled at Biosafety Level II. At this level, laboratory staff must have training in handling pathogenic agents and work under the direction of a competent scientist. Access to the laboratory must be limited and biological safety cabinets must be used for large-volume work or when aerosols are generated.

Low-temperature storage of cells presents unique hazards that necessitate safety precautions. Cryogenic temperatures can result in exposure of personnel to extremely cold conditions, and precautions must be taken to protect personnel during operations in liquid nitrogen freezers. Insulated gloves and long-sleeved laboratory coats or other garb protect the skin from exposure. It is extremely important to wear a full face and neck shield when working in the liquid portion of a liquid nitrogen freezer. As noted previously, improperly sealed glass ampoules may explode when retrieved from liquid nitrogen. To minimize the risk of potential explosions, leave vials retrieved from the liquid phase of the freezer in the vapor phase of the same freezer for a minimum of 24 hours. A face shield that provides neck protection should be mandatory when retrieving vials from liquid nitrogen. The use of Nunc Cryo Flex is also strongly recommended. Special precautions must be taken when working with hazardous biological materials at liquid nitrogen temperatures.

Always thaw and open vials containing hazardous material inside a biological safety cabinet.

Be repaired for exploding and leaking ampoules/vials. Broken ampoules in a liquid nitrogen freezer are a potential source of contamination and contaminants may survive, despite the extremely cold temperatures. When a liquid nitrogen freezer becomes contaminated, the entire unit should be decontaminated after warming to room temperature. When closing down a liquid nitrogen freezer that is not obviously contaminated, remove all material to be retained, warm the unit to room temperature and disinfect it prior to further handling.

6.11 STEP-BY-STEP FOR CULTURED CELLS (NOT STEM CELLS)

- Harvest cells from late log or early stationary growth. Scrape cells from the growth surface if they are anchorage dependent. Centrifuge broth or anchorage independent cultures to obtain a cell pellet, if desired.
- Prepare presterilized DMSO or glycerol in the concentration desired in fresh growth medium. When mixing with a suspension of cells, prepare the cry protective agents in twice the desired final concentration.

- Add the cryoprotectant solution to the cell pellet or mix the solution with the cell suspension. Begin timing the equilibration period.
- Gently dispense the cell suspension into vials.
- Begin cooling the cells after the appropriate equilibration time.
  i. Uncontrolled cooling—place the vials on the bottom of a -60˚C freezer for 90 minutes.
  ii. Semi-controlled cooling—use Mr. Frosty freezing container to freeze the vials in a -70˚C freezer.
  iii. Controlled cooling—use a programmable cooling unit to cool the cells at 1°C per minute to -40˚C.
- Remove the cells from the cooling unit and place them at the appropriate storage temperature.
- To reconstitute, remove a vial from storage and place into a water bath at 37˚C. When completely thawed, gently transfer the entire contents to fresh growth medium.

7

FUTURE OF CRYOPRESERVATION

7.1 AN INTRODUCTON: FUTURE OF CRYOPRESERVATION

The preservative effects of low temperature on biological materials have been long recognised, and cryopreservation is now widely used in biomedicine, including in organ transplantation, regenerative medicine and drug discovery. The lack of organs for transplantation constitutes a major medical challenge, stemming largely from the inability to preserve donated organs until a suitable recipient is found. Here, we described the latest cryopreservation methods and applications. We describe the main challenges—scaling up to large volumes and complex tissues, preventing ice formation and mitigating cryoprotectant toxicity, and discuss advantages and disadvantages of current methods and outline prospects for the future of the field.

Cryopreservation is the storage of biological material at low temperatures. Since ancient times, it has been known that biological material can be preserved longer at low temperatures. Indeed, archaeological findings indicate that as early as 2000 BC, icehouses were used throughout Mesopotamia to store foods. The preservative effect of cold was also a topic of interest to the early experimentalists of the seventeenth century, most notably Boyle, who commented on the ability of ice to preserve human bodies and made several attempts to freeze and revive live animals, discovering species of frogs and fish that could survive encasement in ice.

- Whereas Boyle could only speculate on the nature of cold, the intervening centuries have revealed its preservative effect lies in depriving biological systems of the thermal energy required for normal molecular motion and metabolism, in turn slowing cellular processes and decomposition. The ability to reliably generate the extremely low temperatures required for long-term preservation, typically below – 100 °C, came with the development of cryogenic technologies at the turn of the twentieth century. The modern cryopreservation of living systems, in the sense of biotechnology, can be traced to the discovery of the first effective cryoprotective agents (CPAs), otherwise known as "cryoprotectants", in the 1940s. Notably, Lovelock provided crucial early insights into the origins of cryoinjury and the action of CPAs. Later work by Mazur pioneered the application of quantitative models to describing cell changes during cooling and paved the way towards theoretical approaches to studying cryopreservation.

- The discipline of cryopreservation is now well established as a practical means of storing living cells and tissues and has grown to find applications throughout biology and medicine. As this chapter will highlight, cryopreservation now holds the potential to strongly benefit several areas of medicine by increasing the ease with which therapeutic cells, tissues, and organs can be stored.

7.2 CHALLENGE OF LOW TEMPERATURES

Before discussing the medical applications and opportunities of cryopreservation, we should first consider the physiological effects of low temperatures, and the challenges they present.

Human cells have evolved to operate within an extremely narrow temperature range. Prolonged or extreme exposure to temperatures below the normothermic range of 36–38 °C is frequently harmful, as evidenced by hypothermia and frostbite. Despite this, sub-zero temperatures are not intrinsically pathogenic. The damage associated with low temperatures is, for the most part, the result of uncontrolled transitions between normothermic and low-temperature regimes. Therefore, the successful storage and revival of cells depend critically on the ability to induce and reverse low-temperature states in a controlled manner that minimises or ameliorates transition-related damage. Although this requirement is easy to state, it is in practice the central challenge of cryopreservation. How does the transition to low temperatures

Cryopreservation is an interdisciplinary endeavour between medicine, biology, bioinformatics, chemistry and physics. The main challenges still to overcome are scaling up current methods to larger volumes and complex tissues. The larger the organ, or tissue volumes to be vitrified, correspondingly more time is required to cool and warm the organ. Not only thermal conductivity is an issue here but CPA viscosity limits for perfusion systems play a role. The protocols for cell lines, or even small tissues, such as sperm, eggs or corneas, cannot be replicated in larger human organs, which necessitate toxically high CPA concentration to inhibit ice formation during the longer time spent between the melting temperature and the glass transition temperature, and of course gives more time for toxic insults to accumulate. The best techniques to get around these problems in small tissues use combinations of CPAs to reduce toxic effects of any single agent, using CPAs with weak water interactions to minimise disruption of hydration layers around biomolecules, using CPAs with mutual toxicity neutralisation effects, and damage cells? Although numerous routes exist, much of the damage is driven, either directly or indirectly, by ice formation. In this context, *nucleation* denotes the first step in crystallisation as it initiates the growth of ice crystals. At typical rates of cooling, ice first forms in the extracellular spaces of tissues (Fig. 1). Ice excludes solutes, meaning that the extracellular solute concentration increases as water is removed from the extracellular fluid to form ice crystals. The resultant osmotic gradient across the cell membranes works to drive water out of the cells by osmosis, dehydrating cells and increasing the intracellular solute concentration. This increased solute concentration depresses the intracellular freezing point, preventing intracellular ice formation. This process can damage tissues through two routes: (1) solutes such as salts, which are maintained at safe concentrations under normal conditions, become more concentrated to induce osmotic stress; (2) extracellular ice directly damages tissues, either by puncturing or crushing cell membranes or disrupting extracellular structures. However, the relative significance and interdependence of these two routes are still not entirely resolved. Cryopreservation utilises either slow cooling, in which the sample is frozen at a controlled rate to allow water to flow out of the cell and prevent intracellular ice formation, or vitrification in which a high freezing rate and/or high CPA concentration prevents ice formation in the sample. Ice growth can be further controlled by ice nucleation inhibitors, controlled ice nucleation, ice growth inhibitors or ice recrystallization inhibitors. Devitrification occurs when a vitrified sample is warmed too slowly, resulting in ice formation. Apoptosis inhibitors may also be utilised to prevent cells from dying after cryopreservation from stress-induced apoptosis.

reducing penetrating CPA concentrations by adding non-penetrating CPAs and ice blockers. Nonetheless, little is known about the mechanisms at work. The challenges in cryobiology are not insurmountable. Future research will focus on ever more complex ways to prevent ice formation and mitigate cryoprotectant toxicity; novel cryoprotectants which exert disproportionately large cryoprotective effects compared to their concentration, in silico molecular modelling, and enhanced understanding of the processes that occur during cryopreservation will all be employed. One could envision a universal cryoprotectant solution, suitable for use in a range of tissue types, and physical advancements enabling high cooling and warming rates, or the manipulation of ice formation for large volume vitrification or freezing. Whilst the concepts have been long known, the dedicated field of cryobiology dates back only around 70 years. In that time, it has advanced from freezing spermatozoa using glycerol, to vitrifying tissues, and even small organs using complex multi-component solutions. This is remarkable progress given that cryopreservation is as yet a relatively niche field of study, without garnering much attention in schools or undergraduate courses and utilising a fraction of the funding which is allocated to other causes. As such, there are still many opportunities that lie ahead, from short-term improvements in transplantation biology, to ambitions that may once have been viewed as science fiction, such as the building of organ banks or long-term suspended animation.

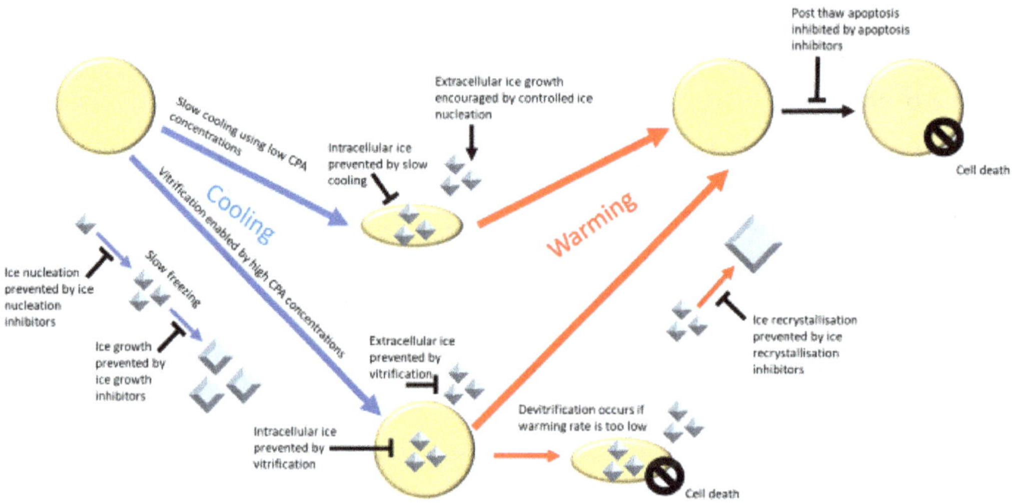

An overview of the mechanisms of cryopreservation and CPAs.

.If a sample is cooled too quickly for water to exit a cell, intracellular ice formation can occur. Intracellular ice crystals are much more damaging than their extracellular counterparts as they are capable of disrupting the cell's internal structures, like lysosomes. Intracellular ice formation is usually, but not always, lethal. Factors affecting lethality include the size of the ice crystals, their intracellular location, their mechanism of growth, their shape (cubic vs. hexagonal) and the warming rate. To reduce the extent of ice-related damage, cryopreservation protocols usually employ CPAs. CPAs act to reduce the aforementioned damage via a number of actions, many of which have not yet been fully identified. Some main effects are those of reducing the concentration of electrolytes [5] and also reducing ice formation by hydrogen bonding with water molecules to prevent them from bonding to ice crystals. CPAs capable of crossing cell membranes are of particular value. The use of such molecules, termed penetrating CPAs (pCPAs), can substantially reduce cryoinjury. The discovery of pCPAs dates back more than 70 years to when the cryoprotective effects of glycerol were discovered. In their landmark study, Polge et al. discovered that glycerol improved the revival of spermatozoa from sub-zero temperatures. Glycerol remains a commonly used pCPA, with other notable examples being propylene glycol (PG), ethylene glycol (EG) and dimethyl sulfoxide (DMSO).

The cellular protection afforded by CPAs has made the cryopreservation of cells and simple tissues a routine procedure; however, there are still several challenges to be surmounted in the preservation of complex tissues and organs. Work to categorise these challenges has identified six broad areas of need. These include:

- Improved control over ice formation;
- Methods to reduce cpa toxicity, to allow their use at higher concentrations;
- The elimination of thermo-mechanical stresses;

- Methods to prevent or reverse ischemic injury prior to or following preservation;
- The prevention of chilling injuries; and
- The development of safe protocols for organ revival.

7.3 MEDICAL NEED FOR CRYOPRESERVATION

7.3.1 Organ and Tissue Preservation

Cryopreservation has multiple and important applications, particularly in medicine. The fact that it can significantly slow down all biochemical reaction kinetics renders cryopreservation highly attractive as a means to preserve organs and therefore facilitate the transplantation process. The lack of organ availability constitutes a major challenge and a significant medical burden for society. According to the World Health Organization (WHO), only 10% of the worldwide need for organ transplantation was met in 2010. The lack of transplantable organs stems partially from a shortage of suitable donated organs, but more importantly from the lack of preservation capability. Although the number of transplanted organs is much lower than what is actually needed worldwide, it was estimated that approximately two thirds of potential donor hearts are discarded. Moreover, up to 20% of potential donor kidneys end up being discarded in the USA, and up to 50% of donated pancreases in the UK, primarily due to the limited time available to find suitable recipients. This extensive waste could be largely prevented by increasing the time through advancing preservation methods, especially cryopreservation, which is currently the only true long-term preservation solution.

The growing shortage of donor organs has become a central problem limiting the effectiveness of transplantation. Advancing preservation technologies can ultimately benefit many thousands of people worldwide by increasing local and global access to transplantation, by improving transplantation outcomes and stimulating progress in related areas such as immune tolerance induction and xenotransplantation. Prolongation of preservation times would enable transportation of organs to wider geographical areas, comprehensive evaluation of donor tissues and better graft characterisation. Notably, the potential public health impact of improved organ banking capability would benefit many areas of medicine by the banking of organs as well as other complex tissues, improving not only transplantation but also tissue engineering, trauma medicine, oncofertility, basic biomedical research and drug discovery.

7.3.2 Pharmaceutical Research

The ability to bank large quantities of tissues and cells is of immense interest for the discovery, development and evaluation of drugs. In particular, drug discovery and testing could benefit from cryopreservation of tissue slices. Tissue slices represent a powerful in vitro tool for studying biological processes as they closely resemble the organ from which they were derived, containing all specific cell types present in the organ. Much effort has been put into optimising the preparation and culturing techniques of various tissue slices such as liver, brain, intestinal and kidney slices. Under optimal conditions, tissue slices can be kept viable for at least 24 h after thawing. The cryopreservation of tissue slices greatly facilitates their use in pharmaco-toxicological research, leading to efficient use of human organ material and a decrease of laboratory animal use.

Human tissues grown from stem cells are also of high interest for the pharmaceutical industry. They can enable large-scale drug screening tests on human tissues, facilitate the identification of adverse effects of drugs and thus revolutionise toxicology screening and drug discovery. Given the costs associated with drug development, improved cryopreservation capability would also constitute an advance for the pharmaceutical industry. For animal welfare, the scenario of using lab-grown tissues additionally constitutes an attractive alternative.

7.4 CRYOPRESERVATION METHODS

7.4.1 Freezing vs. Vitrification

Cryopreservation has become a well-established method for preserving cells and tissues , including sperm, oocytes and ovarian and embryonic tissue which are now cryopreserved on a large scale. Currently, the main approaches for cryopreservation are freezing (i.e. the liquid phase changes to a solid crystalline phase) and vitrification (i.e. the solidification to a glass-like state without ice formation), as illustrated in Fig. 1. The methods of freezing and vitrification are commonly distinguished by whether during sample-cooling ice crystal growth occurs or is suppressed, respectively (Fig. 2). Notably, freezing is applied for a wide range of sample volumes, whilst vitrification usually has a small application size.

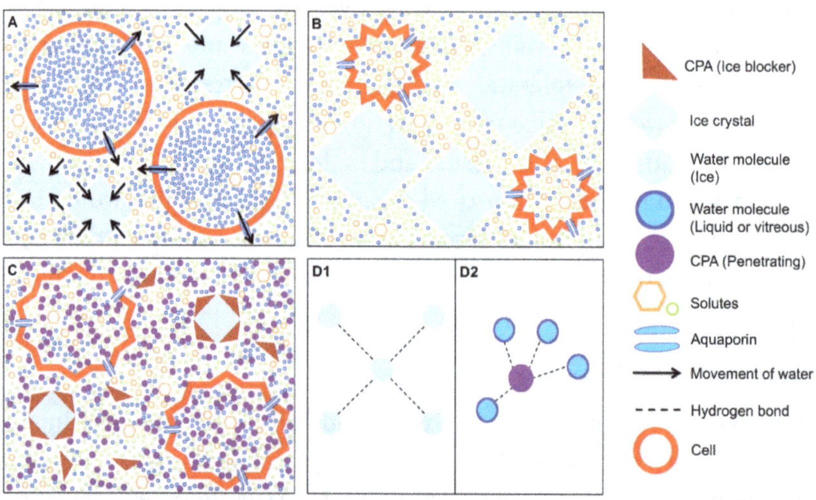

How cryoprotective agents (CPAs) work. A When a sample is cooled, ice first forms in the extracellular space. Ice excludes solutes, so the extracellular solute concentration increases as extracellular water becomes part of the ice crystals. Intracellular water is then drawn out of the cells via osmosis. B In unprotected cells, the intracellular solute concentration increases and causes damage. C Penetrating CPAs permeate the cell membrane and increase the intracellular solute concentration which prevents water loss and dilutes other solutes inside the cell which can cause damage at high concentrations. Ice blockers bind to ice crystals, preventing them from growing, or to nucleators, which prevents heterogenous ice nucleation. D1, D2 Penetrating CPAs interfere with homogenous ice nucleation by colligative interference, which depresses the freezing point of the solution. D1 Water forms a regular ice lattice. D2 A CPA molecule disrupts the hydrogen bonding between water molecules

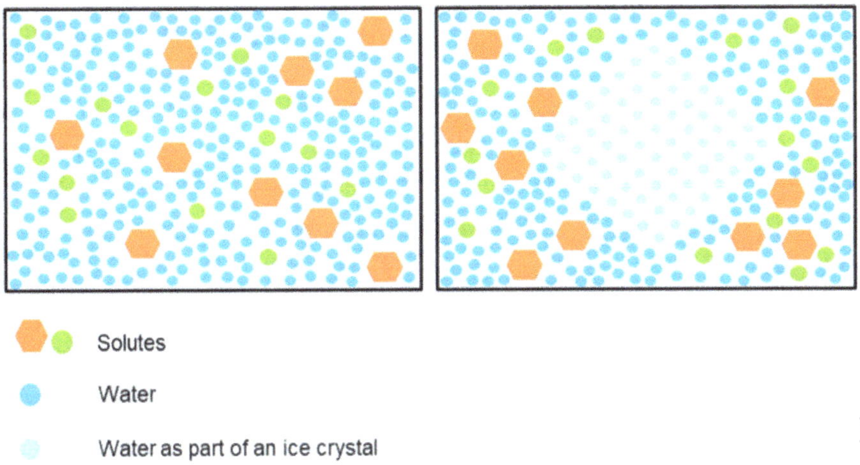

]

A vitrified substance (left) is formed when liquid transitions into a highly viscous, glass-like state that prevents translational molecular motion. In vitrification, molecules remain in the position they were in when the substance was vitrified. This is different to freezing (right), in which ice crystals progressively grow as the temperature is decreased, excluding solutes and thus causing the solute concentration to increase

In freezing, a too low cooling rate can give rise to cryoinjury due to solution effects. On the other hand, the rate needs to be slow enough so that there is sufficient time for the water to exit cells, to prevent intracellular ice formation. Freezing usually involves slow cooling protocols with cooling rates of approximately 1 °C/min. However, different cell types have different optimal cooling rates.

Vitrification is a separate physical process from freezing and occurs at the so-called "glass transition temperature", around – 80 to – 130 °C. In vitrification, samples solidify with no ice crystal formation (Fig. 3). It has been previously shown that the vitrification of small biological samples can be achieved without CPAs using extremely high cooling rates, e.g. with human sperm [40]. In larger living systems where an extremely high cooling rate cannot be achieved, a large percentage of the water in the sample must be replaced by a vitrifying CPA to avoid the damaging effects of ice formation. Vitrification ultimately forms a stable glass-like state, preserving the molecular contents indefinitely.

Importantly, warming after cryopreservation can lead to injury due to ice crystal formation, which can occur both after freezing or vitrification. Along these lines, "ice recrystallization" denotes this damaging process of large ice crystal growth during thawing of frozen cells or tissue. Notably, and somewhat counterintuitively, "devitrification" denotes the process of crystallisation during warming and after vitrification. Hence, devitrification does not mean the process of warming of a vitrified solution, but the formation or recrystallization of ice crystals during rewarming.

In vitrification, CPAs are primarily used to inhibit ice formation by depressing the freezing point of the solution and increasing its viscosity and the glass transition temperature. In this way, CPAs decrease the critical cooling rate for vitrification (Vccr), allowing the sample to solidify before ice formation can occur. Notably, CPAs can also impact on the critical warming rates and so reduce the risk of devitrification. What are the requirements for tissues to vitrify and undergo a transition into the glass-type state? Most importantly, the cooling rate has to be sufficiently high, so that ice formation is prevented. Notably, vitrification requires very high concentrations of CPAs to lower Vccr to practical levels. Unfortunately, the toxic impact on cells caused by these concentrated CPAs is problematic. This problem can to some extent be mitigated, as discussed ahead in the «Cryoprotective agents (CPAs)" section.

Although slow cooling and vitrification are well-established and are used for various applications, they each have their advantages and disadvantages. For instance, the recording of slow cooling protocols is practical which facilitates the demonstration of compliance with standard operating procedures and Good Manufacturing Practice. However, it often requires relatively costly equipment. Moreover, although slow cooling is more time consuming, this apparent disadvantage can entail a more homogeneous distribution of CPAs, which need to diffuse through and permeate the extracellular space to take full effect. Also, the slower and more controlled cooling process in slow cooling is better suited for quantifiable and reproducible protocols. The temperature dynamics within the sample volume are therefore more amenable to modelling in a quantitative manner. However, extracellular ice formed during slow cooling can also have adverse effects. It remains to be investigated how such damage depends on cooling and thawing rates, as well as CPAs.

Rapid cooling to below the glass transition temperature and/or rapid warming from below the glass transition temperature during the warming process may induce fracturing of the glass in which the biological system is embedded. Cell-type or tissue-specific cooling and warming rates can potentially avoid the problems of devitrification and recrystallisation, and further research is required to elaborate on such solutions more information can be found in Fahy and Wowk. Notably, such highly adapted protocols can be problematic with regard to the cryopreservation of organs comprising different cell or tissue types.

Table 1 summarises the key differences between these main approaches. Importantly, there is no gold standard in cryopreservation, and different labs often employ their own cryogenic processing protocols that might substantially differ from other labs' approaches.

Table 1 Comparison of typical key aspects of slow cooling and vitrification

Feature	**Slow cooling**	**Vitrification**
Duration	Long (h)	Short (s)
Usage of CPAs	Low concentrations	High concentrations
Costly components	Equipment	Operation
Controllability/reproducibility/recordability	High	Low
Operation skills required	Low	High
Main risks	Extracellular ice formation, osmotic damage	Intracellular ice formation, CPA toxicity, fracturing, osmotic stress

7.4.2 Directional Freezing and Directional Vitrification

Directional freezing (also called directional solidification) is a method in which a temperature gradient is used to cause the progressive growth of ice along the temperature gradient (Fig. 4). This can be achieved by moving the sample along the gradient at a controlled velocity or by placing the sample between two thermo-conductive blocks. In conventional freezing, the latent heat of fusion released during freezing is conducted through the frozen portion of the sample, this may cause it to melt and can result in repeated freeze-thaw cycles which can cause increased cellular damage. The morphology and rate of ice growth are uncontrolled and non-uniform. In directional freezing, heat is removed by conduction through the unfrozen portion of the sample, preventing damage by freeze-thaw cycles. The directional ice growth results in the formation of lamellae with cells trapped between them, reducing mechanical damage. Directional freezing also allows for faster cooling without inducing intracellular ice formation.

In directional freezing, cold is applied to one side of the sample causing a temperature gradient. Uniform ice lamellae form along this gradient with heat being conducted through the unfrozen part of the sample

Directional freezing has been used to cool a rat's hind limb to −140 °C before submersion in liquid nitrogen (LN2). The hind limbs were then successfully revived and re-transplanted onto the rats. Whilst these limbs were non-functional due to the motor neurons having been severed during amputation, the limbs were otherwise viable. Directional freezing has also been used for the successful cryopreservation of a pig's liver at −20 °C. The liver was later revived and attached to another pig by auxiliary transplantation and demonstrated normal bile production and blood flow.

A related technique is directional vitrification; this is achieved using the same principle as directional freezing, but with a very high cooling rate. Information on directional vitrification is limited; however, this technique may allow for vitrification with cryoprotectant concentrations as low as 17.5% PG.

Both directional freezing and directional vitrification are rich avenues for future research. Further work on directional freezing should aim to further lower the temperature to which organs can be frozen to – 130 °C, the temperature at which storage without degradation can be achieved for centuries or millennia.

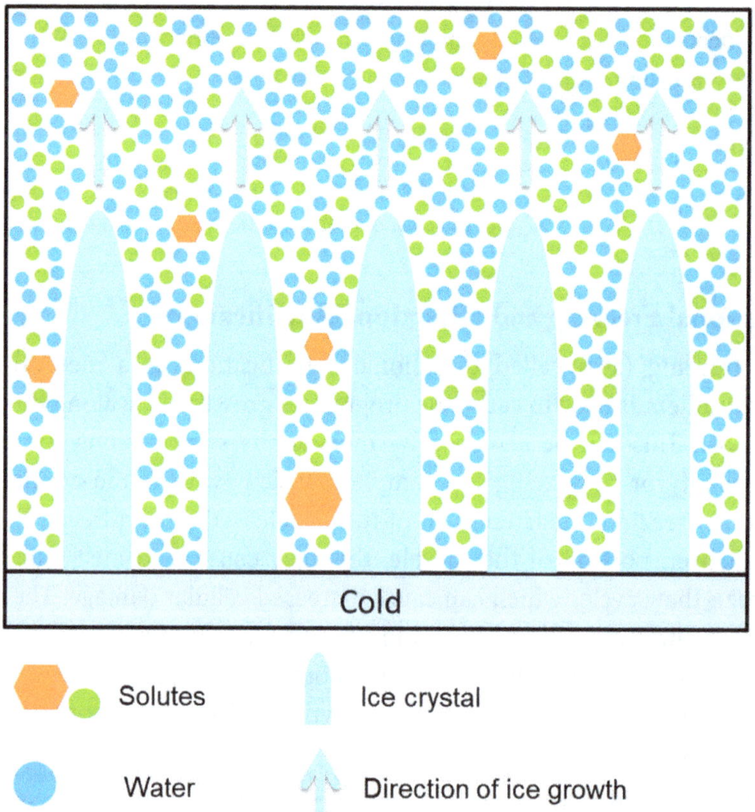

7.5 KEY ASPECTS OF CRYOGENIC PROCESSING PROTOCOLS

7.5.1 Cooling

As mentioned above, during slow cooling, extracellular water freezes first and cells lose water due to osmotic pressure, with more solute outside than inside the cells. During fast cooling, the transmembrane transport of water is not fast enough, and the intracellular water is more likely to freeze as well. It is a complex problem to determine how quickly to best cool cells and tissues: the optimal cooling rate minimises the dangers of injury due to high solute concentrations as well as intracellular ice formation. This "sweet spot" depends on various aspects, including biophysical cell membrane properties, the volume of the sample, the CPA type and the concentration. When cooling a macroscopic sample, e.g. a 2-mL cell suspension, the large heat capacity of water causes the core of the sample to cool more slowly than the surface. The sample therefore usually nucleates in the surface region first because it is closer to the colder cryovial wall. The release of the latent heat will then further increase the temperature lag between rim and core temperatures. For vitrification protocols, the cooling rate needs to be slow enough to achieve a homogeneous temperature distribution, especially around and below the glass transition temperature to not cause cracking.

7.5.2 Warming

It has been established that the quality of a biological sample after cryopreservation depends not only on the method of cooling, but also on the warming of the sample. (Note, we use the word "thawing" only in the context of freezing. In vitrification, no ice crystals are formed, and so the cells or tissues are technically not frozen. Hence, it is more appropriate to use the word "warming"). During warming, ice crystals can grow around an initial nucleus. Both ice nucleation and growth depend on temperature, yet in a different way. The process of ice growth is fastest just below the freezing point and attenuates when the temperature drops. The nucleation reaches the peak well below the freezing point. In other words, the cooled sample first passes through an ice growth zone (but with few nuclei available to seed the growth), then, when the temperature for crystallisation is no longer optimal, enters the zone of nucleation. For this reason, rewarming the sample brings more risk of recrystallisation and devitrification—when the sample enters temperatures optimal for freezing, it is already well nucleated. This is also the reason why the critical rates of cooling and rewarming necessary to outpace the ice formation are different: the former must deal with few nuclei, the latter with many. Notably, multiple small crystals (100–300 nm) are much better tolerated than a few large ones. In this context, it was demonstrated that the addition of a temperature hold during rewarming at the glass transition temperature can significantly alter the impact of physical forces during warming for certain cell types. It remains to be investigated how to optimally choose the dynamics of temperature changes during cooling and warming, in a tissue-specific manner. A novel method for warming has been tested with promising results. Manuchehrabadi et al. demonstrated that inductive heating of magnetic nanoparticles can be employed to improve tissue cryopreservation. Other methods like electromagnetic heating, have also been proposed.

7.5.3 Nucleation Temperature

The initial formation of ice crystals is only possible after a primary nucleation event has occurred. If uncontrolled, such primary nucleation is stochastic. This means, for example, when a specific aqueous sample is cooled at a given cooling rate, it is not deterministic at what time or temperature it will start freezing. The temperature at which primary nucleation occurs within a sample is called the nucleation temperature. It has been shown that the nucleation temperature affects cell survival in a similar way as the cooling rate does, namely, a too high nucleation temperature can be detrimental to cells caused by solution effects, whilst a too low nucleation temperature, close to the spontaneous nucleation temperature, can cause intracellular ice formation. The highest survival of frozen cells was found at intermediate nucleation temperature regimes. It was further found that at optimal intermediate nucleation temperatures the concentration of CPAs could be reduced. In order to reduce the effects of toxic CPAs, the active control of the nucleation temperature has been proven to be beneficial for cell survival after thawing. There are various methods for actively controlling the process of nucleation for slow cooling. Briefly, the nucleation temperature can be controlled by creating locally supercooled conditions, such as via the application of cooled metal rods, expanded gases or Peltier elements. Furthermore, ice-nucleating agents (INAs) exist that can directly induce nucleation events, with examples including silver iodide, ammonium sulfate, long-chain alcohols and nanoparticles; however, as with CPAs, the utility of these is often limited by their toxicity, availability or biocompatibility. Promisingly, high-activity INA proteins have been discovered in multiple frost-tolerant organisms, with the best characterised being those from epiphytic bacteria of the genera *Pseudomonas* and *Erwinia* . The probability of a primary nucleation can be increased by electrofreezing, where a high electrical field is locally applied to a sample. Other physical methods to control nucleation include the use of ultrasound or laser light pulses, reviewed elsewhere.

7.6 CRYOPROTECTIVE AGENTS (CPAs)

CPAs can prevent cryoinjury in multiple ways, including the reduction of solution effects. A good CPA should possess the following properties: it should be water-soluble at low temperatures and to high concentrations, have good penetrance of cell membranes and have low toxicity. The concentration-dependent mechanisms of how pCPAs like DMSO and EG benefit cryopreservation has been investigated through simulations and experiments, which have suggested that at intermediate CPA concentrations, pore formation occurs within the membranes which increase the hydraulic membrane permeability. Some macromolecular CPAs like trehalose, dextran, sucrose or hydroxyethyl starch, which do not permeate cell membranes, are suggested to stabilise proteins by the "preferential exclusion" effect. It was found that they prevent freezing damage best in combination with pCPAs like DMSO. Slow cooling studies revealed that trehalose can alter the ice/cell interaction which can improve post-thaw cell survival. DMSO offers the advantage of enhanced permeability vs. glycerol for many cell types. Currently, DMSO is the most commonly used CPA for the storage of stem cells. However, the infusion of stem cells cryopreserved with DMSO has been associated with toxic reactions such as vomiting, cardiac dysfunction, arrhythmia and others.

Despite these advances in the identification and use of CPAs, CPA toxicity remains the main challenge in cryopreservation. Whilst the mechanisms of CPA toxicity are only partially understood, some mechanisms have been elucidated and classified as either specific or non-specific toxicity. Specific toxicity refers to CPA damage caused through a mechanism unique to that particular CPA. For example, EG is metabolised in the liver to glycolic acid, causing metabolic acidosis. Non-specific toxicity refers to damage that occurs due to properties that CPAs have in common. For example, CPAs form hydrogen bonds with proteins; however, their bonding strength is higher than that of water, with the potential to denature proteins at high concentrations.

Importantly, the toxic effect of CPAs increases with temperature. Loading ramp protocols where CPAs are administered at increasing concentrations as the sample is cooled can greatly improve post-warm viability. Another reason for using a loading ramp is to slowly equilibrate the sample with the CPA to reduce osmotic stress. Indeed, the potentially lethal mechanical stresses on cells due to CPA administration and removal are a major current concern in cryobiology, particularly for vitrification. Moreover, nonequilibrium concentrations of CPAs between regions of the sample or across the membrane can reduce the success of cryopreservation.

7.7 BIOLOGICAL ANTIFREEZE PROTEINS, COMPATIBLE SOLUTES AND ICE BLOCKERS AS ALTERNATIVES

Some animals are known to produce substances that allow them to survive cold temperatures by avoiding ice formation. These substances can be either cryoprotectants (e.g. glycerol) or antifreeze proteins. The latter can bind to ice and hinder its growth; however, the use of such proteins is limited by their high costs. Synthetic alternatives are used in research. These include polyvinyl alcohol (PVA) as an ice growth inhibitor, which binds to and prevents ice crystals from getting larger, and polyglycerol (PGL) as an ice nucleation inhibitor, which prevents ice from nucleating in the first place. They are effective even in small quantities and widely used. Compatible solutes are small organic osmolytes like hydroxyectoine, ectoine and L-proline which are produced by microorganisms under stress. These have been used to reduce the amount of pCPAs in freezing protocols. There may be potential benefits in using compatible solutes for vitrification methods.

7.8 VITRIFICATION SOLUTIONS

Vitrification solutions are used to decrease V_{ccr}, allowing larger samples to be vitrified. Pure water can be vitrified if it is cooled fast enough, but to cause this, extremely high cooling rates (in the order of 10^6 °C/s) are needed. To counter this problem, high CPA concentrations can be used to lower V_{ccr} to a level achievable in practice by plunging a sample into LN_2. However, samples with a high volume to surface area ratio exhibit a lower internal cooling rate. To preserve larger tissues with a higher volume to surface area ratio, vitrification solutions need to be made more concentrated, or otherwise adapted to further lower the V_{ccr}. This often comes at the cost of increased toxicity. As such, the development of vitrification solutions is dependent on finding new CPAs or combinations of CPAs that decrease V_{ccr} whilst having lower levels of toxicity. A number of advances in vitrification solutions have been made in recent years, as follows.

7.8.1 Neutralising Toxicity

The simplest vitrification solutions are high concentrations of single CPAs such as EG or DMSO, which are toxic at high concentrations. It has been found, however, that different CPAs can be combined in a single solution to reduce toxicity. By using a solution of different CPAs, it is possible to avoid using any one CPA at a concentration that would cause specific toxicity. Additionally, certain CPAs can directly neutralise the toxicity of other CPAs, for example, DMSO neutralises the toxicity of formamide, allowing for a higher total concentration of CPAs to be used with reduced toxicity. This allows for a higher total CPA concentration for the same level of toxicity.

7.8.2 Ethylene Glycol

It was found that using ethylene glycol in vitrification solutions in place of PG reduces the non-specific toxicity of the solution. This is because EG forms weaker hydrogen bonds than propylene glycol, meaning that macromolecules are protected by increased hydration from the greater number of remaining water molecules.

7.8.3 Non-Penetrating CPAs (npCPAs)

Ice can form more readily in the extracellular space than in the intracellular space because the intracellular space has a higher concentration of solutes. For this reason, a lower concentration of cryoprotectants is needed inside the cell. High-molecular weight polymers that do not penetrate cell membranes, such as polyvinylpyrrolidone (PVP), can be used to mimic the effects of intracellular solutes in the extracellular space, meaning that a lower concentration of the more toxic penetrating cryoprotective agents (pCPAs) is needed for vitrification. npCPAs also reduce chilling injury as the resulting solution is hypotonic.

7.8.4 Synthetic Ice Blockers

Ice blockers are another kind of npCPA which bind directly to ice and to heterogeneous ice nucleators. These are the synthetic equivalent to anti-freeze proteins]. Non-penetrating ice blockers such as polyvinyl alcohol (PVA) are a cheaper and more effective alternative to animal-derived antifreeze proteins. Adding ice blockers in small amounts inhibits heterogeneous ice nucleation.

7.8.5 Methoxylation

M22 is an advanced proprietary vitrification solution created by the company twenty-first Century Medicine (21CM). The solution, described, contains the methoxylated polyol 3-methoxy-1,2-propanediol, which has increased cellular permeability and decreased viscosity. This decreased viscosity is important for vitrifying larger tissues and organs because it increases the ability of the CPA to perfuse into tissues. Additionally, 3-methoxy-1,2-propanediol increases the glass transition temperature of tissues and decreases the critical warming and cooling rates of the solution.

7.8.6 Hypertonicity

Chilling injury can be reduced by making hypertonic vitrification solutions, which have an increased concentration of non-penetrating components. Whilst the reasons for the protective effect of hypertonicity are poorly understood, it has been suggested that it reduces damage due to the thermal contraction of cell membranes.

7.9 CURRENT APPLICATIONS OF CRYOPRESERVATION

7.9.1 Organ Preservation and Banking

Many thousands of lives worldwide would benefit if replacing organs and tissues on demand became a reality. Some of the approaches to extend organ lifetime are ex vivo continuous perfusion at normothermic temperatures (35–37 °C) and a combination of perfusion with hypothermic temperatures (4–10 °C), which is routinely used for kidney transplants and can extend lung survival times from 8 to 21 h. The addition of CPAs and suppression of ice formation (supercooling) in rat livers enabled perfusion on even lower temperatures (– 6 °C), extending storage times to 4 days. However, so far, there are no effective methods to reliably preserve solid organs beyond 3–12 h. Not only days or weeks but even hours of additional preservation time would greatly increase the number of possible transplantations.

Kidneys and hearts have been the most widely studied organs, but neither has been consistently recovered after cooling to temperatures lower than – 45 °C. Nevertheless, sporadic survival of kidneys has been claimed after cooling to lower temperatures. Along those lines, Fahy et al. reported success in vitrifying a rabbit kidney at – 130 °C which was rewarmed using a special conductive warming technique combined with perfusion. After warming, the kidney was transplanted into a recipient rabbit that lived for 48 days with a working kidney before being sacrificed. More recently, Marco-Jimenez et al. succeed in creating a long-term biobank of kidney precursors as a potentially unlimited source of organs for transplantation as well as cryopreserving renal primordia able to developed morphologically normal glomeruli after warming and allotransplantation.

7.9. 2 Tissue and Cell Banking

7.9.2.1 Corneal tissue

According to the WHO, 10 million people worldwide require surgery to prevent corneal blindness as a result of trachoma, with a further 4.9 million suffering from total blindness due to corneal scarring. There is a considerable discrepancy between the supply and demand of transplantable corneas. Even though currently the main problem in most countries is the insufficient availability of transplantable corneas, corneal cryopreservation would contribute to an amelioration of the situation. Moreover, corneal tissue engineering is a highly researched topic and has the potential to increase the supply. Corneal cryopreservation would offer an unlimited storage time and eliminate the time-dependent deterioration. Human corneas for transplant are commonly stored at 2–8 °C for up to 14 days, or by organ culture, which allows up to 4 weeks of storage. The first cryopreservation technique developed for rabbit, cat and human corneas used 7.5% DMSO and 10% sucrose. Although cryopreserved corneas have been successfully transplanted, significant endothelial damage occurred as a result of freezing. Cells in monolayers, such as corneal endothelium, are more susceptible to freezing injury compared to isolated cells frozen in suspension. In human corneas, endothelial cells have a very limited proliferative capacity, and cells lost during cryopreservation are not subsequently replaced by mitotic division. Well-preserved corneal endothelium is therefore crucial to the successful outcome of full-thickness corneal transplantation. A number of attempts have been made to improve the reliability of corneal cryopreservation. Rabbit and human corneas have been vitrified using 6.8 M propane-1,2-diol and shown to retain endothelial function. Although this technique showed some success, it requires exposure to very high concentrations of solutes and is very time-consuming. Recently, a cryopreservation protocol involving interrupted slow cooling at 1 °C/min to a plunge sub-zero temperature (−35 °C for porcine and −45 °C for human) before storage in LN_2, and using a combination of a pCPA (5% DMSO) and a npCPA (6% HES), has been successfully applied to cryopreserve corneal endothelial cells. So far, however, cryopreserved human corneas have not been adopted by eye banks for routine use due to their significantly lower performance compared with fresh corneas.

7.9.2.2 Retinal Pigmented Epithelium (RPE)

Transplantation of RPE may have a potential clinical application for the treatment of RPE-specific retinal degeneration such as age-related macular degeneration (AMD). AMD is the most common cause of blindness in the developed world, affecting one in three people by 75 years of age] with an estimated prevalence of ~600,000 significantly visually impaired people in the UK and over 8 million worldwide. The development of a suitable storage method for RPE is necessary for the establishment of future RPE replacement therapy. The feasibility of an RPE storage bank has been investigated by Durlu and Tamai who performed transplantation of viable cryopreserved RPE cells into the subretinal spaces of adult albino rabbits and 23-day-old rats. Successful transplantation was confirmed postoperatively by light and electron microscopy. In rabbits, xenografted RPE cells residing on Bruch's membrane of the host retina were disclosed without any morphologic difference between the fresh and cryopreserved RPE cells in situ at day 25 following transplantation. In rats, subretinal injection of cryopreserved RPE cells partially rescued photoreceptor cells locally at the transplanted area as observed 3 months postoperatively. Together, their data confirmed that cryopreserved RPE cells can be used for RPE transplantation. Although changes in viability, altered cellular functions and impaired biological activities caused by cryopreservation are usually transient, and cells resume their activities with time or after repeated passage, some irreversible changes have also been reported. For example, Basu et al. found an increase in abnormal chromosomes in cultured bovine RPE cells after freezing, which further increased after cell passage. Another study investigating RPE cell lines suggested that inappropriate freezing conditions may not only decrease cell viability but also induce replicative senescence. Valtink et al. also demonstrated that RPE cells can be cryogenically stored and kept available for transplantation for a prolonged period of time.

7.9.2.3 Retinal Müller Glial Cells Cryopreservation

Müller cells are the dominant macroglia of the vertebrate retina. In addition to their various important functional roles in retinal physiology, they are also involved in retinal pathology, for example, the cells display altered membrane properties in cases of proliferative vitreoretinopathy. The use of human tissue is limited by reasons of availability, thus developing a method for long-term storage of these cells is desirable. Biedermann et al. described a cryopreservation method in which dissociated Müller cells are stored in 10% DMSO. They showed that the main electrophysiological properties were not altered by this method and that voltage- and ligand-gated currents can be recorded from cryopreserved cells even after 2 years of storage.

7.9.2.4 Nerves and Brain Tissue

The replacement of damaged or missing cells is an emerging treatment strategy for a number of neurological disorders. The cryopreservation of neural cells and tissues has not been applied with sufficient success to enable it to be incorporated into routine clinical practice. Isolated brain tissue has a short shelf-life and cells are usually used for experiments immediately after isolation or refrigerated for a short period of time before use. Storage at 4 °C allows good survival of neural tissue up to a maximum of 8 days; therefore, developing long-term storage procedures is of vital importance. Cryopreservation of primary neural cells would also greatly facilitate basic neuroscience research. The ability to cryopreserve brain tissue would also facilitate pooling of tissue from multiple donors, allowing plenty of time for necessary testing to ensure safety and function of the cells prior to transplantation. Successful cryopreservation would be beneficial for neural tissue engineering too, allowing long-term storage for purposes of inventory control, quality control and product distribution.

The first attempts to achieve nerve tissue cryopreservation were published in 1953, by Luyet and Gonzales, who demonstrated that chicken brain tissue survived freezing to – 196 °C, following exposure to EG and rapid cooling. Since then, different studies have suggested methodologies to cryopreserve neurons and neural stem cells as summarised by Paynter. Many studies have measured the qualitative outcome of viability or cell membrane integrity; however, only a few authors have characterised cryopreserved cultures in deeper detail. Post-thaw viability (based on membrane integrity assays) does not necessarily reflect the quality of the cryopreservation procedure and proper functionality tests must be carried out. For example, Higgins et al. provided a qualitative evaluation of neuritic tree formation whilst Quasthoff et al. included a parametric analysis of morphological and functional development of the cultures.

In addition, cryopreservation of organised adult cerebral tissue slices is of potential interest for pharmaceutical neuropsychiatric drug evaluation and development. Pichugin et al. showed that the viability and the structure of mature organised, and complex neural networks can be well preserved by vitrification. Cryopreservation of single neural cell suspension has also been described, like neural progenitor cells and neurally differentiated pluripotent stem cells. Moreover, cryopreservation of brain spheroid cultures is another emerging topic. For example, Ma et al. compared the survival rate and viability after cryopreservation between three different states of neural stem cells (NSC): single-cell suspension, NSC spheres with diameters of 30–50 μm and 80–100 μm. NSC spheres with a diameter of 80–100 μm achieved the best survival rate of 82.9%, and the NSCs still sustain the multi-differentiation potentiality. Overall, efficient cryopreservation of neural tissue would be beneficial not only to potential clinical use but also for basic science and pharmaco-toxicology studies.

7.9.2.5 Tissue-Engineered Constructs Banking

Cryopreservation of tissue-engineered products by maintaining their structure and function is a prerequisite for large-scale clinical applications. Since the engineered tissue substitutes are undergoing clinical trials, and foreseeing a growing demand for cultured cells and tissues, the tissue engineering community is becoming increasingly worried about the challenge of providing sufficient amounts of these products to the market. Cryopreservation is the only established mechanism for long-term preservation of cells, tissues and organs, as well as engineered tissues; therefore, it is a key step for the improvement of tissue engineering.

7.9.2.6 Cryopreservation of Retinal Organoids

An optimised vitrification method enables en bloc cryopreservation of stratified neural retina (NR) of human origin. They developed an efficient cryopreservation method for storing stratified NR epithelia derived from human ESCs and found that a vitrification method with pre-treatment in 10% DMSO + 5% EG + 10% sucrose on ice was quite effective for NR cryopreservation. More recent findings show that both human iPSC-derived retinal organoids and dissociated retinal cells can be easily cryopreserved whilst retaining their phenotypic characteristics and the preservation of photoreceptor precursors.

7.9.3 Cryopreservation of Bioartificial Liver

Hepatocyte microbeads can temporarily replace the function of damaged hepatocytes in acute liver failure. Hence, intraperitoneal transplantation of hepatocyte microbeads is an attractive option for the management of acute liver failure. Interestingly, alginate-encapsulated hepatocyte spheroids were successfully cryopreserved by Massie et al.. More recently, Jitraruch et al. developed an optimised cryopreservation protocol that could improve the outcome of cryopreserved hepatocyte microbeads for future clinical use.

7.9.4 Cryopreservation of Decellularised Oesophagi For Tissue Engineering

Efficient storage could potentially allow a timely use of decellularised oesophagi, essential for oesophageal tissue engineering and clinical translation. Recently, Urbani et al. published a protocol for long-term storage of decellularised oesophageal scaffolds for tissue engineering purposes which includes slow cooling with cryoprotectant solution in LN_2 vapour.

7.9.5 Cryopreservation of Tissue-Engineered Skeletal Muscle

The tissue-engineered skeletal muscle plays an important role in the field of regenerative medicine and in emerging areas such as soft robotics, organ-on-a-chip disease models and drug testing [167]. Grant et al. published an optimised protocol in which the tissue-engineered skeletal muscle was frozen; tissue made from either differentiated myotubes or their undifferentiated myoblast precursors were investigated. After thawing the tissue maintained cell viability, and moreover, an optimised protocol in which the skeletal muscle was frozen undifferentiated, showed a threefold increase in force production as compared to unfrozen muscle.

7.9.6 Cryopreservation of Engineered Neural Tissue

The ability to preserve engineered neural tissue without disrupting cellular and extracellular components and structures is important for clinical translation and commercialization. Of note, Day et al. studied the effect of cryogenic preservation on engineered neural tissue.

7.9.7 Cryopreservation of Tissue-Engineered Pancreatic Substitute

The use of encapsulated insulin-secreting cells represents a promising approach towards the treatment of insulin-dependent diabetes. Mukherjee et al. investigated cryopreservation of a model tissue-engineered pancreatic substitute by two ice-free cryopreservation (vitrification) solutions in comparison with a conventional freezing protocol and found vitrification to be a promising preservation procedure for this encapsulated cell system.

7.9.8 Cryopreservation of Tissue-Engineered Bone

Yin et al. examined the feasibility of cryopreservation by vitrification of tissue-engineered bone composed of osteo-induced canine bone marrow mesenchymal stem cells and partially demineralized bone matrix scaffold and showed maintenance of cellular viability and osteogenic function after cryopreservation. Recently, Tam et al. compared the effects of two potential preservation methods on the survival, quality and function of human tissue-engineered bone grafts from iPSCs. They found that storage at – 80 °C resulted in cell death and structural alteration of the extracellular matrix, whilst hypothermic storage at 4 °C did not significantly affect tissue viability and integrity.

7.9.9 Cryopreservation of Tissue-Engineered Skin

Tissue-engineered epidermal membranes are useful for clinical wound healing. Chen et al. showed that when transplanted into nude mice, trehalose-cryopreserved artificial skin repaired skin defects in a similar manner to a non-cryopreserved control. Moreover, they showed that trehalose-cryopreserved artificial skin resulted in engraftment and wound closure that was significantly enhanced compared with that of DMSO-cryopreserved epidermal membranes indicating that the use of trehalose improves cryopreservation of tissue-engineered epithelial sheets. Cryopreservation of tissue-engineered human dermal replacement also plays an important role in skin tissue engineering and skin banking. An optimal cryopreservation protocol consisting of a cooling rate at 1 °C/min in 10% DMSO with the viability of studied dermal equivalent treated by this protocol being 75% of that of fresh control.

7.10 FERTILITY CRYOPRESERVATION

Fertility preservation (FP) is an important and rapidly growing field of reproductive medicine. Since 1978, when the first birth following in vitro fertilisation (IVF) was reported, at least 8 million babies have been produced with the help of medically assisted reproduction. Cryopreservation techniques play a vital role in this treatment revolution, allowing the long-term storage of gametes and embryos without degradation of their quality for later use. With a growing number of women postponing childbearing into their mid-thirties and beyond, the need for FP treatment has never been greater.

Beside elective FP available for healthy individuals, numerous medical conditions that compromise fertility require special attention in terms of FP. Particularly important is the need for FP in cancer patients. Most cancer treatments like chemotherapy, radiotherapy or a combination of both are highly toxic to the gonads and can cause infertility in both male and female patients. Non-malignant pathologies such as autoimmune diseases and their treatment, benign haematological diseases (most prominently β-thalassemia), gender dysphoria prior to starting hormone therapy, gynaecological or urologic and some genetic disorders (Turner's syndrome, Fragile X syndrome, Klinefelter syndrome) could also lead to impaired fertility.

Continuous improvements in cancer treatment and early diagnosis significantly increased long-term survival rates in cancer patients. Cancer survival statistics are particularly impressive in children, with 80% of prepubertal cancer patients (0–14 years) likely to survive the disease [180, 181]. Survival rates in adolescent and young adult cancer patients (15–39 years) also improved dramatically in the last few decades. Such an improvement in survival rates of cancer patients together with advancements in assisted reproductive technologies greatly increased the demand for FP for the childhood, adolescent and young adult patients.

7.10.1 Methods for Male Fertility Preservation

Successful semen cryopreservation was achieved first, in the 1950s, thanks to their relative abundance for experimentation and small size. Although slow cooling protocols can induce significant genetic damage, leading to low post-thaw mobility rate, freezing techniques for semen cryopreservation remain the most commonly used worldwide. Comparative studies between slow cooling and vitrification showed that the use of vitrification techniques is less damaging for DNA, slightly improves post-warm motility rates and is more cost and time-efficient. Several studies confirmed that long-term storage of sperm specimens (up to 40 years) did not affect the post-warm fertilisation of sperm. However, a number of questions regarding the impact of specific cryopreservation protocols and suitability of donors are unresolved, and the research topic remains very active and clinically relevant.

For prepubertal boys who have not undergone spermatogenesis yet, and therefore not producing mature sperm, testicular tissue biopsy and spermatogonial stem cell cryopreservation is the only available option for fertility preservation for boys who are undergoing fertility reducing treatment such as chemotherapy. Germ cells are removed and frozen before treatment and are transplanted back into testes after treatment to restore spermatogenesis.

7.10.2 Methods for Female Fertility Preservation

Embryo and oocyte cryopreservation are both well-established FP techniques. Embryo cryopreservation is the process of freezing and storing embryos at either cleavage or blastocyst stage. This technique has been proven as a safe and effective technology of FP over the last four decades. Since the first birth after embryo cryopreservation reported in 1983, more than half a million live births were achieved using this technique. Embryo cryopreservation represents the gold standard for FP due to high pregnancy rates. Importantly, human embryos preserved by vitrification showed higher survival rates post-warm than embryos preserved using slow cooling.

Oocyte cryopreservation is the most widely used FP technique in the world for delaying childbirth. Oocyte cryopreservation is technically more challenging than embryo cryopreservation due to oocyte high water content that makes it more prone to cryoinjury. Therefore, oocyte cryopreservation involves a higher risk of damage during the preservation process and results in lower overall pregnancy rates compared to embryo cryopreservation. Some of the adverse effects of oocyte cryopreservation as summarised by are hardening of the zona pellucida, egg's meiotic spindle, cytoskeleton and cortical granular damage by ice crystals. Lower survival rates were reported after freezing of oocytes compared to vitrification as well as lower fertilisation rates. When compared to natural conception, the incidence of congenital anomalies in born children does not differ significantly after cryopreservation.

Ovarian tissue cryopreservation is not a conventional FP technique but is the only available option for prepubertal girls and women who cannot delay the start of cancer therapy. It involves harvesting and freezing of ovarian tissue allowing cryopreservation of oocytes within the primordial follicles presented in the ovarian cortex. After thawing, ovarian tissue can be either transplanted back into a cancer survivor, or immature oocytes can be harvested and matured in vitro. The first live birth after autotransplantation of previously frozen human ovarian tissue was reported in 2004 with the number of live births exceeded 130 as of June 2017. The majority of these pregnancies occurred after natural conception, which is the main advantage of the ovarian tissue cryopreservation compared to oocyte and embryo cryopreservation. Besides the restoration of fertility, the endocrine function of ovarian tissue was also successfully restored with the main duration of about 4–5 years after transplantation, with graft functioning up to several years, making this procedure an option to delay menopause.

7.11 STEM CELL CRYOPRESERVATION

The commercial and clinical application of stem cells relies on cryopreservation as the only available long-term storage option. Stem cells represent highly promising resources for application in cell therapy and regenerative medicine, drug discovery, toxicology and developmental biology research. Stem cell banking offers the opportunity to cryogenically preserve stem cells at their most potent state for later use. Many stem cells come from once-in-a-lifetime tissues, like umbilical cord blood; therefore, all over the world, stem cell banks are established in order to preserve these cells for their potential clinical application or future use in basic or translational research.

7.11.1 Umbilical Cord Blood (UCB) Cryopreservation

Whilst UCB was treated as a medical waste in the past, nowadays, it is used to treat various diseases such as blood cancers as a valuable haematopoietic stem cell source. Since the first successful transplantation of cryopreserved UCB to treat an infant with Fanconi anaemia was performed in 1988, the field of UCB banking and transplantation has grown exponentially and a global network of public and private biobanks of UCB was established. By 2013, over 600,000 UCB units have been stored for transplantation worldwide, and more than 30,000 UCB transplantations have been performed.

7.11.2 Human Pluripotent Stem Cells Cryopreservation

Due to their unlimited propagation and differentiation abilities, human pluripotent stem cells including ESCs and iPSCs are also attractive types of stem cells for banking. Furthermore, iPSCs has opened up the potential for personalised cell therapies, and over the past decade, several initiatives have been established to collect and generate a large amount of human iPSCs for scientific research. Many iPSC banks were established worldwide including the European Bank for induced pluripotent Stem Cells (EBiSC). Other available stem cell sources for banking include bone marrow, umbilical cord tissue and adipose tissue. Dental stem cells are also an easily accessible source of multipotent adult stem cells which makes them a suitable alternative for bone marrow-derived mesenchymal stem cells. For that reason, dental tissues have become an attractive source of mesenchymal stem cells and dental stem cell banks are formed worldwide.

7.11.3 Haematopoietic Stem Cells Cryopreservation

Haematopoietic progenitor cells (HPCs) are a highly relevant from a clinical point of view. Notably, allogeneic haematopoietic stem cell transplantation (HSCT) remains the only curative treatment for many blood disorders and is the most established stem cell treatment to date Indeed, over the past 50 years, HSCTs have been performed in over 400,000 individuals. Along those lines, banking haematopoietic progenitor cells (HPCs) from the bone marrow, peripheral blood and umbilical cord blood is highly important for facilitating the finding of matched donors for patients needing HPC transplants. Cryopreservation of HPCs is the standard mode of handling for HSCT but has certain disadvantages as compared to the usage of non-cryopreserved HPCs. Hence, further research is needed to improve the cryopreservation of this type of stem cells.

7.12 FUTURE RESEARCH AND APPLICATIONS

7.12.1 Emergency Medicine

Besides the medical applications described above that are already being explored, improved capabilities for artificially lowering the temperature of biological tissue could have a significant impact in many other fields. For instance, the possibility to gain more time is crucial in emergency medicine. Indeed, moderate states of hypothermia are common in clinical practice. Along those lines, therapeutic hypothermia has been associated with beneficial outcomes in patients following out-of-hospital cardiac arrest. Moreover, animal studies demonstrate improved neurological outcome. Such capabilities would be also advantageous in warzones, where availabilities of equipment and medical staff are restricted, and so the extension of transfer time would have a big impact. Furthermore, cryopreservation of red blood cells is commonly used in settings where their availability is limited or unpredictable. Despite these existent use cases, further research is required, e.g. to establish critical parameters in hypothermic emergency treatment, such as the duration, temperature and the substances to be administered.

7.12.2 Space Travel

Improved technologies for lowering the body temperature would also be of major benefit for space travel. Currently, some of the most limiting factors for manned, long-duration space travel (e.g. to Mars) have to do with the requirements for keeping space travellers alive: beyond the availability of resources needed for normal physiology and metabolism, there are also significant health challenges posed by the long-term exposure to interplanetary radiation and zero-gravity, as well as the psychological stresses from confined space and associated social dynamics. Low-temperature biostasis or cryopreservation could render these issues negligible. Along those lines, the European Space Agency (ESA) and the US National Aeronautics and Space Administration (NASA) have both investigated the feasibility of suspended animation for deep-space travel. Notably, it has been shown that rats (which are non-hibernators) can be synthetically induced into a hibernation-type state, namely torpor. Torpor is a state of reduced metabolic rate and decreased body temperature. Intriguingly, it can protect from ionising radiation, further highlighting the potential benefits of this line of research for space travel.

7.13 QUANTITATIVE METHODS

With the advances in computing power in the last decades, mathematical and computational approaches for analysing and modelling processes have gained substantial attraction in many biological fields (e.g. for omics, brain dynamics or synthetic biology research). Also in cryopreservation, quantitative methods have been useful to formalise complex physical dynamics. Along those lines, the seminal work of Mazur established that there is an optimal cooling rate for different cell types, which gave rise to mathematical equations describing the relationship between cell type-specific parameters and the optimal cooling rate or the addition and removal of CPAs.

Importantly, computer models can be applied at various spatial scales and to varying detail: for instance, use of molecular dynamics simulations to assess the cryoprotective effect of different CPA concentrations and temperatures. Such modelling can help understand the impact of CPAs on a molecular level, and so potentially guide rational CPA design. Also making use of computational toolsets, but following a complementary approach, employ a computational model to study the volumetric properties of thermo-mechanical stress on a much larger spatial scale. Overall, it is anticipated that in the future, such computational methods will play an important role in the design of cryogenic processing protocols for medical applications, similar to current methods using computer-aided design tools, and so constitute a valuable tool in cryopreservation, which currently is mainly an experimental discipline.

8

MAMMALIAN OOCYTE CRYOPRESERVATION

8.1 SCIENCE OF CRYOBIOLOGY

The demand for effective bio-preservation methods in the medical community continues to increase with advances in transplantation and transfusion medicine. In reproductive medicine, pre-implantation embryo cryopreservation has become an integral component of overall patient care, increasing the success rate per oocyte retrieval cycle. Oocyte cryopreservation is becoming increasingly important due to legal restrictions on the creation and transplantation of supernumerary pre-implantation embryos as well as ethical considerations surrounding the cryopreservation of pre-implantation embryos. The current chapter will attempt to provide a broad overview of cryobiology, and refer to the reproductive biology literature when appropriate.

8.2 ANATOMY OF CRYOPRESERVATION

Cryopreservation is the successful preservation of the normal function of cells or tissues by a reduction in temperature below which normal chemical reactions take place. It is not the long-term storage of cells at these temperatures that is damaging, but the progression to these temperatures and back to normothermia that results in cryoinjury. Cryopreservation nearly always entails the use of one or more compounds that confer protection to cells during freezing. These socalled cryoprotectants (CPA) are typically very simple, low molecular weight mlecules with high water solubility and low toxicity. One feature that is common among these compounds is their ability to interact with water *via* hydrogen bonding. Application of CPA is done (in most cases) simply by incubating the cells in solutions in which these compounds are dissolved. After this exposure, the cells are cooled to a low sub-zero temperature (specimens are typically held at the temperature of liquid nitrogen; -196 °C). At the appropriate time, the specimen is warmed, washed free of the CPA, and used in whatever manner is deemed appropriate. While this seems like a 4 relatively straightforward procedure, many types of injuries can result from any one of the steps; thus numerous lethal effects need to be avoided.

8.3 THE EFFECTS OF WATER PRECIPITATION (AS ICE) DURING COOLING ICE NUCLEATION, CRYSTALLIZATION, VITRIFICATION, AND DEVITRIFICATION

Figure 1shows a phase diagram for a generic aqueous solution. The major physical transitions which occur as a result of cooling are described with such a diagram. The temperature at which these transitions occur depends upon the concentration of solutes in the solution. The curve labeled Tm describes the melting temperature of the solution (which is also the equilibrium freezing temperature). The dashed extensions of the curves represent extrapolations across hypothetical boundaries. As the solute concentration of the solution increases (moving along the X-axis from left to right), the melting temperature of the solution decreases. This is a well known result of the colligative effects of the solutes. The curve labeled Th characterizes thehomogenous nucleation temperature. Homogenous nucleation is defined as the nucleation of ice crystals in the absence of nucleating agents. Aqueous solutions will usually crystallize at temperatures much higher than this due to impurities which act as effective nucleators. A solution containing pure water will remain liquid down to -39 °C, at which point the entire solution will crystallize (the point where Th intersects the Y-axis). This temperature decreases with increasing concentration of solutes.

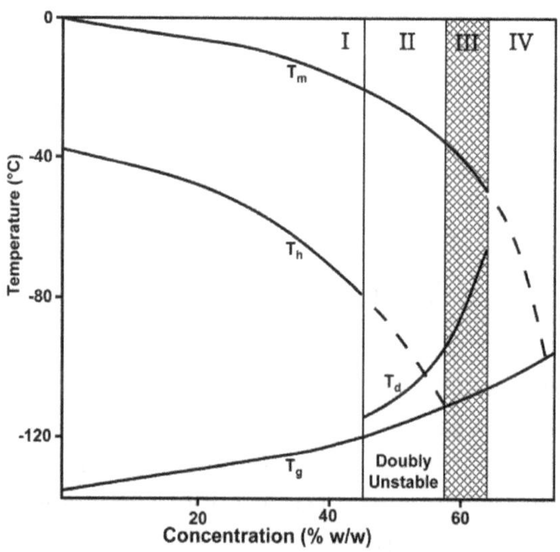

Fig. 1: A phase diagram for a hypothetical solution is shown.

The concentration and temperature dependent physical transitions, including melting (Tm), homogenous nucleation (Th), glass formation (Tg), and devitrification (Td), are described by the respective curves. The use of such diagrams allows the calculation of variables which are important considerations for freezing injury.

The reason why water can remain liquid below its melting point is because the creation of a crystal entails the creation of a liquid-crystal interface with an associated interfacial free energy. The size of a thermodynamically-stable crystal (i.e. one that will continue to grow by the addition of water molecules) is dependent upon temperature (smaller crystals are more stable at lower temperatures). So as the temperature is lowered, the probability of formation of a stable crystal increases until Th, where the probability is 1.

The curve labeled Tg represents the glass transition temperature. At this temperature, liquid solutions will transition to a stable glass (vitrify) and remain vitreous upon further cooling. In region I of this chart (solutions with concentrations ≤ ~45 wt% (weight/weight) in this example), achieving true vitrification is practically impossible. Nucleation (both homogenous and heterogeneous) is essentially unavoidable (at least with practical cooling rates). In region II, the difference between the Th and Tg curves is small enough that vitrification can be achieved with practical cooling rates.

However, a solution which does attain a vitreous state in this region is thermodynamically unstable. In the regions marked III and IV, vitrification is easily achievable. Notice that the Th curve actually intersects Tg at the transition between regions II and III. Heterogeneous ice formation will not occur at temperatures below the glass-transition temperature.

If a sample with a composition described by regions II and III is cooled fast enough to vitrify, ice may still form during warming due to nucleation and ice growth. The temperature at which this happens is described by Td, the devitrification temperature. Essentially, between Tg and Tm (melting temperature), a solution free of ice is in a metastable state. Because the solution is no longer a glass (the glass "melts" to form an unfrozen liquid with the molecules having translational mobility), nucleation and ice growth can occur during warming.

Why is this so? Consider the following two facts: 1) The probability of ice nucleation increases as temperature decreases (notice that as temperature decreases, the sample will get closer to Th; 2) crystal growth, however, being a kinetic phenomenon, will be faster at higher temperatures. Now, imagine the following scenario. A sample with a solute at a concentration of 40 wt% is cooled very quickly so that nucleation occurs only at the temperature around Th (~ -70 °C). However, at -70 °C, ice crystal growth is very slow (at least relative to higher temperatures). Thus, crystal growth is minimal, especially considering that cooling is still taking place. However, during warming, once the sample gets above Tg (~ -120 °C), ice crystals can grow from the nucleation sites. However, it will still be slow, but will increase as the sample warms. Because there were no ice nuclei above -70 °C during cooling, ice growth at higher temperatures could not happen. However, ice nuclei are present during warming and crystal growth can occur until Tm. Therefore, if nucleation does occur during cooling, warming must be very fast to avoid crystal growth at temperatures below Tm. The striking differences between the critical cooling rate (cooling rate necessary to achieve vitrification) and the critical warming rate (defined as the warming rate necessary to avoid more than 0.2% crystallization during warming) are illustrated by the analysis of Baudot and Odagescu. According to their calculations, for a

40 % (w/w) solution of ethylene glycol in water, the critical cooling rate is 569 °C/min, but the critical warming rate is 1.08 x 1010 °C/min.

8.4 SOLUTE CONCENTRATION AS A RESULT OF ICE CRYSTALLIZATION, THE ASSOCIATED OSMOTIC EFFECTS, AND CELL DEATH AT SUPRA-OPTIMAL COOLING RATES

As water precipitates from a solution as ice during cooling, only water molecules comprise the ice crystals. As a result, all other components (salts, *etc.*) become concentrated in the remaining solution. As the solution concentration increases, the chemical potential of the water in the solution decreases. Water will continue to crystallize until the chemical potential of the water in the liquid phase equals the chemical potential of the water in the solid phase. In other words, the remaining solution will reach its equilibrium freezing point (the curve defined by Tm). Therefore, the concentration of the remaining liquid phase can also be determined from a phase diagram.

For example, assume that Figure 1 represents the phase diagram of a sodium chloride-water binary solution. If you start with an isotonic saline solution (0.9 wt%) and cool it to -20 °C, ice will form until the remaining solution is at its equilibrium melting point. In this example, the remaining solution will attain a concentration of ~ 45 wt% (note the point where the Tm curve reaches -20 °C). In this hypothetical example, the unfrozen solution would be roughly 14 M sodium chloride (compared to 0.15 M initially). In reality, sodium chloride will only concentrate ~ 4 M at -20 °C, which is still an impressive concentration effect.

When cells are frozen in suspension, the cells are sequestered in the channels of concentrated unfrozen medium. The high concentration of the remaining unfrozen solution establishes an osmotic gradient across the cell membrane, and as a result, water will flow out of the cell *via* exosmosis Below a cell's equilibrium freezing point, the cytoplasm is in a supercooled state. If the sample is cooled slowly enough, exosmosis occurs to a sufficient degree to keep the cytoplasm in a near-equilibrium state with the extracellular solution. Such a situation will preclude intracellular ice formation (IIF). On the other hand, if the cooling rate is relatively rapid, water cannot leave the cell fast enough to maintain a near-equilibrium state with the extracellular solution, and at some point equilibrium will be re-established by ice formation (Figure 2). This situation is described schematically in Figure 3. The formation of intracellular ice is usually (but not necessarily) fatal to cells. Direct cryomicroscopic observation of intracellular ice formation in mouse oocytes (similar to that which is seen in Figure 2) and the correlation to cell survival were some of the most convincing data to support the assertion that ice formation was the lethal.

The rate at which water flows out of a cell is dictated by the cell membrane water permeability. The permeability of cells to water is dependent upon several factors including temperature and the presence of CPA. For example, in the absence of cryoprotectant, human sperm water permeability is 1.84 µm/min/atm at 22 °C, but is reduced to 1.23, 0.84, 0.77 and 0.74 μm/min/atm in the presence of propylene glycol (PG), dimethylsulfoxide (DMSO), glycerol, and ethylene glycol (EG), respectively. Furthermore, water permeability can vary greatly across cell types. For example, water permeability for human erythrocytes is an order of magnitude higher than the value for human oocytes. Because intracellular ice formation is dependent upon the degree of supercooling (hence the water content), the rate of cooling which results in intracellular ice formation differs widely across cell types. The theory of cell death due to intracellular ice formation resulting from the interaction of cooling velocity and water loss outlined above was developed quantitatively. The practical benefit of this theory comes from its potential to predict optimal cryopreservation procedures.

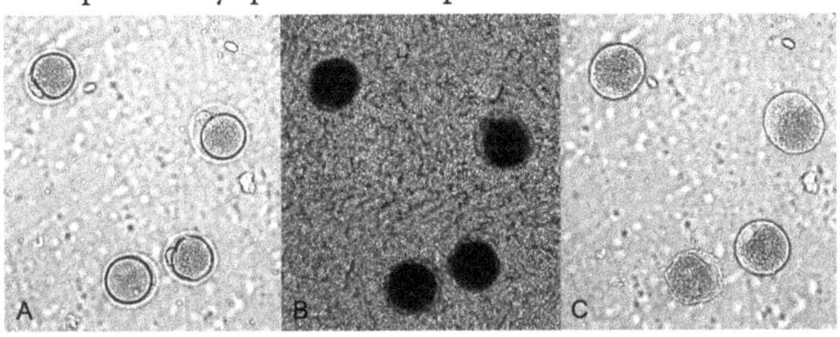

Fig. 2: Photomicrographs of intracellular ice formation in mouse oocytes cooled at 100 °C/min in an isotonic solution is shown.

In Panel A, intact oocytes are shown prior to ice crystal formation; note the well-defined oolemma within the zona pellucida. As cooling proceeds, ice forms in the extracellular solution and eventually, the intracellular solution. The darker background in Panel B is due to ice, and the "blackening" of the oocytes indicates that intracellular ice formation has occurred. When the solution is warmed and the ice melts (Panel C), the cell membrane within the zonapellucida is no longer visible, indicating cell lysis. The speckled appearance of the background in Panels A and C is due to atmospheric water precipitation on the cold glass surface of the cell chamber, and not due to ice crystals.

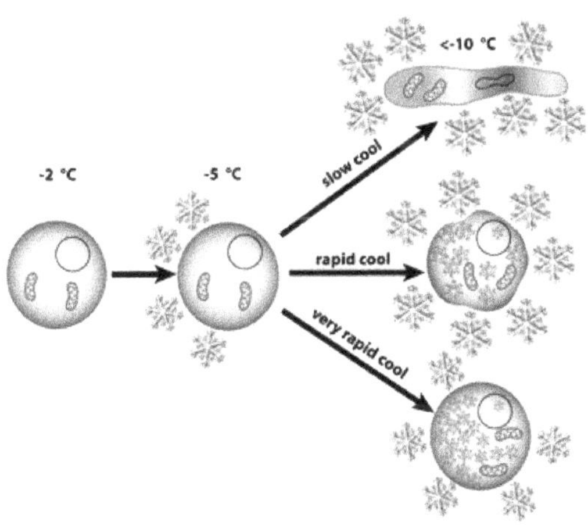

Fig. 3: The cooling-rate-dependent fate of intracellular water resulting from extracellular ice formation is shown.

As the temperature of the solution is cooled below the equilibrium freezing point, ice will form in the extracellular solution. As a result, water is driven out of the cell by an osmotic gradient across the cell membrane. If cooling is slow (upper cell, right side), sufficient water leaves the cell and intracellular ice formation does not occur. If the cooling rate is faster, ice will form inside the cell, and the amount of ice that forms and the size of the crystals will depend upon the cooling rate. Intermediate cooling rates (middle cell, right side) result in partial cell dehydration, larger crystals, and less intracellular ice. Very rapid cooling rates result in virtually no cell dehydration, a greater amount of ice formation, but smaller crystals. If cells cooled very quickly (lower cell, right side) and warmed slowly, the small crystals will aggregate to form larger crystals (a process known as recrystallization). This will be more damaging to the cell compared to very rapid warming.

Knew the degree of cell dehydration necessary to prevent damaging ice formation and the membrane permeability, it would be possible to predict the cooling rate which would prevent intracellular ice formation and the temperature at which cooling could stop and the sample could safely be transferred to liquid nitrogen. See the original description of mammalian embryo cryopreservation for a relevant example. Ice formation in the cytoplasm of cells is not necessarily damaging. Studies over the years have investigated the correlation between the morphology of cytoplasmic ice, cooling and warming rates, and survival (Bank, 1973; Shimada and Asahina, 1975). These studies showed that larger ice crystal size correlated positively with cell death. As discussed above for bulk solutions, devitrification and recrystallization can occur to cell water/ice during warming resulting in the growth of large ice crystals. Recrystallization is the phenomenon whereby small ice crystals fuse to become larger crystals which are more stable at a given temperature. According to a more recent study, the formation of small ice crystals may actually be beneficial to cell survival. Such a result is likely due to the reduced level of osmotic stress.

Attempts have been made to explain the mechanism(s) which cause intracellular ice formation, and to date several theories have been put fourth. As explained in more detail by Mazur (Mazur, 2004), theories of intracellular ice formation must account for several experimental facts: (1) in order for ice formation to occur in cells above ~ -30 °C, extracellular ice must be present, and the proximity of the cells and ice is important; (2) extracellular ice is not a necessary precondition for intracellular ice below ~ -30 °C; (3) intracellular ice formation usually happens immediately if the cells and the surroundings are supercooled -15 to -20 °C and extracellular ice is rapidly initiated; (4) if extracellular ice forms near the cell's equilibrium freezing point, intracellular ice usually does not form; and (5) the nucleation temperature decreases substantially if the extracellular solute concentration increases.

Several lines of evidence suggest that intracellular ice formation can be triggered from more than one mechanism. It is generally agreed that ice formation below -30 °C in the absence of extracellular ice is due to the presence of intracellular nucleators (or as a result of homogenous nucleation at lower temperatures). The fact that intracellular ice formation above this temperature requires the presence of extracellular ice strongly suggests that the extracellular ice is acting to nucleate the intracellular ice. As extracellular ice does not nucleate intracellular ice at low degrees of supercooling, an intact plasma membrane effectively blocks the passage of ice into the cell. However, the plasma membrane is implicated mechanistically in the major theories put fourth to explain the initiation of intracellular ice formation.

Mazur hypothesized that ice crystals can grow through membranes *via* protein pores like aquaporins. While evidence exists that ice can grow through channels which connect cells (*i.e.* gap junctions, the pore size in these channels is much larger than those in aquaporins, making ice growth more likely. Mazur and colleagues are currently using genetic engineering techniques in oocytes as a means to test this hypothesis directly (*cf.* (Mazur *et al.*, 2005a; 2005b; Guenther *et al.*, 2006) for results from initial experiments). Toner and colleagues have suggested that ice interaction with the plasma membrane causes a structural change to the inner membrane surface, resulting in an increase in the efficiency of ice nucleation. Muldrew and McGann have put forth a different mechanism altogether which suggests that ice grows through the membrane after the formation of a lesion as a result of the osmotic pressure gradient and resultant water efflux. A similar argument regarding the formation of a membrane lesion as a prelude to intracellular ice formation had been proposed by Steponkus and colleagues several years earlier.

Despite the debate as to the nature of intracellular ice formation initiation, the evidence that IIF is a very damaging (and usually lethal) event is overwhelming. Hence, preventing ice formation in the cytoplasm (at least to a large degree) during cryopreservation is critical.

8.5 CELL DEATH AT SUB-OPTIMAL COOLING RATES AND THE ROLE OF CRYOPROTECTANTS IN MITIGATING CELL DAMAGE

Given that the probability of IIF decreases proportionally with cooling rate, an obvious question is: why not just cool cells at very slow rates to prevent damage from intracellular ice? The simple answer is because IIF is not the only cause of cell damage during freezing, and some of the other causes are more detrimental at lower cooling rates. When cell survival is plotted as a function of cooling rate, an inverted "U-shaped" curve is generated (Figure 4). The peak of this curve represents the optimal cooling velocity, cell viability being lost at both higher and lower cooling rates. The shape of this curve has been interpreted to suggest at least 2 mechanisms of cell damage, each of which is oppositely dependent upon cooling rate. As described above, damage due to IIF can explain the loss of cell viability at cooling velocities higher than optimal (the right side of the curve). The mechanisms responsible for the loss of viability at lower than optimal cooling rates are varied, and we still do not have a comprehensive understanding of the nature of these injuries. Nevertheless, a general idea has emerged as to the nature of some injuries which occur at slow cooling rates. One of the principal mechanisms of cell injury at suboptimal rates of cooling is directly tied to the nature of CPA.

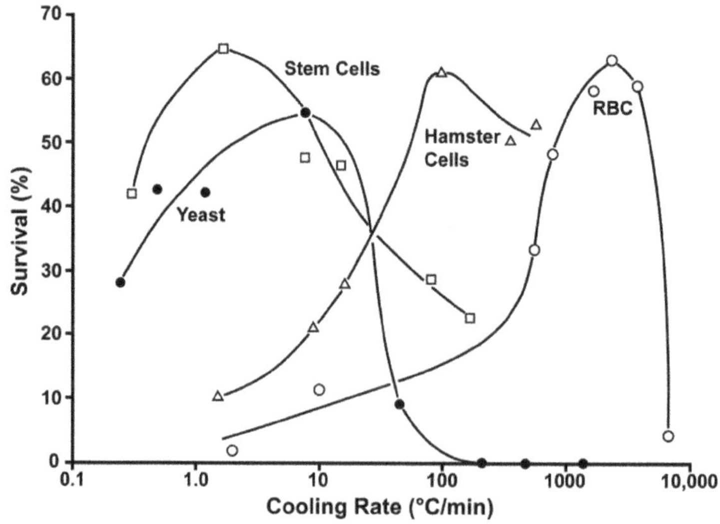

Fig. 4: The relationship between cooling rates and cell survival, for different cell-types, is shown. Higher water permeability allows faster cooling rates to be applied without a high probability of cell death due to intracellular ice formation. The specific optimal rate is also dependent upon other factors such as the presence of CPA and the warming rate.

8.6 CRYOPROTECTANTS: BENEFICIAL EFFECTS

Cryoprotectants are defined in a functional manner as compounds that allow a higher degree of cell survival during freezing in their presence than in their absence. Although our knowledge of the mode of action of CPA is still incomplete, it is likely that the effects of CPA are multi-factorial, and CPA of different classes (*e.g.* alcohols, sugars, diols, amides, large polymers) may act by different mechanisms. One of the earliest theories of the mechanism of action of CPA was developed from a series of experiments investigating the protective action of glycerol on erythrocytes. James Lovelock, a physical chemist by training, studied the effects of salt concentration on hemolysis. In his initial experiment he investigated the relationship between the salt concentrations in partially frozen saline solutions which caused cell damage with the degree of cell damage when cells were exposed to the same salt concentrations without freezing. He determined that the degree of damage could be explained by the increase in salt concentration due to ice precipitation (Figure 5).

Since that time, others have repeated these experiments and confirmed Lovelock's original findings. Lovelock proceeded to show that when cells are frozen in solutions containing glycerol, the temperature at which hemolysis began was progressively lower as the amount of glycerol was increased. As discussed above, solutes depress the equilibrium freezing point of a solution. By adding glycerol to a cryopreservation solution, the amount of water that freezes at any given temperature will be reduced. As a consequence, the final concentration of the salts in the remaining solution will also be reduced. Lovelock's experimental results supported the conclusion that the colligative depression of the freezing point and concomitant reduction in salt concentration explained the protective mechanism by which glycerol exerted its effect (Figure 6). Hemolysis always began at the same concentration of sodium chloride (~0.8 M). Several other modes of action have been proposed for CPA. One effect includes interacting with water molecules and altering the water structure in a solution, and reducing the ability of water to join the ice phase. Polymers can also facilitate vitrification upon cooling and reduce the concentration of permeating CPA necessary to attain a glassy state. CPAs have also been shown to either directly interact with or be preferentially excluded from biosurfaces (*e.g.* the surface of lipid bilayers or proteins). The apparent opposite nature of these modes of interaction seems to suggest opposite effects.

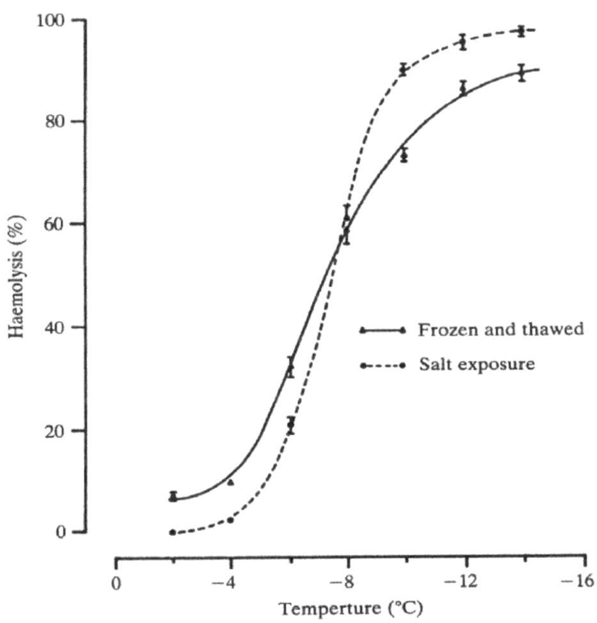

Fig. 5: The correlation between hemolysis and salt concentration for cells frozen or only exposed to salt is shown. The striking correlation lead Lovelock to conclude that the concentration of salt resulting from ice formation was a primary mechanism of cryodamage. Others have argued that the effect of the salt concentration on cell volume was the true cause of cell damage.

However, each mode of interaction can be beneficial to the stability of these structures. In addition, Rudolph and Crowe have shown that trehalose and proline can prevent freezing-induced fusion of lipid vesicles.

Perhaps it is not surprising that many organisms living in climates where freezing temperatures are encountered have evolved to include the metabolic production of

CPA as a survival strategy. As discussed by Erica Benson and reviewed by Ken Diller the cryoprotective properties of sugars and glycerol in plants were described by Nikolay Maximov in the early 20th century. The farsighted nature of his conclusions is remarkable considering what has been learned about CPA and their mechanisms since that time. A great deal of research has been conducted to understand the response of various members of the animal kingdom to freezing temperatures. The metabolic production of CPA is also a common strategy in these organisms. Inhibition of freezing at high sub-zero temperatures is one strategy among arthropods and fish, and is accomplished by regulative supercooling. In many instances thefreezing point of physiological solutions is regulated by thermal hysteresis proteins. The tertiary structure of these proteins allows them .

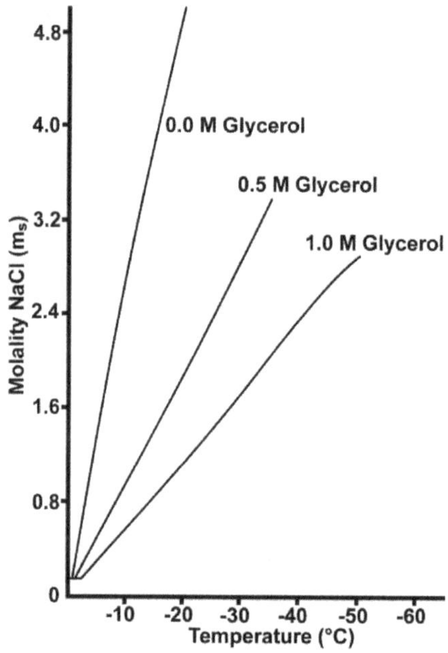

Fig. 6: The presence of CPA reduces the salt concentration at a given subzero temperature due to a colligative reduction in the freezing point of the solution (shown here for a glycerol and sodium chloride solution mixture).

Notice that, as the concentration of glycerol increases, the salt concentration is significantly reduced. To directly interact with ice crystals due to polar residues along the protein backbone. These proteins lower the freezing point of water without significantly altering the melting point (this phenomenon, the separation of the temperature of ice growth and the melting point of a solution has been termed thermal hysteresis. Thus, their mode of action is not colligative. According to a general model for their activity, these proteins bind to ice crystals and alter the radius of curvature of the growing crystal, which reduces the temperature at which it is thermodynamically favorable for additional water molecules to join the crystalline phase for more detail on an early model of the mechanism of fish hysteresis proteins, and for a recent review on the mechanism of action). Overall, these proteins restrict ice growth when the environmental temperature is slightly below the equilibrium freezing temperature of the body fluids.

As an alternate strategy, organisms across many phylogenetic groups have developed mechanisms of regulating ice formation *in situ*. In naturally freezetolerant organisms, avoiding the formation of intracellular ice is managed by actively promoting and regulating the formation of extracellular ice. This allows freeze-induced dehydration of the cells and prevents ice from forming in the cytoplasm. As a coupled strategy, mechanisms to avoid the damaging consequences of cellular dehydration and ischemia that accompany freezing have also evolved.

Ken and Janet Storey, in a review on the subject, described four requirements for the successful freeze tolerance in animals: (1) ice must be confined to extracellular spaces and damage from ice crystals must be minimized; (2) the rate of freezing must be slow and controlled; (3) cell volume reduction beyond a minimum tolerable volume must be avoided; and (4) mechanisms must be present to prevent damage from resulting ischemia. These requirements are often met through behavioral and physiological adaptations. For example, slow, controlled temperature change is often facilitated by the chosen hybernaculum of the organisms. Controlled ice nucleation can be performed by specific ice nucleating proteins in the blood. Cryoprotectant synthesis in some organisms is initiated by freezing) (see the review by Storey and Storey for more details). Membrane adaptations to cold have also been described. As improved cryopreservation methods are sought, it is likely that attention paid to nature's laboratory will provide insights into appropriate means to avoid cryodamage.

8.7 CRYOPROTECTANTS: DETRIMENTAL EFFECTS

As their name implies, CPA are beneficial during freezing. Their use, however, is not necessarily benign. Since the time of Lovelock's original work, it has been pointed out and experimentally confirmed that the correlation between the freezing damage in the presence of glycerol and the associated increase in salt concentration is strongest at low levels of hemolysis. In addition, as the concentration of glycerol is increased, the concentration of salt that causes a given degree of hemolysis decreases, suggesting that high concentrations of glycerol contribute to cell damage during freezing (Figure 7). This suggests that high glycerol concentrations (particularly as a result of ice precipitation) contribute to the damage of cells frozen slowly. Similar results have also been shown for dimethylsulfoxide.

Injury from CPA is not limited to those which occur during freezing. Exposing cells to solutions containing CPA prior to cooling can be damaging due to an osmotic effect. Almost all of the commonly used permeating CPA have lower plasma membrane permeability coefficients compared to that of water. This relationship results in cells experiencing osmotically-driven volume excursions during cryoprotectant addition to and removal from the cell during the course of a cryopreservation procedure. It has been shown in numerous cell types that damage to cells can occur as a result from volume excursions alone. Furthermore, many studies have demonstrated a beneficial effect of prolonging cryoprotectant addition and/or removal a process which reduces the associated volume excursions.

The rate of water movement across the plasma membrane is determined by several factors and can be described by equation 1:

$$\frac{dV_w}{dt} = -L_p ART\left(M_s^e + M_n^e - \frac{n_s^i}{V_w} - \frac{n_n^i}{V_w}\right) \quad (1)$$

where *dVw/dt* represents the change in the cell water volume over time, *Lp* and *A* represent the cell membrane hydraulic conductivity and surface area, respectively, *R* and *T* represent the gas constant and temperature, *M* represents molal concentration, *n* represents the number of moles and *v* represents the cell volume.The letters *e, i, s*, and *n* in the super- and subscripts represent the extra- and intracellular compartments, and permeating (*s*) and non-permeating (*n*) solutes respectively. A concentration gradient of permeating CPA will also result in movement of these compounds across the cell membrane. The rate of change in intracellular cryoprotectant resulting from such a gradient can also be described by an ordinary differential equation (equation 2):

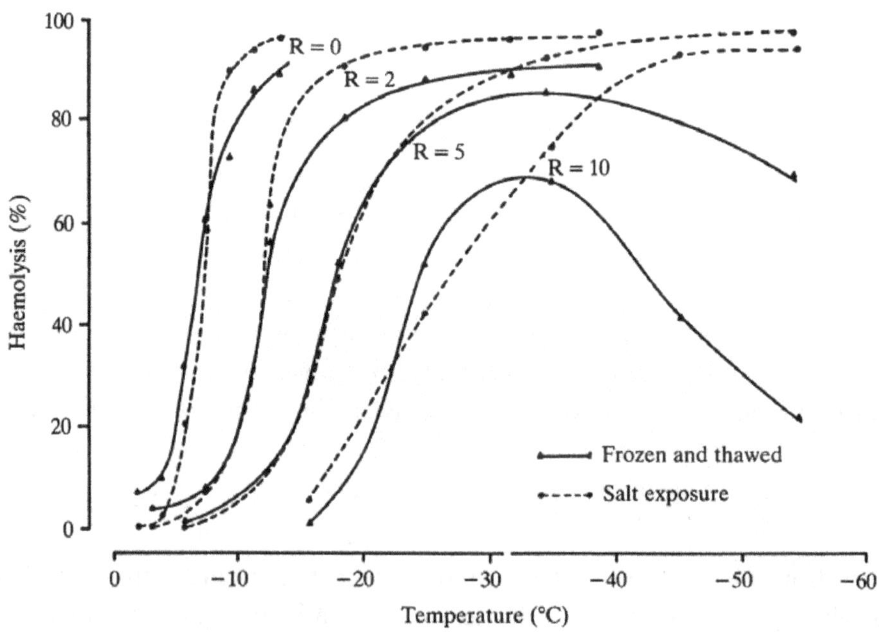

Fig. 7: The correlation between hemolysis and salt concentration is greatest at low levels of cell damage. As the ratio of glycerol to sodium chloride (R) increases, the correlation becomes weaker, particularly at high levels of hemolysis.

$$\frac{dn_s^i}{dt} = P_s A\left(M_s^e - \frac{n_s^i}{V_w}\right) \quad (2)$$

where *dn/dt* represents the change in the number of moles of intracellular cryoprotectant over time, *Ps* represents the membrane permeability, and the remaining variables are equivalent to those in equation 1. Here we have shown the so-called 2-parameter membrane transport model (*Lp* and *Ps* are the phenomenological parameters defining the permeability of the cell membrane to water and cryoprotectant). A 3-parameter model incorporating an interaction coefficient (σ) was proposed by Kedem and Katchalsky on the basis of irreversible thermodynamics for membranes where water and solute move through a common pathway. It has been argued that the interaction coefficient is not applicable to biological membranes as water and CPA usually travel through independent pathways. Furthermore, being phenomenological in nature, a 3-parameter model is less parsimonious than a 2- parameter model.

Osmotic damage is often ascribed to the associated volume reductions. Cell volume response can be controlled during cryoprotectant addition and removal by modifying the procedures for adding and removing these compounds. As a result, cryoprotectant addition and removal can be accomplished in a manner that prevents injury due to excessive volume excursions. Because the volume response of cells can be modeled on a computer when the parameters in equations 1 and 2 for the cells are known, one can proactively predict optimal methods for this process.

True chemical toxicity is also a concern associated with the use of CPA. This is particularly true for vitrification methods as very high concentrations of these compounds are necessary to achieve and maintain a vitreous state (a solid state free of crystalline structures). The precise nature of the toxic effects of CPA remains, to a large degree, uncertain.

Repolymerization after treatments is common, but the particular organization of the polymers often does not represent those of untreated oocytes. Frequently, toxicity is argued to be a significant cause of cell death in oocyte cryopreservation studies. However, rarely is the chemical effect isolated from the osmotic effect in such experiments. In a previously unpublished experiment in our laboratory, the osmotic effects associated with exposure to 2.5 M PG and EG were controlled when assessing the effects of exposing mouse oocytes to CPA by including a treatment simulating the volume excursions associated with cryoprotectant addition and removal. The results suggested that the damage associated from exposure to PG was not a result of the osmotic effects, but a true chemical effect. Exposure to the same concentration of EG was not detrimental to mouse oocyte survival (Figure 8).

It was mentioned earlier that most CPA have hydrogen bonding capability, and altering water structure is one of the mechanisms by which CPA are hypothesized to function. It is also recognized that the toxic concentration differs for different CPA. In a more recent report, Fahy and colleagues have determined that a compositional variable they call *qv** is directly associated with the toxic properties of a cryoprotectant when toxicity is non-specific. The proposal these authors make is that *qv** is related to the degree of hydration of a cryoprotectant.

It also show that the total concentration of CPA was not as strong of a predictor of toxicity as was the ratio of the molarity of water (MW) in the solution to the molarity of polar groups in the solution (MPG; *i.e.* $qv^* = MW/\ MPG$).

Polar groups were defined as hydroxyl groups (-OH), sulfoxide groups (- S=O), carbonyl groups (-C=O), and amino groups (-NH2) on the cryoprotectant.

In their experiment, they initially tested the toxicity of various CPA with differing *qv** indices using rabbit renal cortical slices and examined the K+/Na+ ratio after 31 exposure. As *qv** increased in the range from 2 to 6, the K+/Na+ ratio decreased from ~ 80% to ~ 10% relative to the controls.

The significance of the polar groups is such that they account for the interaction with water molecules, and compounds with a lower *qv** can interact with fewer water molecules. Such compounds do not form a vitreous state (*i.e* form a glass) as easily as those with a higher *qv**. Hence, weak glass forming CPA are less toxic. Using this new information, the investigators were able to predict and confirm that substitution of PG (a very good glass former) with EG (a very poor glass former) in a previously developed vitrification solution (VS41A) would be asuperior vitrification solution using rabbit renal cortical slices and mouse oocytes. In previous reports, others have shown that the protective potential of CPA was correlated to the molarity of potential hydrogen bonding groups. As discussed below, reducing cryoprotectant toxicity is one approach to improving vitrification methods.

8.7.1 Cell Death at Suboptimal Cooling Rates

Many of the factors which might contribute to cell damage as a result of freezing are interdependent. For example, ice crystal formation may have deleterious mechanical effects on cells in suspension or in tissues and the amount of ice formed and the crystal structure is dependent upon cooling rate, warming rate, and the presence of CPA. Not surprisingly, the existence of an optimal cooling rate for a cell is an oversimplification. The cooling rate at which cell survival is highest is dependent upon other factors such as cryoprotectant concentration and warming rate.

Intracellular CPA have noticeable effects on IIF. In general, in the presence of permeable cryoprotectant, cell water crystallizes at slower cooling rates than in its absence. This effect is likely a result of several factors. One includes the reduction in the water permeability of cell membranes in the presence of cryoprotectant as mentioned above. A second is likely a result of lowering the freezing point of the cytoplasm which causes a general reduction in the temperature at which a given driving force for water efflux is present. Because of the temperature dependence of water permeability, less water can move out of the cell in a given amount of time under such circumstances.

The generic term "solution effects" has been coined to collectively describe the various forms of injury to cells cooled slowly enough to preclude damaging IIF. This term reflects a notion that the damage results from the solution conditions created by ice formation as described above. Meryman and colleagues suggested that damage was a physical and not a biological event, resulting either from the osmotic dehydration of the cells and the resulting stress placed upon the cell membrane due to cell volume reduction, or a direct osmotic effect on the membrane itself. In earlier work, Meryman described a hypothesis of cryoinjury based upon the cell reaching a minimum critical volume. This later work supports this theory.

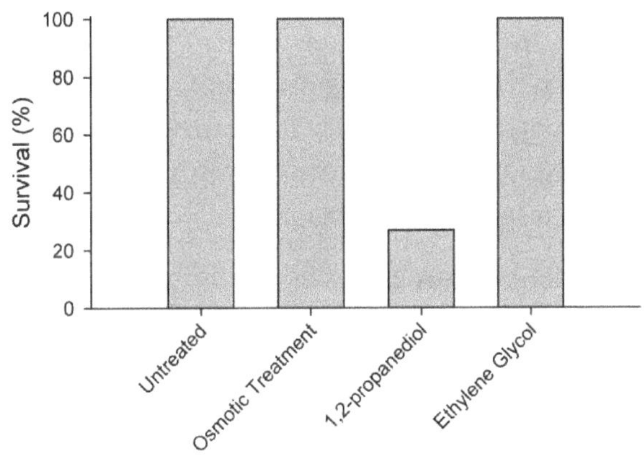

Fig. 8: The effect on cell viability of exposing mouse oocytes to 1,2- propanediol or ethylene glycol (2.5M final with 0.3M sucrose) is shown. A solution containing 0.5M sucrose which simulates the osmotically-driven volume excursions of the other treatments was included as a control. Neither stress nor ethylene glycol exposure had an effect on oocyte viability after 6-hour incubation. Exposure to 1,2-propanediol resulted in a dramatic loss in cell viability (cell lysis).

Another interesting hypothesis has been put forth in relation to the formation of ice. In a series of studies, Mazur and colleagues investigated the effects of the fraction of the solution which remained unfrozen on cell damage. Their data showed a strong correlation between survival and the unfrozen fraction when the unfrozen fraction was low (5-15%). They proposed that as the unfrozen fraction is reduced, the cells are damaged by mechanical effects of the ice and/or close apposition with other cells. When the unfrozen fraction is increased, damage was less strongly dependent on that variable and more on the salt concentration until the effects of the unfrozen fraction were lost. When the unfrozen fraction is not a damaging mechanism, the loss was attributed to the osmotic effect of the solute (both during exposure and dilution).

The interpretation of these data came under scrutiny as pointed out that the unfrozen fraction variable was confounded by the treatments used to change the unfrozen fraction. Such treatments would systematically alter the volume excursions which the cells would undergo during the experiment, and this difference could also result in the outcome seen.

Perhaps the most important message to get from this particular debate is the difficulty in designing experiments to isolate the effect of a single variable on cell damage during freezing when numerous potential variables are interdependent for an elaboration). Another good example of this is the challenge to the explanation for slow-cooling injury resulting from increased salt concentration. As discussed above, concentrated solutes cause exosmosis and result in a reduction in cell volume. Thus, either high salt concentration or volume reduction could explain the damage (it could also be an interaction of the 2 factors). The minimum volume hypothesis was strongly supported by the results of Williams and Shaw with erythrocytes following up on earlier work by Meryman.

In more recent years, the molecular mechanisms of cryodamage, particularly the induction of apoptosis, have been investigated. These authors have suggested that the trigger for apoptosis is not necessarily an immediate effect of the cryopreservation stresses, but can be delayed for several hours. Clearly, we are far from understanding all of the mechanisms which result in cryodamage.

8.7.2 Chilling Injury / Cold Shock

Even in the absence of ice, sub-physiological temperatures can have profound effects upon cells. Injury from cooling cells is often differentiated by the degree to which the rate of cooling causes the specific event. Injuries from rapid cooling are usually categorized as cold shock injuries. These types of injuries occur quickly after cooling, and are generally independent of the warming rate. In the context of cryopreservation, a significant body of literature has been produced which describes the effects of cold shock on cell membranes, particularly for spermatozoa.

A description of the liquid crystalline model of cell membranes can be found in a standard cell biology text. In general, amphipathic lipid molecules form a bilayer structure with various proteins being integrated throughout. At physiologic temperatures, there is a general fluidity such that molecular mobility is high and many of the proteins and lipids are free to diffuse laterally within the bilayer (opposite faces of the bilayer are not identical, and moving from one face of the bilayer to the other is energetically unfavorable).

The structure that lipids can take in solution is more diverse than this, however. A lamellar (*i.e.* bilayer) structure is common, but micelles, inverted micelles (micelles within the bilayer), hexagonal-II, and cubic phase structures can occur . The particular arrangements lipids take is dependent upon factors such as water activity, temperature, pH, salt concentration, and interaction with other molecules (*e.g.* proteins).

When membranes are cooled, they exhibit thermotropic behavior; that is to say they tend to undergo phase transitions. As membranes are cooled, the lipids tend to transition from a liquid-like state to a gel-like state, with the molecules being arranged in an orderly, crystalline fashion with a characteristic hexagonal arrangement. Due to the complexity of biological membranes, a transition is not like a crystallization event in a simple solution (*i.e.* a rapid precipitation), but more like a (relatively) slow lateral separation of membrane lipids into distinct domains (see Figure 1.9). Nevertheless, this transition is a distinct change from the usual lipid arrangement, and can have significant effects on membrane function.

The temperature at which this transition occurs is dependent upon several factors, including the length of the hydrocarbon chain in the lipid group, the presence and location of cis-unsaturated bonds (transition temperatures decrease as the position of the bond moves away from the polar group and toward the middle of the chain), and the concentration and valence of cations in the solution. An increase in the concentration of polyvalent cations increases the phase transition temperature, whereas monovalent cations increase lipid fluidity and decrease the phase transition temperature. The presence of cholesterol in a membrane can also affect phase transition behavior by (1) altering the ability of lipid species to transition to a gel-like configuration; and (2) increasing the disorder of the gel phase.

The propensity to develop a lamellar or hexagonal-II structure varies across lipid species. Different species tend to aggregate into domains during the phase change, and the creation of inverted micelles (hexagonal-II structures) within a bilayer can occur as a result (Figure9) . Rearrangements such as these can result in alteration to the selective permeability of membranes and loss of cell homeostasis.

Changes in the biochemistry of spermatozoa as a result of cold shock have been examined. A reduction in anaerobic glycolysis and respiration, ATP levels, Cytochrome C loss from the mitochondria, and release of numerous intracellular enzymes have all been described (reviewed in Furthermore, changes in the distribution of intracellular ions have also been noted.

Numerous compounds have been shown to confer protection to spermatozoa from cold shock. Protective agents include glycerol, phosphatidylserine, egg yolk, lecithin, milk, and albumin. The low density lipoprotein fraction of egg yolk is particularly effective at preventing cold shock injury , with phosphatidylcholine being a particularly active component.

The results from Quinn and colleagues suggest that the effect is a result of interactions with the surface of the membrane, and not as a result of components intercalating within the lipid bilayer. The mechanisms of these compounds are not fully understood, but one model for the effect of adhering CPA on phase separations of membranes has been put fourth and is shown schematically in Figure 10. Studies during recent years have focused on the genetic regulation of the response to cold shock. While it seems a bit ironic, members of the heat-shock protein family are often upregulated as a response to cold stress. Other proteins that have been referred to as cold-shock proteins are also upregulated as a response to cold temperatures. Compared to that which is known in prokaryotes, our knowledge of the genetic response to cold temperatures in eukaryotic cells is just developing. For further review of these topics, see and references therein.

8.8 VITRIFICATION AS AN ALTERNATE TO EQUILIBRIUM FREEZING

Vitrification can be employed as an alternative to equilibrium freezing. The obvious benefit of this approach is that the damage due to IIF can be completely avoided. Unfortunately, other kinds of damage are much more likely to occur when using this method. Recalling the initial discussion of phase transitions in solutions during cooling, it was mentioned that with a sufficiently high solute concentration, ice formation could be avoided altogether. The easiest way to achieve vitrification would be to use a solution that has a concentration as indicated in section IV of Figure 1. Water in solutions with such a composition will not crystallize nor de-vitrify even at cooling rates applicable to equilibrium freezing methods. Unfortunately, the toxicity of such solutions is too high to make them practical. Similarly, solutions with concentrations around 60 weight % (Region III) are often too toxic to be useful, although they can also be cooled slowly without crystallization. Various strategies have been described to counter the potential toxicity of solutions, and include: (1) the use of a combination of solutes, each of which is below a concentration that is very toxic, yet in combination will facilitate vitrification; (2) the substitution of polymers in the extracellular medium for the smaller permeating agents; (3) the application of hydrostatic pressure; (4) the use of compounds which counteract the toxicity of other agents (*e.g.* acetamide with dimethylsulfoxide); and (5) reducing the time for which and/or the temperature at which the biomaterial is exposed to high concentrations of CPA. Hydrostatic pressure has an effect by shifting the Th (lower) and Tg curves (higher) such that their intersection point occurs at lower concentrations of solutes. As discussed above, when solution concentrations are reduced, the likelihood of devitrification during warming increases. Hence, the warming rate is an especially important consideration when designing vitrification strategies. In the first report of successful vitrification of mammalian embryos, Rall and Fahy used a combination of these strategies (1, 4, and 5) to overcome the toxicity associated with the vitrification solution (VS1). Mouse embryos (8-cell) could survive exposure to the solution for up to 15 min at 4 °C; survival dropped precipitously with increasing time. Furthermore, loss of viability occurred when samples were cooled slowly compared to rapidly when the concentration of VS1 used was reduced by 25%, and when the warming rate was reduced to 10 °C/min from 300 or 2500 °C/min.

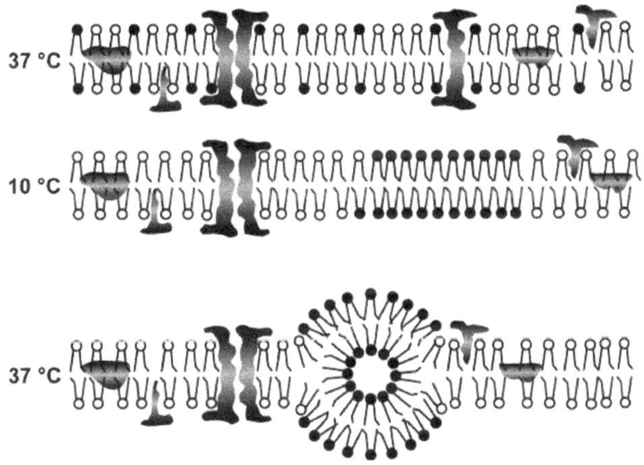

Fig. 9: A model of temperature-induced phase changes to membranes is described in this figure.

In the upper portion of the figure, a typical biomembrane is shown, with various integral membrane proteins and lipid species. As the temperature is reduced from physiologic to hypothermic (10 °C in this instance), lateral redistribution of the various molecules occurs, with lamellar-forming lipid species (represented with white polar groups) and exagonal-II-forming lipid species (represented with black polar groups) separating into distinct domains. Upon warming, an inverted micelle structure is created by the hexagonal-II-forming lipids. Such a configuration could result in a significant disruption of the membrane selective permeability, and the possibility of membrane failure and cell death.

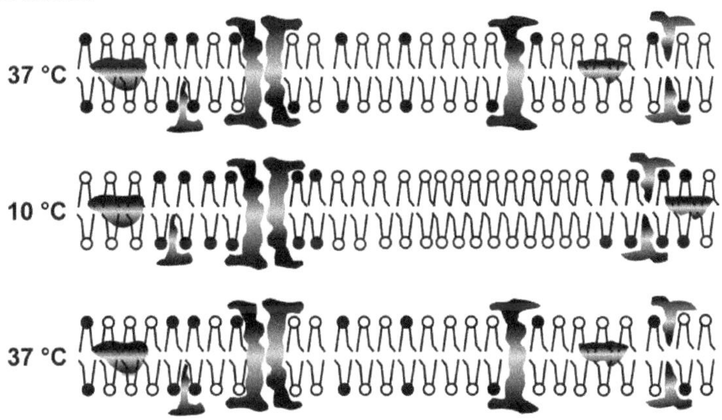

Fig. 10: A model of the effect of CPA on preserving biomembrane stability.

In the presence of CPA (not shown schematically), the hexagonal-II -forming lipid species preferentially associate with membrane proteins during cooling and only lamellar-forming lipid species segregate into distinct domains. When the cell is warmed, the hexagonal-II-lipid – protein interactions prevent a non-lamellar transition, promoting the return to a normal bilayer configuration.

These results support the idea that vitrification at moderate solute concentrations can be precarious. This region is labeled "doubly unstable" on the chart, as the use of solutions in this range are likely to be highly nucleated upon cooling, and prone to devitrification and recrystallization upon warming. Reducing the solute concentrations in combination with higher cooling rates is a strategy being currently employed for vitrification of mammalian oocytes.

However, modification of the cooling and warming rate is not the only strategy that might lead to improvements. As has been discussed throughout this chapter, several types of cryoinjury exist, and the development of optimal cryopreservation strategies will require that all of these factors are taken into account, along with the specific cryobiological properties of the cells under study.

8.9 MAMMALIAN OOCYTE DEVELOPMENT: GENERAL AND COMPARATIVE ASPECTS

In mammals, ovarian follicles begin to form during the early stages of fetal development, for humans, it has been estimated that, at birth, the ovary contains between 250,000 and 500,000 follicles. Most of these follicles are in the primordial stage of development, characterized by low metabolism and having small oocyte surrounded by a single layer of squamous, pregranulosa cells. The entire structure is encapsulated by a basement membrane. Entry of these quiescent follicles into the growth phase occurs at slow rates over time, with the majority remaining in the primordial state for months to years. The regulation of this transition remains ill defined, although factors such as the *Kit* receptor and its ligand (Kitl) and anti-Müllerian hormone (AMH) have been implicated as possible contributors. When these follicles begin to develop, the first sign of a change occurs in the somatic cells; They transition from having a squamous to a cuboidal morphology and are positively labeled with [3H] thymidine; at this stage the structure is termed a primary follicle (~80 – 100 μm in diameter in humans). The somatic (granulosa) cells then undergo clonal expansion, resulting in significant proliferation and an increase in the number of cells which surround the oocyte.

Over the course of several weeks, additional changes occur to the developing follicle. Proliferation of the granulosa cells continues, the zona pellucida begins to form, and a second layer of somatic cells (theca) develops outside of the granulosa cells. At this stage, the diameter of the follicle in humans attains a size ranging from approximately 100 – 150 μm.

The early stages of follicular development appear to be directed by factors contained within the ovary (Rajkovic *et al.*, 2006); however, later stages of follicular development (early-antral stages and beyond) require pituitary hormones (*i.e.* follicle-stimulating hormone;. As granulosa cell proliferation continues, the follicle increases in size and fluid-filled cavities begin to appear among the cells. Over time, these cavities coalesce to form a single antrum, and in humans, the follicle attains a size of about 2 mm. This process of cell division and antral expansion continue over the course of another few weeks, and, if selected to continue growth, the preovulatory follicle in humans eventually approaches 2 cm in size. Final oocyte maturation is triggered by the mid-cycle gonadotropin surge, culminating in ovulation.

The mature oocyte is one of the largest cells in mammals, with a diameter of approximately 70 μm in mice and 120 μm in humans, cows, and pigs. In most vertebrates, oocytes progress to the second meiotic division stage near the time of ovulation, and maintain a meiotic arrest until fertilization (Figure 12). As such, the chromosomes are maintained in a condensed state and positioned on the metaphase plate of the meiotic spindle. While in general oocytes from mammals are similar, some differences do exist that are thought to be important determinants to cryosurvival. As mentioned above, mouse oocytes are smaller in size than oocytes from humans, cows, and pigs (although they all are approximately spherical in shape). The larger oocytes have a smaller surface-tovolume ratio, which makes the rate of water flux under osmotic influences slower, all else being equal. This difference could influence the probability of IIF (*cf.*, as for a given cooling rate, more water will remain trapped inside of the cell over time (again, assuming the membrane permeability properties are equal). Another notable difference in oocytes across mammalian taxa is the amount of cytoplasmic lipid inside of the cell (see Figure 13). Removal of cytoplasmic lipids from porcine embryos rendered them less sensitive to cooling, and presumably the presence of lipid in the cytoplasm of oocytes also makes them chilling sensitive.

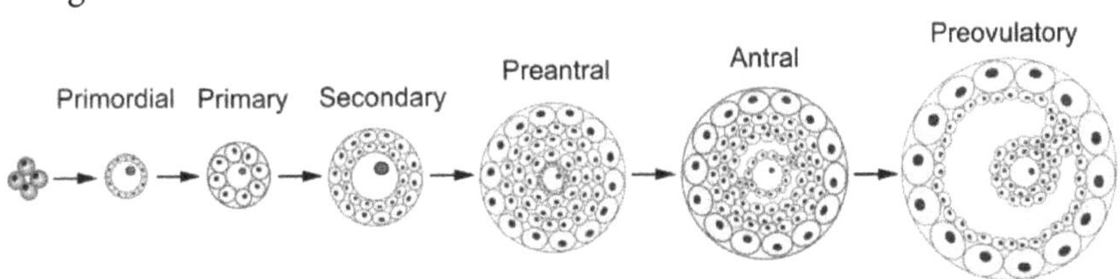

Fig. 11: A diagram showing the development of a hypothetical ovarian follicle is shown. Primordial germ cells initially enter the gonadal ridge and eventually develop into primordial follicles. After re-initiation of development, the follicle progresses through several developmental stages (primary, secondary, preantral, and antral) with proliferation of the inner granulosa cells and outer theca cells. If the follicle is selected to continue development, it progresses to the preovulatory follicle stage, containing a large fluid-filled cavity (antrum) with the oocyte surrounded by cumulus granulosa cells.

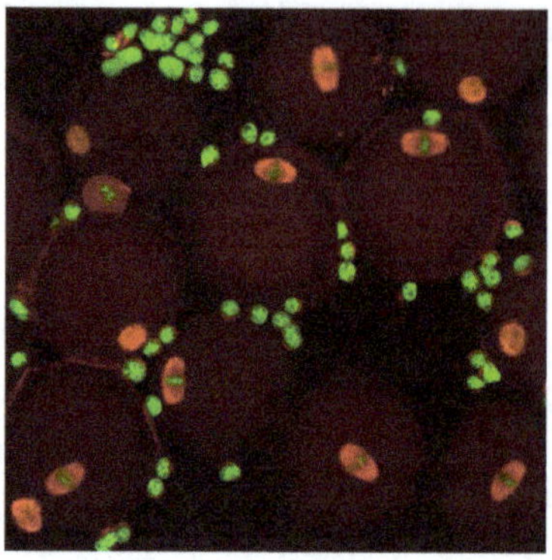

Fig. 12: Immunocytochemical staining of the MII spindle in mouse oocytes (red) and counterstaining of the chromosomes (green). The smaller green spots on the edge of some of the oocytes represent nuclear staining of granulosa cells that remained attached to the oocytes after collection.

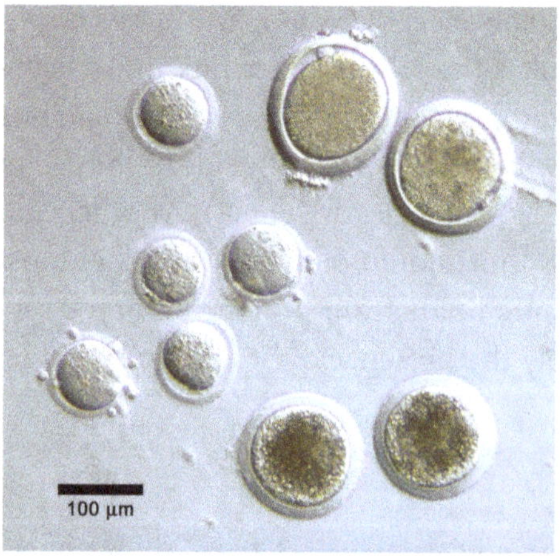

Fig. 13: Photomicrographs of oocytes from mice (smallest; n=5), cows (n=2, upper right), and pigs (n=2; lower right) are shown. Notice the difference in darkness of the cytoplasm in these examples, particularly in the cow and pig oocyte compared to the oocytes from the mouse. A human oocyte has a lipid concentration similar to a mouse. The high concentration of lipids is one proposed means that renders oocytes (and embryos) from pigs and cows more sensitive to cooling compared to oocytes from mice. See text for more details.

The shape and orientation of the meiotic spindle also differs between human, cow, pig, and mouse oocytes. The spindle in mouse oocytes is barrel-shaped, with a pole-to-pole distance longer than the distance across the metaphase plate (see figure 13). However, spindles from cow and pig oocytes are nearly spherical. Human oocyte spindles are intermediate to these two shapes. The MII spindle of mouse oocytes is positioned tangentially to the oolemma at metaphase arrest, with the metaphase plate being orthogonal to the oolemma. The spindle from bovine, porcine, and human oocytes is positioned such that the metaphase plate is parallel to the oolemma (or orthogonal relative to the spindle in mouse oocytes (see figure 14). Another interesting biological difference between oocytes from mice (and other rodents) and most other mammals is the inheritance pattern of the centrosome. In most mammals, centrosomal proteins are biparentally inherited and the centrioles are contributed to the oocyte by the fertilizing spermatozoon. In mice, however, centrosomes are contributed wholly by the oocyte with centrioles being generated *de novo* around the blastocyst stage The metaphase spindle in mammals has been shown to be easily disrupted by reduced temperatures and cryoprotectants . One significant difference in the spindle properties that is relevant to cryobiology is that the spindle in mouse oocytes is easily repaired after such damage (Pickering and Johnson, 1987); this is not the case in oocytes from other mammals, however. In general, a higher proportion of mouse oocytes can be cryopreserved more easily compared

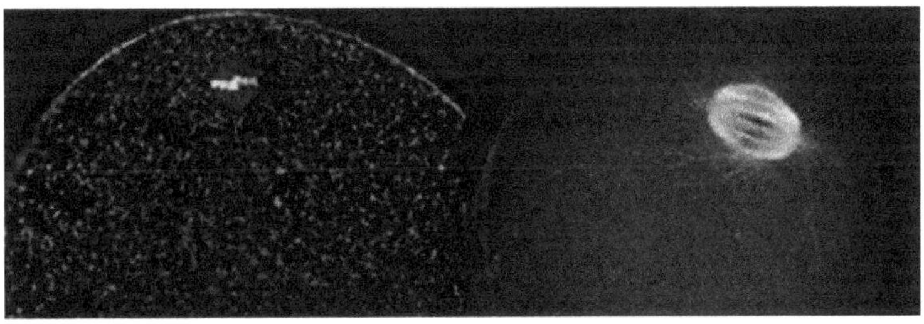

Fig. 14: Immunostained MII spindles in a human oocyte (left) and a mouse oocyte (right) are shown. Notice the difference in size and orientation of the spindles in the human oocyte compared to the mouse oocyte. to oocytes from other mammalian taxa . This difference has lent credibility to the hypothesis that disruption of the MII spindle is a primary mechanism for cryoinjury in oocytes from most mammals.

8.10 SIGNIFICANCE OF MAMMALIAN OOCYTE CRYOPRESERVATION

Although adequate means for the cryopreservation of preimplantation embryos are available across many taxa, methods for cryopreserving oocytes for these species are less successful. Mature oocytes have several unique characteristics, such as the large intracellular water volume, cortical granules, and meiotic spindle, which make them vulnerable to damage during cryopreservation. To date, many studies have been published describing attempts to cryopreserve oocytes from mammals using an empirical approach without taking into consideration the fundamental cryobiological properties of the cells. It is my thesis that developing a comprehensive understanding of the fundamental cryobiology of oocytes will yield insight into developing efficacious means to cryopreserve these cells.

8.10.1 Role of Oocyte Cryopreservation in Human Reproductive Medicine

For many individuals, having a family defines their sense of self and place in their community, and the desire to have children is one of the strongest emotional drives in humans. Unfortunately, many couples (estimated to be as high as 40%) of reproductive age would be infertile without medical intervention. For many individuals, the experience of infertility is a serious life crisis. Women often describe infertility as being the worst crisis they have ever faced; having symptoms of depression and anxiety that are similar in women facing heart disease, cancer, and incurable infections. Emotions such as anger, grief, frustration, and guilt are often felt by persons experiencing infertility.

The Society for Assisted Reproductive Technology and The American Society for Reproductive Medicine in conjunction with the Centers for Disease Control and Prevention compile annual statistics on human assisted reproductive technology (ART) activities performed in the United States. In 2002, nearly 86,000 cycles of ART treatments were initiated. Just over 14,000 transfers using cryopreserved embryos were performed during that year with approximately 3500 deliveries reported from those cryopreserved embryos and a live birth rate per transfer ranging from 28 % for women under the age of 35 to 17 % for women between the ages of 41-42 (National center for chronic disease prevention and health promotion, 2004). Live birth rates per transfer of nonfrozen embryos ranged from 43 % to 15 % for these respective age groups. A similar report of the work completed in 2002 from 25 European countries has recently been produced (Andersen *et al.*, 2006). This report documented nearly 325,000 cycles initiated from 631 clinics. In 13 countries for which all clinics reported their data, more than 175,000 cycles were reported, and the clinical pregnancy rate for *in vitro* fertilization (IVF) and intracytoplasmic sperm injection (ICSI) was approximately 30 %. Deliveries *per* transfer ranged from ~10 % to >30 % across clinics.

When IVF is utilized as a treatment option, a series of interventions are utilized. Initially, the woman's menstrual cycle is artificially controlled through the administration of hormones to increase the number of oocytes available for fertilization. In the majority of cases, more oocytes are collected than initially used; yet due to the lack of adequate methods for preserving oocytes, it is common for the entire cohort to be subjected to IVF. This usually results in the creation of supernumerary embryos. Commonly, the remaining embryos that look healthy are cryopreserved so future transfers can be made in the event that the first transfer fails to produce a child. As a result of this practice, the number of human preimplantation embryos currently in cryostorage is estimated to be on the order of several hundred thousand.

Unfortunately, there are significant social concerns when it comes to human embryo cryopreservation. These issues have lead to increases in societal regulation and concomitant increases in costs. Having successful methods to cryopreserve oocytes would help to alleviate these concerns by allowing the creation of only the numbers of embryos deemed appropriate for transfer, yet would maintain the potential for future attempts at pregnancy should the initial one fail. Such an approach has beenmandated by law in Italy, leading to a significant increase in the numbers of oocytes being subjected to cryopreservation cycles.

Additionally, having successful methods available to cryopreserve human oocytes would allow the establishment of oocyte banks to assist those female patients who cannot produce viable oocytes. Oocyte cryopreservation would not only obviate cycle synchronization between donor and recipient, a method which is not always successful, but it would allow sufficient time for seroconversion of the donor in the event of a recent infection with a transmittable disease. Such practices are routine in semen banks and are facilitated by the ability to cryopreserve human semen.

Another area in human medicine where oocyte cryopreservation could prove beneficial is cancer therapy. With the use of aggressive chemo- and radiotherapy, as well as bone marrow transplantation, greater than 90 percent of many cancers of children and young women can be cured. One unfortunate iatrogenic side effect of these treatments is the loss of gonadal function and subsequent infertility. Although promising, ovarian tissue freezing is still experimental in nature. Additionally, *in vitro* maturation techniques for ovarian follicles are presently experimental. Having a means to preserve mature oocytes could preserve the fertility of young women undergoing such therapy. Oocyte cryopreservation is also important given that the average age of childbearing continues to increase and embryo cryopreservation is not desirable for many women without partners.

8.10.2 Role of Oocyte Cryopreservation in the Use of Animal Models for Human Health and Disease

Cryopreservation of reproductive cells dates as early as 1776, with the report by Spallazani of sperm freezing. The use of slow cooling rates as well as CPA has led to improvements in overall post-thaw viability in cellular cryopreservation. Successful freezing of mouse embryos was first demonstrated in 1972. Continued research has improved post-thaw viability significantly, with embryo cryopreservation often yielding post-thaw survival rates greater than 90 percent. Oocyte cryopreservation, on the other hand, is much less successful.

Two technological advances occurred in the 1980's that had enormous implication in the ability to study physiology and disease in mammals on a genetic level. The successful isolation of mouse embryonic stem cells reported in 1981 and the development of targeted gene disruption strategies, pioneered by two groups allow investigators to alter the genomes of mice with the precision of a single base-pair. Since these techniques have become routine, hundreds of genetically-modified mouse strains have been created.

Given their value as models for biological systems, it is desirable to preserve these strains indefinitely for future studies. Historically, this has been performed by the maintenance of breeding colonies. Expenses for this maintenance not only include food and bedding, but also building maintenance and disease monitoring and control. Therefore a substantial cost is being borne by the research community, thus society, for the care of these animals (Sharp and Mobraaten, 1997). Cryopreservation of germplasm offers a way to reduce these costs by allowing the maintenance of "frozen" colonies as a backup in the event of a disease outbreak or other emergency, or as a full replacement of breeding colonies that are not currently in use but whose maintenance is desired for future research. Additionally, a frozen colony derived from founder animals would help alleviate inbreeding depression and genetic drift, thus maintaining overall genetic vigor while minimizing the number of animals necessary for maintenance of a live colony. While oocyte cryopreservation may not offer the most efficient means to preserve such rare genotypes, it may play a role in special cases such as when sperm cryopreservation in inefficient.

Genetic modification of animals such as sheep, cattle and swine also offers benefits to progress in biomedicine, including the creation of pharmaceuticals, and may even allow the production of tissues and organs for xenotransplantation. Furthermore, in some instances, these organisms are preferred models for human physiology and pathology. Genetic engineering technology can be applied to livestock species for the creation of transgenic and knockout organisms for physiological studies, similar to that described for mouse models above.

Due to the lack of embryonic stem cell technology in livestock species, the use of nuclear transfer (NT) technology is relied upon for the creation of targeted genetic modifications in animals such as sheep, cattle and swine. At the present time, NT technology is relatively inefficient, and relies on the ready access to mature oocytes. Having an efficient means to cryopreserve oocytes would ensure a steady supply of these cells for NT procedures, and could alleviate inefficiencies associated with seasonal variability of oocyte quality. The value of a ready supply of oocytes for research is evident by the development of commercial operations that sell oocytes. Under the current system, oocytes have to be shipped and used shortly after acquisition from abattoir-derived ovaries. Having effective means to cryopreserve oocytes might improve operational efficiencies by loosening this restriction, and also allow stockpiling of oocytes during the best season as just described.

8.10.3 Role of Oocyte Cryopreservation in the Management of Domestic Livestock

Initially with the use of artificial insemination (AI) with frozen-thawed semen, and continuing with other methods of assisted reproductive technologies (ART), increases in the average value of domestic animals progressed much more rapidly than would have been the case using natural mating. This is due to the much broader dissemination of reproduction events from elite animals than would be feasible without ART. Harvesting oocytes can offer significant additional increases in offspring potential of females. One estimate suggests that up to 300 offspring can be produced from a single animal. By using *in vitro* produced (IVP) embryos, significant improvements in reducing the generational interval can be attained. This technology would also allow the rescue of gametes from otherwise infertile animals, as well as the use of abattoir-derived oocytes that are increasingly being used in research. Although maximizing reproductive potential of individual animals can have detrimental inbreeding effects, it is expected that with careful management this can be held to acceptable levels. Genetic engineering technology also has the potential to benefit food and other farm-related products, The maximum potential of such improvements are unrealizable without efficient methods to preserve oocytes. The viability of mature oocytes is measured on the scale of hours. Even though methods exist to mature bovine oocytes *in vitro*, this only prolongs their potential by an additional 24 hours. Additionally, the creation of highly valuable transgenic animals as discussed above will make successful cryopreservation critical. For one thing it will allow international distribution of these genotypes. Secondly, it will facilitate the creation of repositories in the event of a catastrophe such as disease outbreak or natural disaster. Any event that has the potential to destroy the living animals must be considered, and preparations for such events must be made. While application of embryo cryopreservation has the potential to address these issues to a large degree, having cryopreserved gametes will allow more flexibility in future decision making. For example, it would allow planned genetic recombination through fertilization, allowing the creation of cross-bred animals.

8.10.4 The Sensitivity of MII Mammalian Oocytes to Various Steps in the Cryopreservation Process - Focus on Oocytes from Mice, Cattle, Pigs, and Humans Sensitivity of MII Oocytes to Cooling Below Physiologic Temperatures without Intracellular Ice Formation Effects on Development

Many cell types are adversely affected by cooling below the temperatures at which they normally function. The rate of cooling can play a significant role in determining the viability of cells after cooling. As described earlier, cold shock is a term defined as an adverse effect of abrupt cooling to a lower temperature. It has been shown that cooling cells slowly to temperatures at which they are susceptible to cold shock injury can be innocuous. The term chilling injury is used when an adverse effect occurs irrespective of the cooling rate. Mammalian oocytes are susceptible to chilling injury or cold shock. However, oocytes from some mammals are less sensitive than others, and this difference lends some insight into possible mechanisms which result in the loss of viability after cooling. Investigations into injury from cooling oocytes (and early embryos) were pioneered by M.C. Chang in the middle of the 20th century using a rabbit model. In his first report, he showed that oocytes could undergo cleavage divisions after having been stored at 10 or 0 °C for 24 hours (~80% and 40% respectively). In the second report, however, investigation into further development after storage at either temperature for 6-7 hours or 30-31 hours showed that only 3-11% developed normally for 20 – 26 days (average gestational length for rabbits is approximately 31 days.

Injury to oocytes, as a result of cooling below physiologic temperatures, from cattle, mice, humans and pigs has also been investigated, to various degrees. The earliest investigation on the effects of cooling oocytes from these groups appeared in 1959, with the report on mouse oocytes by Sherman and Lin. They showed that mouse oocytes can tolerate cooling to 0 °C (at 0.2 °C/min or 1.0 °C/min) with immediate re-warming. Viability was unaffected as measured by morphological assessment and supravital staining. Furthermore, the proportion that survived IVF and embryo transfer was not different in comparison to untreated oocytes (gestation was allowed to proceed for 19 days; gestation for mice is 19-20 days. A lower temperature (-10 °C in the presence of 5% glycerol) or longer storage times proved deleterious, with very low survival after -10 °C exposure, or no survival after 24 hour exposure at any sub-physiological temperature. About 50 % of the oocytes were able to survive storage at 5 °C for 6 hours relative to those abruptly cooled to and then immediately warmed from this temperature. In 1995, two additional reports appeared which described similar experiments. The robust nature of mouse oocytes to cooling to 0 °C at relatively slow rates (the precise rate was not given in the report, but the cells were cooled from room temperature in a ¼ cc straw to near-zero by placing the straw into an ice bath) was confirmed by Bouquet and colleagues), but somewhat contradicted by Hunter and colleagues (Hunter *et al.*, 1995). In the later study, both rapid cooling (> 1000 °C/min) and slow cooling (3 °C/min) to 0 °C were investigated. The proportion undergoing fertilization and further development to hatching blastocysts was recorded. Both rates of cooling affected the proportion of viable cells, with fertilization being reduced from 95 % to 75 % and 83 % for the rapid and slow cooling groups, respectively. Further development was also reduced to 67 and 69 % for the respective groups compared to 92 % for the controls. It has been shown that oocytes from cattle are extremely sensitive to cold temperatures, with very short exposures at some temperatures having dramatic effects on developmental potential. Arav and colleagues, in a study focusing on the effects of cooling bovine oocytes to 13 or 4 °C, showed that polyspermy was more prevalent in oocytes rapidly cooled to these temperatures (20 – 25 % polyspermy in cooled oocytes *vs.* 12% in controls). Further development was not reported. Martino *et al.* reported on the effects of cooling *in vitro* matured bovine oocytes to 20, 10, or 0 °C for 30 min, and the effects of holding oocytes at 0 °C for various times up to 30 min, on *in vitro* development.

All cooling was conducted at rates of about 120 °C/min. While cooling to 20 °C showed no effect on fertilization and development, cooling to 10 or 0 °C was damaging, with a reduction in blastocyst development to 8 and 3 %, respectively. They also showed that polyspermy increased as a result of cooling. They proceeded to analyze the kinetics of injury at 0 °C, and demonstrated that cooling for short periods of time (as short as 30 seconds) had a noticeable effect on development; the proportion of oocytes developing to blastocysts was reduced by 50% after 30 seconds, and nearly all oocytes failed to develop to blastocysts after holding for 30 min. These results were confirmed a short time later by Azambuja *et al.*, where they also showed that cooling *in vitro* matured bovine oocytes to 10 or 0 °C for 5, 10, or 20 min resulted in the loss of the developmental potential of nearly all of the oocytes. They went on to show that changes to the zona pellucida as a result of cooling could account for some of the developmental failure, with approximately 30 % and 38 % fewer of the oocytes being cooled to 10 and 0 °C, respectively, undergoing fertilization due to effects on the zona (zona free oocytes had a higher probability of sperm penetration).

Only limited basic research into the effects on development of cooling human and porcine oocytes has been performed. In 1992, Bernard and colleagues published a report on the effects of cooling mature human oocytes to 0 °C at one of 2 rates – 3 or 1000 °C/

min and holding for 30 min prior to warming (Bernard *et al.*, 1992). They showed that the proportion of oocytes that became fertilized after cooling at either rate was high, and similar to the untreated control oocytes (≥ 70 % for all groups). Development for an additional 24 hours of 17 of the embryos was also similar between groups. Only one of 16 oocytes used asparthenogenetic controls underwent cleavage. Restrictions on research with human embryos precluded the assessment of further development in this study.

Cooling *in vitro* matured porcine oocytes to high sub-physiological temperatures for 1 hour was shown to be slightly detrimental to fertilization and embryo development. The proportion of oocytes undergoing fertilization was reduced to 57, 46, and 43 % after having been cooled to 30, 25, and 20 °C (compared to 62 % fertilization when the oocytes were maintained at 38.5 °C) at slow rates of cooling (cooling within ¼ cc straws and transfer to a water bath). The proportion of oocytes undergoing cleavage was 45, 49, and 36 % at those temperatures compared to 46 % for non-cooled oocytes.

8.10.5 Effects on the Oolemma

Cooling can directly affect the plasma membrane of cells, resulting in a transition from a liquid-crystalline phase to a gel phase caused by lowering the thermal energy of the phospholipid molecules. The result is an altering the interactions between the lipids, and hence the physical properties of the membrane. The temperature of this transition is affected by several factors, including the length of the hydrocarbon chain on the particular species, the presence of unsaturated hydrocarbon chains and the degree of unsaturation, cooling rate, type and concentration of integral membrane proteins, and the composition of the solution surrounding the cell, including the presence of CPA. The phase transition temperature of MII oocytes has been investigated from some mammalian species. In 1996, Arav and colleagues documented this phenomenon in bovine oocytes, showing that a very broad transition occurs, centered around 10 °C (the temperature of transition ranged between ~20 and 0 °C). The presence of cholesterol in a lipid bilayer is known to broaden a phase transition such as that described for bovine oocytes and may account for this property. These results are in contrast to a more narrow transition for human oocytes centered at ~17 °C.

8.10.6 Effects on the Metaphase II Spindle

Many research studies investigating the effects of cooling mature mammalian oocytes have been directed at understanding the effects on the metaphase II (MII) spindle. The first report appeared in 1980 using mouse oocytes. It was shown that cooling on ice for 15, 30, 45, or 60 min resulted in progressive depolymerization of the microtubules of the spindle. The majority of the spindle fibers were depolymerized after the shortest treatment, with the continuous, or pole-to-pole fibers being completely depolymerized, and the kinetochore fibers being mostly depolymerized. Nearly all of the kinetochore microtubules had depolymerized by 30 min, and longer treatments completed the depolymerization process. Rapid warming to different temperatures (30 min at 18, 25, or 37 °C) resulted in different degrees of repolymerization and normal spindle structure, depending upon the temperature. No repolymerization occurred at 18 °C, and little occurred at 37 °C, but complete repolymerization occurred at 25 °C. They went on to show that progressive warming (warming in an 18 °C water bath, and transferring to a 25 °C bath, and finally to a 37 °C bath) resulted in rebuilding of a normal spindle as well.

Cooling also has deleterious effects on the MII spindle of oocytes from other mammals. However, one of the major differences is that the ability of the spindle to undergo normal repair (by which is meant the morphology of the spindle appears the same after warming as before cooling) is reduced. Less than half of human oocytes had normal spindles after cooling to room temperature for either 10 or 30 min and warming to 37 °C for either 1 or 4 hours in one study. Several other studies have investigated the effects of cooling human oocytes to various temperatures for various times and observed similar effects. In the study by Zenzes and colleagues, they showed that evidence for microtubule depolymerization could be observed in as little as 1 minute at 0 °C, although only slight effects were discernable. However, after 2 min it was more noticeable, 4-5 min caused substantial disruption, and 10 min resulted in complete microtubule depolymerization. Most of these studies have shown that the chromosomes do not disperse far from the spindle area, but clumping and some displacement off of the metaphase plate is frequent. Similar effects have been shown with bovine and porcine oocytes as well, suggesting one possible mechanism for the significant difference of the cooling sensitivity of oocytes from mice and other mammalian species as discussed above.

8.10.7 Effects on Cortical Granules Exocytosis

Cooling oocytes has also been shown to affect the zona pellucida, presumably as a result of cortical granule exocytosis. Johnson and colleagues showed that rapid cooling to 4 °C resulted in a reduced proportion of oocytes undergoing fertilization after sperm incubation, and this result could be 8.10.8 eliminated by removal of the zona. As mentioned above, Azambuja and colleagues also showed cooling affects the bovine zona pellucida.

8.10.8 Sensitivity of MII Oocytes to Intracellular Ice Formation

Freezing of intracellular water has a strong correlation with cell survival during cooling, as described above. Due to their large size and spherical shape (hence low surface-to-volume ratio), mammalian oocytes are particularly susceptible to IIF due to the likelihood of a large volume of water being trapped inside the cells during cooling (please see Figure 1.2). Oocytes were used as a model cell in some of the earliest experiments which documented the relationship between IIF and cell death. In this report by Leibo and colleagues, oocytes were cooled at various rates in the presence of 1.0 M DMSO to temperatures as low as -80 °C under a cryomicroscope to observe the incidence if IIF (evidenced by a sudden darkening, or "flashing" of the cell) and the temperatures at which it occurred.

The IIF temperature was shown to be slightly related to the rate of cooling, and occurred at a surprisingly low temperature, around -40 °C, on average. The cumulative survival (as determined by the inverse of the incidence of ice formation) was shown to strongly correlate with cooling rate. Rates at or above ~5 °C resulted in IIF in all cells examined, while cooling at 1.3 °C resulted in only 13 % of oocytes exhibiting evidence of IIF. Some of the cells could be cooled to temperatures near -60 °C before exhibiting evidence of IIF.

Toner and colleagues extended this line of investigation to look at IIF parameters for mouse oocytes in the absence of permeating CPA. They found that the temperature at which IIF occurred was much higher (median temperature of IIF was ~ -12 °C). This temperature was also shown to be dependent upon the tonicity of the extracellular solution (hence the concentration of water in the cytoplasm). The median temperature of IIF was approximately -8, -12, -17, -27, -29, and -32 at extracellular osmolalities of 200, 285, 510, 735, 820, and 1035 mOsm, respectively. Again, there was not a strong cooling-rate dependence of the IIF temperature, the cooling rate for which 100 % of oocytes exhibited IIF was about 5 °C/min in an isotonic solution. One important conclusion the authors reached is that at least 2 mechanisms of ice formation appeared to be operating. As the temperature of the suspending medium was lowered, the cumulative fraction of oocytes containing ice increased rapidly after reaching ~ -31 °C. The authors speculated that heterogeneous ice nucleation was responsible for the differences in ice formation at temperatures above and below -31 °C, the former being catalyzed by extracellular ice and the later by intracellular nucleating agents. Other important findings in this investigation included a difference in the cumulative fraction of oocytes exhibiting IIF at constant temperatures (lower temperatures tended to result in a higher proportion of oocytes freezing internally), and that the presence of extracellular ice was absolutely necessary for the formation of intracellular ice at temperatures above -20 °C. When cells were supercooled to temperatures around this point, and then ice spontaneously formed outside of the cells, ice would immediately form inside the cells as well.

More recently, Mazur and colleagues investigated IIF temperatures in mouse oocytes in the presence of EG and glycerol. Similar to DMSO, EG and glycerol depressed the nucleation temperature of the oocytes well below that of oocytes in an isotonic solution without permeating CPA. Freezing oocytes in solutions containing one molar CPA resulted in an average flash temperature of -41 and -37 for glycerol and EG, respectively (the temperature was -14 °C for cells in solutions of phosphate-buffered saline (PBS) alone). Interestingly, the presence of the CPA seemed to mitigate the effects of extracellular ice on membrane integrity, such that in a solution of PBS, about half of the oocytes appeared to have lost membrane integrity as a result of the external ice (the lack of "flashing" of the oocytes when plasma membrane integrity was lost was interpreted to be a direct effect of the ice on plasma membrane integrity, such as a mechanical effect and not a nucleating effect). The results from this study also did not support the conclusion of an effect of cooling rate on IIF temperature when cooling rates ranged from 5 to 50 °C/min.

The analysis of the solution compositions was extended to temperatures below freezing to determine the fraction of unfrozen solution at the flashing temperatures. The results showed that in the presence of only PBS or a solution with glycerol, flashing took place when approximately 94 % of the water in the solution was converted to ice. This percent was independent of the initial concentration of glycerol (the values were virtually identical at 0, 0.5, and 1.0 M glycerol). In comparison, in the presence of EG, flashing occurred when between 84 to 91 % of the water was converted to ice.

IIF was also investigated in bovine oocytes by Chandrasekaran and colleagues and Ruffing and colleagues. In the absence of CPA, Chandrasekaran *et al.* showed that the average temperature for bovine oocyte IIF occurred at -11.2 °C when cooled at 16 °C/min. The report by Ruffing *et al.* used *in vivo* matured oocytes for one experiment where the cells were only exposed to an isotonic solution. *In vitro* matured oocytes were also utilized to investigate IIF after equilibration in 1.5 M EG, PG, or glycerol. The cells were cooled at one of 3 rates (4, 8, or 16 °C/min) from 5 to -60 °C. For *in vivo* matured oocytes in isotonic medium, a total of 45, 75, and 93% of the oocytes exhibited IIF at the 3 cooling rates. The median temperature for ice formation was -13.3 and -14.2 °C for the cooling rates of 8 and 16 °C/min, respectively. The median temperature of IIF for *in vitro* matured oocytes in the presence of EG, glycerol, or PG was significantly reduced. Specifically, IIF occurred in half of the cells by -45.6, -46.5, and -34.4 °C when cooled at 4 °C/min; -39.2, -33.5, and -28.1 °C when cooled at 8 °C/min; and - 27.8, -20.7, and -19.4 °C when cooled at 16 °C/min in the presence of the respective CPA.

8.10.11 Sensitivity of MII Oocytes to Anisosmotic Conditions

As discussed above, cryopreservation usually requires the exposure of cells to one or more cryoprotective compounds. Base solutions used for freezing cells are approximately isosmotic with serum. Therefore, the addition of CPA to such solutions makes the overall solutions hyperosmotic. Exposing cells to these solutions results in osmotically-driven volume excursions which can be detrimental to cell viability. However, the degree of volume change associated with exposure to solutions containing permeating CPA can be controlled in a manner which reduces the absolute volume excursions and can prevent the cells from exceeding their osmotic tolerance limits. This means of cell damage is particularly acute for oocytes, as these cells are spherical, having a minimum surface-to-volume ratio. In order to minimize the deleterious effects of osmotic volume excursions, the tolerance of cells to changes in volume must be defined. The results of studies investigating the tolerance of mammalian oocytes from some species to anisosmotic solutions have been reported in the literature. In 1997, MII mouse oocytes from an outbred stock (ICR) were exposed to isotonic or hypotonic concentrations of PB1 medium, the later made by dilution of an isotonic PB1 solution with double-distilled water. The concentrations of the hypotonic solutions were 0.5, 0.3, 0.25, and 0.2 X isotonic (150, 92, 77, and 62 mOsm, respectively). After exposure and return to an isotonic culture medium, the oocytes were allowed to recover for 30 to 60 min before morphological assessment. Only the oocytes having been exposed to the 150 mOsm solution were not affected, on average. Oocytes exposed to the other hypotonic solutions showed a loss of viability directly related to the solution concentration (~ 90, 60, and 20 % viable cells, respectively). Oocytes having previously been vitrified showed a similar pattern when immediately exposed to the hypotonic solutions, with a slightly higher proportion of oocytes undergoing lysis after the treatments. A recovery period in culture of 1.5 – 2 hours before exposure to the hypotonic solutions eliminated this effect.

The effect of anisosmotic exposure on MII bovine oocytes has also been investigated. Hypo- and hypertonic solutions of sodium chloride (final osmolalities of 75, 150, 600, 1200, 2400, and 4800 mOsm) were used as the treatments. In this experiment, the oocytes were not denuded of cumulus cells prior to the treatment. All of the treatments affected the viability of the oocytes as measured by post-IVF development. The proportion of oocytes able to develop to the blastocyst stage decreased as the solution concentrations diverged from isosmotic. The proportion of oocytes developing to blastocysts (normalized to the controls) was approximately 37, 68, 60, 50, 48, and 19 % for the treatments, respectively. These results show that even brief exposure (10 min) to anisosmotic solutions can lead to a significant loss of developmental competence of bovine oocytes. Only limited information on the effect of anisosmotic exposure to human oocytes has been reported to date. Exposure to anisosmotic solutions ranging in concentration from 0.25 to 1.5 M had no effect on the immediate survival of MII human oocytes as measured by a vital dye assay.

8.10.12 Sensitivity of MII Oocytes to Cryoprotectant Exposure Effects on Cellular Components

The effect on mammalian oocytes of exposure to some permeating CPA has also been investigated. Johnson and Pickering showed that exposure to DMSO changed the polymerization pattern of nearly all of the microtubules in the MII spindle of mouse oocytes. Exposure to a 1.5 M concentration at 37 °C for 10 to 20 min resulted in the poles of the spindle showing an astral pattern of microtubules, compared to the normal anastral pattern seen in untreated oocytes. Furthermore, other cytoplasmic asters of microtubules formed, focused around pericentroliar material not associated with the spindle. This pattern was also seen after transfer to 4 °C. In both cases, the overall shape of the spindle was preserved. Exposure to 4 °C without DMSO was associated with rapid depolymerization of the spindle, as expected. Prolonged exposure (60 min, with or without transfer to 4 °C) resulted in disappearance of the spindle microtubules, with the tubulin monomers becoming incorporated into the cortical asters. This effect was also associated with chromosomal dispersal from the metaphase plate. Restoration of the spindle after DMSO removal occurred in some oocytes (between 56 to 60 %, depending upon the treatment) after incubation at 37 °C for 30 min, but none of the spindles were restored in oocytes that remained at 4 °C. More oocytes (89 %) were able to restore a normal spindle apparatus if the DMSO was added in increments instead of in one step, with no cooling below 37 °C, suggesting that the osmotic effects of the cryoprotectant addition and removal play a factor in spindle stability. Changes to the actin cytoskeleton in mouse oocytes were also investigated after a direct, 30 minute exposure to DMSO at concentrations of 0.25, 0.5, 1.0 and 1.5 M (Vincent, Pickering, Johnson and Quick, 1990). Only 1.0 and 1.5 M concentrations affected the cortical actin meshwork, with irregular patterns shown after the treatment. The effects on the actin meshwork appeared to be fully reversible after a 30 minute wash at 37 °C. In this study, chromosome scattering was also evident due to the treatment, and not always reversible. In both of these studies, exposing mouse oocytes to DMSO at 4 °C resulted in far less disruption to the spindle or actin meshwork in mouse oocytes. These authors also showed that exposure to DMSO resulted in hardening of the zona pellucida surrounding mouse oocytes and that exposure to this compound at 4 °C reduced the proportion of oocytes affected. They also showed that a stepwise procedure for addition and removal was more detrimental to the zona hardening effect, and increasing osmotic stress (due to exposure to increasing concentrations of a solution containing sucrose up to 1.0 M) had little effect on the zona. Cortical granule exocytosis resulting from exposure to DMSO in mouse oocytes was confirmed and extended to human oocytes, and was also shown to be caused by PG . As a logical conclusion to these experiments, it was shown that exposure to DMSO at 4 °C improved fertilization compared to exposure at 37 °C. This result was subsequently extended to human oocytes. Depolymerization of the spindle microtubules in mouse oocytes also results from exposure to PG, but more strongly at lower concentrations (0.5 or 1.0 M). Higher concentrations (1.5 and 2.0 M) resulted in less disruption. All of these concentrations resulted in minor disruption of the cortical actin band, with the exception of the actin-rich region just above the metaphase spindle. Similar effects on the spindle and actin cytoskeleton were shown in bovine oocytes after exposure to 1.5 M EG, but to a lesser degree. The proportion of oocytes displaying a normal spindle was reduced from 75 – 80 % in control oocytes to ~ 65, 55, and 50 % after exposure for 20 min, 1 hour, and 3 hours respectively. The proportion of oocytes exhibiting normal actin staining patterns was reduced from ~ 85 % to ~ 65 – 70 % with the same incubation times. The EG addition occurred in a single step, and the removal proceeded in a stepwise fashion at ~ 25 °C in this study. Rho and colleagues demonstrated an effect of exposure to 3.2 M EG and 2.36 M DMSO on microtubules (only 64% of the cells had a normal spindle pattern vs. 92 % for controls) and mitochondria membrane potential (74 % normal vs. 82 % for controls) in bovine oocytes. In a different study, bovine oocytes were exposed in a single step to DAP213 (2 M DMSO, 1 M acetamide, and 3 M PG) for 1.5 or 5 min, then transferred to an isotonic solution supplemented with 0.5 M sucrose for 5 min and then to an isotonic solution (. In this study, the expanded cumulus cells had been partially removed prior to the treatment. The oocytes were subsequently processed for IVF or transmission electron microscopy (TEM). Ultrastructural changes were affected to a minor degree by the 1.5 minute treatment, but more so by the 5 minute treatment. Specifically, a higher proportion of oocytes exhibited a loss of cytoplasmic cortical granules (90 % vs. 8 % for the controls) and an increase in the number of cytoplasmic vesicles (53 % vs. 16 % for the controls). While the proportion of oocytes with abnormal microvilli and mitochondria was doubled in the 5 minute treatment group, the results did not reach statistical significance.

8.10.13 Effects on Development

There is conflicting evidence for an effect of exposure to DMSO on mouse oocyte developmental potential. The study by Carroll *et al.* showed that exposure to a concentration of 1.5 M caused a small but significant loss of developmental potential to the blastocyst stage and Bouquet *et al.* showed a significant effect on development to 2-cell embryos; although freezing and thawing had a more pronounced effect. It showed that there was no effect of 1.5 M DMSO no effect of an exposure to a concentration of 3.5 M on development to blastocysts, although the cell number was reduced in the treated group compared to controls, and the number of polyploid embryos was increased, albeit slightly in each case. In each study, the B6CBF1 strain was used. One difference is that in the study by Van der Elst *et al*, the oocytes were allowed to recover in culture for 1 hour prior to IVF. A recovery time was also shown to be beneficial to development showed that exposure to 6 M DMSO for 110 seconds caused a significant reduction in the oocytes ability to progress to the first meiotic metaphase after Sr++ activation (29% progressed normally) compared to an exposure for 90 seconds (65% normal). However, the number of embryos developing to the 2- cell stage did not differ from controls when the incubation was 110 seconds. O'Neill *et al.* also showed that exposure to 6 M DMSO decreased the proportion of oocytes that were able to develop to blastocysts (58 *vs.* 65%;). However, the addition of polyethylene glycol to the medium (1mg/ml) increased the developmental potential to 58 %, which was not statistically different than the controls. Men *et al.* showed no effect of 1.5 M PG (in the presence of 0.2 M sucrose) but a slight negative effect of 3.5 M DMSO (in the presence of 0.5 M sucrose) on oocyte fertilization using Kunming mice. Conflicting results have also been presented regarding the tolerance of mouse oocytes to PG. In the report by Gook *et al.*, a very high proportion (68 %) of mouse oocytes degenerated in culture after the cumulusenclosed mouse oocytes were exposed to 1.5 M PG. However, only a minor effect was seen in a different report. Exposing mouse oocytes to increasing concentrations of EG (> 4 M for more than 5 min) resulted in membrane blebbing and disruption of the actin cytoskeleton (lower concentrations had minimal effect).

A concentration of 8 M EG was tolerated only if the exposure time was less than 1 minute. In a different report, fertilization rates were high after exposure to 7 M EG in a 1-step or 2-step procedure (67 and 66 % cleavage *vs.* 75 % for untreated controls), yet the development rate to blastocysts was significantly reduced (31 and 45 % *vs.* 64 % for controls.

Numerous studies have investigated the effects of cryoprotectant exposure on bovine oocytes. This has particularly been of concern for bovine oocytes since methods to cryopreserve these cells using low cryoprotectant concentrations and slow cooling have been abandoned in favor of vitrification methods, due to the extreme sensitivity of these cells to chilling. Papis and colleagues exposed bovine oocytes to 2, 4, or 6 % EG (0.36, 0.71, and 1.07 M) and showed that only the concentration of 6% had a noticeable effect on the cleavage rate (77 % *vs.* 94 % for controls;. Martino *et al.* exposed bovine oocytes to solutions containing EG at concentrations of 4 or 5.5 M in a single-step, and determined that there was a significant decrease in development as a result (10 84 to 15 % development to blastocysts compared to 42 % for untreated oocytes;. Rho and colleagues demonstrated a significant decrease in developmental potential from exposure to a vitrification solution containing 3.2 M EG and 2.36 M DMSO (12.5 % blastocysts rate *vs.* 24.7 for untreated oocytes). Fuku and colleagues showed that exposure to DAP213 for 1.5 min had no significant effect on fertilization, cleavage or development to the blastocysts stage. A 5 minute exposure, however, resulted in a significant reduction compared to control oocytes as measured by fertilization (56 *vs.* 79 %), and blastocysts development (4 *vs.* 13 %). The study by Rojas *et al.* also showed that a stepwise exposure to EG at a final oncentration of 40% (~7.1 M) had a serious effect on the *in vitro* development of porcine oocytes. Only 11.2 % of the oocytes exposed to the cryoprotectant solutions developed to blastocysts compared to 31 % for untreated oocytes. Shi *et al.* showed that only 3.8 % of porcine oocytes exposed to a solution containing to 20 % EG (~3.6 M) and 20 % DMSO (~2.8 M) with 18 % ficoll and 0.3 M sucrose developed to blastocysts compared to 26 % of untreated oocytes. Only 11 % of oocytes exposed to a solution containing 2.4 M EG and 1.77 M DMSO with 0.5 M sucrose developed to blastocysts compared to 23 % for controls. Compared to these significant reductions in developmental potential, Somfai and colleagues showed no effect of exposure to a solution of 35 % EG (~6.3 M) with 5 % polyvinylpyrrolidone and 0.3 M trehalose after an exposure to 4 % (0.7 M) EG (25 % blastocysts development compared to 27.5 % for controls). It should be noted that experiments aimed at determining the effect of CPA rarely take into consideration the possibility of an osmotic effect independent of the chemical toxic effects of the exposure making the true cause of the injury from exposure to such solutions uncertain.

8.11 PROGRESS ON CRYOPRESERVATION OF OOCYTES FROM CATTLE, PIGS, MICE, AND HUMANS HUMAN OOCYTES

The first report of a pregnancy and subsequent delivery of a baby derived from a frozen and thawed oocyte appeared in 1986 and 1988, respectively. In the first report, Chen described unsuccessful attempts to freeze oocytes using the method developed for human embryos. A modified method was developed using a mouse model that provided high survival, and was subsequently applied to oocytes from IVF patients. This specifics of this method included exposure of the oocytes to precooled DMSO (1.5 M concentration) in a

PBS solution with 10 % fetal bovine serum for 15 min, and gradual cooling to -7 °C (rate not described). After extracellular ice induction (seeding), the temperature was lowered to -36 °C at a rate of -0.5 °C per minute. The sample was subsequently plunged into liquid nitrogen. Warming was rapid (immediate immersion into a 37 °C water bath; ~2500 °C per minute) and the cryoprotectant was diluted by adding a 4x volume of PBS. Of the 40 oocytes frozen with this method, 32 (80%) survived the freezing and thawing, and 25 of 30 were successfully fertilized. Eighteen of those underwent cleavage divisions. One of the 4 patients receiving embryos became pregnant (as described in the initial report), and the later report described the full-term delivery.

Several other reports appeared in the late 1980's describing additional attempts to cryopreserve human oocytes. One notable feature of these reports (and many other reports on oocyte and embryo cryobiology) is the lack of fundamental experiments designed to characterize the cryobiology of human oocytes. Instead, simple changes to a standard equilibrium protocol were made and outcomes were assessed. Changes included altering the addition and removal of the cryoprotectant (stepwise and at room temperature; and assessing the effects of PG *vs.* DMSO. In general, the outcomes of these early reports were poor and highly variable. For example, immediate survival of 136 oocytes recovered after thawing was 32% , 58% of those survived underwent fertilization, 2 pregnancies ensued, but neither went to term. In the report by, 28 % of the oocytes frozen in DMSO survived, and only half of those fertilized. Oocytes frozen in PG tended to survive better (32 %) and 75 % of those fertilized. The second report of a live birth came in 1987 using an equilibrium method with DMSO as the permeating CPA. While concerns over possible damage to the meiotic spindle during cryopreservation of oocytes slowed progress in this field during the early years, a few reports in the mid-1990's offered evidence to refute this notion. Deborah Gook and colleagues published a series of reports describing the outcome of equilibrium cryopreservation of human oocytes using PG. The first report documented a higher survival of oocytes (64 %) compared to earlier studies. Perhaps more importantly, this report suggested that damage to the spindle apparatus may not be as prevalent as originally thought. Of oocytes untreated, exposed only to the CPA, and those frozen, 81 %, 63 %, and 61 % had normal spindles, respectively, as assessed by conventional epifluorescence microscopy. In a second report, a karyotype analysis was performed on blastocysts having been derived from frozen and thawed oocytes. All 4 which had a chromosome spread that could be analyzed showed a normal karyotype. It should be noted that only those oocytes undergoing normal fertilization and proceeding to blastocysts were assessed, and of the oocytes not undergoing normal fertilization, all had abnormal spindles. This suggests the possibility that aneuploid embryos may be selected against during early embryo culture, biasing these results.

This topic continues to generate controversy, as some reports provide evidence that spindle damage occurs in the majority of oocytes frozen while other reports suggest that this is not the case. Using live cell imaging technology, Bianchi and colleagues showed that microtubule density does increase during post-thaw culture, but does not return to the same intensity noted prior to cryopreservation. Given the inherent challenges associated with research on human oocytes and embryos, including small numbers and restrictions on destroying human embryos, it will be difficult to resolve this controversy. Nevertheless, the fact that a significant proportion of transferred embryos perish *in vitro* and *in utero* supports the hypothesis that a high number of aneuploid embryos are generated from frozen and thawed human oocytes.

In two reports appearing in 1995, ICSI was tested against IVF to determine if ICSI would improve development. Indeed, this was the outcome. These reports seemed to have a significant influence on the practice of human oocyte cryopreservation, for the use of DMSO was wholly abandoned for PG after this time. It should also be noted that at this time, the use of low concentrations of sucrose (0.1 M) in the freezing medium became routine, and likely contributed to the increased post-thaw morphological survival of oocytes with this method (although survival was not always consistently high. Another important note is that in the reports, the investigators showed that a proportion of oocytes lysed in culture after thawing and CPA removal prior to sperm injection. The first live birth using an equilibrium method with PG and sucrose as CPA appeared in 1997. In this report, 33 % of the oocytes survived freezing morphologically (4 of 12), 2 fertilized normally, one underwent cleavage and was transferred on day 2. These investigators also utilized sucrose at a concentration of 0.2 M in the freezing medium.

Possibly the most important contribution, from a basic research standpoint that has resulted in improved post-thaw viability appeared in 2001, where Fabbri and colleagues systematically compared the effects of different sucrose concentrations on oocyte survival with the equilibrium method using 1.5 M PG. Increasing the concentration of sucrose from 0.1 M to 0.3M caused a marked improvement in survival, from ~33 % to ~82 %. Due to these results, most reports have utilized a 0.3 M sucrose concentration as a cryoprotectant during human oocyte cryopreservation.

Since 2001, at least 10 reports have appeared describing the results of freezing human oocytes using a standard equilibrium method with 1.5 M PG and sucrose (See Table 2). In most cases, sucrose has been used at a concentration of 0.3 M. This and other subtle differences may explain at least some of the variation in success seen in these reports (see notes in Table 2). Some of these reports show very high survival and early development rates, and this has lead some individuals to emphatically defend human oocyte cryopreservation and suggest that the clinical results are nearly equivalent to the use of fresh embryos. Other investigators have been more cautious in their interpretation, and in two recent reports with a very large number of cycles, the results were still rather poor, especially when compared to the use of fresh embryos.

Vitrification has also been utilized as a means to cryopreserve human oocytes in recent years, albeit less often that the traditional equilibrium method. In 1999, Kuleshova *et al.* described a birth resulting from an oocyte vitrified with a solution containing ethylene glycol and sucrose. For this work, the investigators utilized an open-pulled straw (OPS) for the procedure. Since this time, other reports on human oocyte vitrification have been published (Table 2b), with all of these reports utilizing open container systems to achieve socalled "ultra-rapid cooling". On average, the results from the vitrification trials have been better than the trials using slow-cooling (as measured by the number of oocytes to achieve a pregnancy). However, the results are still limited in number and general conclusions are difficult to reach at this time. Overall, progress on human oocyte cryopreservation has been significant during the past few years, as evident by the number of reports appearing in the literature. Furthermore, changes in legislative mandates in recent times (*e.g.* have pushed some clinics toward the necessity of freezing human oocytes, increasing the demand for a clinically robust procedure. Progress on investigations into new methods will continue to be slow, however, as dictated by the difficulty in conducting controlled studies with human oocytes.

8.12 MOUSE OOCYTES

The mouse is the first mammal for which MII oocytes were successfully frozen and stored in liquid nitrogen, and subsequently developed into live offspring. This report appeared in 1977 by the same investigator that contributed to the first successful method for mammalian embryo cryopreservation. The procedure used was somewhat similar to the one used for mouse embryos. It included exposing the cells (with or without the cumulus mass) to 1.5 M DMSO at 0 °C, slow cooling (< 1 °C/min) to -65 °C, and slow warming (average of 8 °C/min). The authors fertilized the oocytes either in vitro or in vivo for this study. Two strains of mice were used as oocyte donors, MF1 outbred and B6CBAF1 hybrid mice. For oocytes having been fertilized in vitro, some oocytes from both MF1 and B6CBAF1 oocytes were transferred to pseudopregnant foster dams at the 2-cell stage. However, only oocytes from the B6CBAF1 strain were cultured to later stages in vitro before being transferred (culture methods had yet to be developed to overcome the 2-cell block in embryos from many strains. Overall, the proportion of morphologically normal oocytes after freezing and thawing ranged from 65 to 75 %. For the results for development to the 2-cell stage, freezing had a significant effect as did strain (B6CBAF1 oocytes cleaved at a higher rate), yet the presence or absence of the cumulus mass had no effect. A significant loss of embryos occurred in utero from the oocytes having undergone fertilization in vivo. It was estimated that only about 5 % of the oocytes frozen with cumulus cells and 15 % without developed to day E14. For oocytes fertilized in vitro post-thaw, about 16 % from the MF1 strain and 38 % from the B6CBAF1 strain developed to day E14 For oocytes fertilized in vitro post-thaw, about 16 % from the MF1 strain and 38 % from the B6CBAF1 strain developed to day E14 and live young after having been transferred at the 2-cell stage. Approximately the same proportion (36 %) developed to term having been transferred at the blastocyst stage (control = 25 %). B6CBAF1 oocytes tended to have a higher developmental capacity to term compared to MF1 oocytes. Stage of transfer for the B6CBAF1 oocytes did not seem to have an effect on outcome. Fertilization was in the range of 50 – 80 % and possibly could have been higher had the authors used serum in the freezing medium (as discussed above). Since this time, dozens of papers have appeared in the literature describing results of mouse oocyte cryopreservation (see Table 3 for a summary of the more important developments).

8.13 DIFFERENT CRYOPROTECTANTS OR SOLUTION COMPOSITIONS

Todorow and colleagues directly compared the effects of DMSO, PG, or a combination of the two (all at a final concentration of 1.5 M with 0.1 M sucrose) in a slow-cooling procedure. These authors tested a stepwise addition (5 equal molar steps) and a 5-step removal as well as cooling to either - 35 or -80 °C at a cooling rate of 0.3 °C /min. The combination of the two compounds gave the best results, and cooling to -80 °C was also superior to -35°C. They did not see a significant difference in stepwise or single-step CPA addition and removal.

Surrey and Quinn also compared DMSO and PG in a slow-cooling method (Surrey and Quinn, 1990). The temperature at which the controlled cooling was terminated differed between the two compounds (-40 °C *vs.* -30 °C, respectively). In addition, 0.2 M sucrose was present with the PG, but no sucrose was present with DMSO. Even though survival was lower in the PG group (33 *vs.* 24 %), fertilization (50 *vs.* 71 %) and blastocysts formation rates (30 *vs.* 60 %) were higher. Thus, the overall proportion of oocytes developing to blastocysts was higher in the group frozen with PG compared to DMSO (5 *vs.* 10 %). Unfortunately, the confounded nature of the study makes it difficult to determine the true cause of the differences between the methods.

Stachecki and colleagues published a series of reports, , investigating a novel approach to slow-cooling mouse oocytes; namely, modifying the freezing solution to replace sodium ions with choline ions. The rationale for this approach was the hypothesis that the high concentrations of sodium ions that result from the precipitation of water as ice during freezing are responsible for the loss of cell viability. In the first report, they demonstrated a significant preservation of developmental potential after replacing nearly all of the sodium ions with choline. When the sodium concentration was reduced from 167 mM to 30 mM (with a concomitant increase in choline concentration), 99 % of the cells recovered from the freezing procedure with intact membranes, and 35 % of the total number of oocytes frozen were able to develop to blastocysts. In a second report, they tested several variables of the freezing procedure on developmental potential, including thawing rate, plunge temperature after slow-cooling, and the composition of the holding medium used prior to freezing (choline chloride-based medium *vs.* NaCl-based medium). In their best results using a choline chloride holding and freezing medium along with a moderate rate of warming, they demonstrated that 58 % of the oocytes frozen could develop to blastocysts. In a later report they went further to show that a high proportion of oocytes could develop to term after embryo transfer. Of recipients becoming pregnant, 48 % of the transferred embryos developed to fetuses on E10 – 13 and 27 % of the total number of oocytes frozen could develop to term. The logic of replacing sodium chloride with choline chloride stems from the hypothesis that the increased concentration of salt resulting from water precipitation as ice during cooling is a primary mechanism of cryoinjury when equilibrium freezing methods are utilized, as was described earlier. However, the authors also suggest that choline could interact directly with the membrane and assist in its stabilization during freezing.

Studies such as those described above show how the solution composition can have a great influence on the outcome of attempts to cryopreserve cells. Solution properties can have differing influences, from the ability to depress the rise in extracellular solute concentration concurrent with ice precipitation to effects on the hydration level of cells, and direct chemical effects. The last series of studies discussed is particularly insightful in this regard, as the replacement of a common inorganic ion with an organic ion was shown to have a beneficial effect, and the authors suggest the resulting benefit was due to a reduction in the Na+ concentration. However, one possibility these authors seem to dismiss is an intracellular role for choline. Choline is a metabolic precursor of betaine, a compound commonly used as an organic osmolyte for counteracting the effects of osmotic stress. It is possible that choline entry into the cell could act as an intracellular osmolyte during freezing, and the fact that these authors saw an increased protective role when the cells were incubated for longer periods in the choline chloride-based medium prior to freezing lends support to this hypothesis. Regardless of the specific mechanism, these results show how significant the composition of the cryopreservation solution can be on the ability to achieve successful cryopreservation.

8.13.1 Rapid Cooling without Vitrification

A small number of experimental reports have been published describing attempts to rapidly freeze mouse oocytes in moderate concentrations of either EG or DMSO. Sathananthan and colleagues reported on freezing mouse oocytes with 3.5 M DMSO or PG with 0.5 M sucrose (Sathananthan *et al.*, 1988). The time for exposure to the CPA was approximately two min and the straws were then plunged into liquid nitrogen. There was no dilution procedure postthaw that included an osmotic buffer, however. Of the oocytes frozen, only 33- 34% survived morphologically. Of these, 15 % frozen in DMSO and 7 % frozen in PG developed to blastocysts. Ultrastructural analysis was performed with the thawed oocytes within 1 hour of thawing in one group, and another group was fixed at 0 °C immediately after thawing. All oocytes fixed immediately at 0 °C showed signs of severe ultrastructural damage including spindle disorganization and membrane disruption. Various levels of disruption were apparent, but to a lesser degree in the oocytes thawed to 37 °C and held at approximately 22 °C prior to fixation.

Van der Elst and colleagues published two similar reports with such a method (Van der Elst *et al.*, 1993; Van der Elst *et al.*, 1998). In both repots, the authors froze the solution containing the oocytes by rapidly quenching the straws in liquid nitrogen after exposure to 3.5 M DMSO and 0.5 M sucrose. Survival rates were very similar. The fertilization rate of the frozen oocytes was similar (59 % and 69 % respectively). The development to the blastocyst stage of oocytes reaching the 2-cell stage was better in the report in 1998 compared to 1993 (82 % *vs.* 49 %). In the 1993 report, the authors compared ploidy in the resulting embryos and determined that exposure to the CPA solution resulted in a small reduction in the proportion of embryos having diploid nuclei (78 % *vs.* 81 % for CPA exposure and control, and 74 % *vs.* 91 % for frozen and thawed oocytes *vs.* control). Also in the 1993 report, the total number of nuclei was reduced in the frozen group compared to the controls (38 *vs.* 70) and the proportion of blastocysts that developed to d17 viable foetuses was reduced to one-third of the unfrozen controls (11 % *vs.* 32 %). In the 1998 report, they also showed that the total cell number per blastocyst was reduced in the frozen group (40 *vs.* 54) and that the average number of cells in the inner cell mass (14) was less in the frozen group compared to the other treatments. In 1994, Rayos and colleagues reported on rapid freezing of mouse oocytes using 3 M EG and 0.25 M sucrose or trehalose. They tested the effects of different incubation times in the CPA prior to putting the straws into liquid nitrogen (5, 10, 20, and 40 min). Overall, the sugar treatment had no effect. The five minute incubation was the only treatment that resulted in a reduced survival compared to the other times (52 % *vs.* 67, 77, and 77 %, respectively). The blastocyst development rate was also reduced with the five minute group compared to the others (26 % *vs.* 38, 44, and 43 %; the controlrate was 88 %). Fetal development was also reduced due to freezing (20 % *vs.* 49 % for controls).

This particular method, rapid freezing in moderate concentrations of CPA, at first appears to be a fair compromise between the potential toxic effects of high concentration of solutes used in true vitrification procedures and the potential for cooling damage of equilibrium methods. However, the process itself is somewhat precarious. Using a moderate concentration of cryoprotectant may allow true vitrification during rapid cooling, but the likelihood of devitrification and recrystallization during warming is very high. Critical warming rates are usually many orders of magnitude higher than critical cooling rates (the rates needed to avoid ice crystal formation during warming and cooling, respectively and at the cooling and warming rates attainable with ¼ cc straws, solutions with CPA concentrations in the range used in these studies are very likely to undergo devitrification. Even with higher concentrations of CPA, devitrification can be a serious challenge, resulting in high variability between experimental runs.

8.13.2 Vitrification

Reports of vitrification of mouse oocytes appeared in the literature shortly after the successful vitrification of mouse embryos by Rall and Fahy (Rall and Fahy, 1985). In 1988,

Kola and colleagues vitrified cumulus-intact oocytes using Vitrification Solution 1 (VS1: 2.5 M DMSO, 2.36 M acetamide, 1.19 M PG, 5.4 % polyethylene glycol) and compared the results to a traditional slow-cooling procedure with 1.5 M DMSO to -80 °C before storage in liquid nitrogen. Exposure to the VS1 solution alone resulted in morphological changes to the oocytes and reduction in their developmental potential. The fertilization potential of the slowly-frozen oocytes was not affected a great deal, but further development was impaired even more than the oocytes having been vitrified.

The incidence of aneuploidy was slightly higher in the vitrification group compared to the untreated oocytes (2-3X higher); an increase was also seen in the group cooled with slow freezing, but less than the vitrified group. In 1989, Nakagata and colleagues reported very high survival of oocytes vitrified using VS1 medium when the total time for exposure to the solution was 10 seconds or less and the dilution was rapid (immediate transfer to a 10X volume of 0.3M sucrose). Eighty-eight percent of the thawed oocytes were morphologically normal, and the fertilization rate was very high with control oocytes as well as those only exposed to the modified VS1 medium and vitrified (97 %, 86 %, and 82 %, respectively). Cleavage for these groups after IVF was also high (98, 84, and 78 %), as was the proportion of live young (60, 43, and 46 %). These high survival rates were not repeatable by another group using this method, however (Shaw *et al.* reported that the use of Nakagata's original procedure resulted in 54 % morphological survival in their hands). In this report, Shaw *et al*, used a 90 % concentration of the original VS1 recipe, and made modifications to the CPA addition and removal method original used by Rall and Fahy (1985). They found that extending the CPA dilution procedures (without sugars as osmotic buffers) could improve the overall viability. When the oocytes were only exposed to the vitrification solution and the dilution procedure was extended, reducing the osmotic effects, the morphological survival improved (96 % *vs.* 77 % for the original method). The proportion that developed to the blastocyst stage from the 2-cell stage was also greater (95 % *vs.* 51 %). These results were better, despite the oocytes being in contact with the dilute vitrification solutions for longer periods of time. When sucrose was added as an osmotic buffer and the exposure to solutions containing the permeating CPA was eliminated during the dilution procedure, the results were also improved. The proportion of oocytes developing to the blastocyst stage having been vitrified by the original method where the CPA was removed stepwise without the presence of sucrose was more than doubled by using a 1-step dilution with sucrose as an osmotic buffer (24 % blastocyst development *vs.* 55 % with sucrose dilution). The effect of this procedure when exposure to the cryoprotectant without cooling was tested was also dramatic (31 % *vs.* 72 % blastocysts development, respectively). The morphological results support the conclusion of a protective effect of sucrose on the membrane after vitrification, for when the longer dilution procedure without sucrose was employed, fewer oocytes survived morphologically (45 %) compared to when sucrose was employed (79 %). A report that appeared the same year also showed that changes to the CPA dilution procedure after vitrification of mouse oocytes in VS1 could also result in high survival and developmental rates. In this study, the authors also performed embryo transfer and showed that the overall term development rate was approximately 25 % after vitrification. Other investigators have shown the beneficial effects of extended dilution procedures on post-thaw oocyte viability after exposure to a high concentration of EG (7 M), with and without vitrification.

The following year, Shaw and colleagues also reported on the effects of brief exposures of mouse oocytes to 90 % VS1 solutions and the subsequent viability. They investigated exposing oocytes directly to the solution for 10, 15, 30, 60, or 300 seconds. While exposures up to 1 minute were tolerated well (>85 % morphologically normal oocytes), a 5-minute exposure reduced the number of morphologically normal oocytes to 48 %. Fertilization and cleavage rates were also very high for the treatments lasting ≤ 60-seconds, and blastocyst development was good as well for the 10, 15, and 30-second groups (≥77 %; the group exposed for 60 seconds had a blastocyst developmental rate of 45 %). The results for the groups having been vitrified were markedly lower, with rates of blastocyst development of 22, 32, 12, 10, and 0 for the respective treatments; untreated oocytes developed to blastocysts at a rate of 87 %. The development rates were highly variable within the treatments, and the authors attribute this, at least in part, to devitrification of the medium during thawing, even though the thawing took place in a 37 °C water bath. These results suggest that the cells are able to tolerate high osmotic stress for brief periods of time. The authors mention that there was very little evidence of any volume recovery of the cells for the first 60 seconds of exposure, suggesting that only an infinitesimal volume of cryoprotectant penetrated the oocytes during this time.

The time of exposure to a 6.0 M DMSO solution has also been shown to be important to mouse oocytes. These authors used a stepwise exposure to DMSO instead of a direct addition to the final solution. With the short exposure time, the percent of viable fetuses was shown to be similar between the untreated control, DMSO exposure, and vitrified groups.

Results such as these have lead to the development of certain practices for vitrifying oocytes and embryos from mammals, incorporating very brief times of exposure to the CPA. Some authors have argued that restricting the times of exposure to only a few seconds is difficult in practice, and not an optimal method for practical application as there is little room for error. Considering that the 1991 reports by Shaw *et al.* and Kono *et al.* Have shown that oocytes can tolerate exposures to less concentrated solutions of VS1 for longer periods of time, and similar results have been reported with DMSO, there appears no need to resort to the extreme measure of direct exposure to the final concentration of the vitrification solution for brief periods of time as first described by Nakagata (1989). Since this time, numerous other reports have appeared in the literature describing successful mouse oocyte vitrification with varying modifications, including the use of modified complex vitrification solutions and vitrified sperm, ethylene glycol, the addition of various macromolecules, and devices. Table 3 summarized the results of embryo development from several reports where mouse oocyte cryopreservation was investigated.

In general, rapid cooling methods, including vitrification procedures, are becoming more commonly used to cryopreserve mammalian oocytes. As discussed above, the idea that short exposure times are more beneficial when using high concentrations of CPA has become established. However, as results such as those of Shaw and colleagues and Kono and colleagues have shown, when a reduction in the osmotic consequences of cryoprotectant exposure are incorporated into the method, short exposure times may not be necessary. This is a very important conclusion, as ensuring that cryoprotectant exposure occurs for only a very brief time is difficult in practice. Furthermore, the likelihood of IIF can be higher than expected if devitrification is considered, and solutions are frequently used that are likely to undergo devitrification during warming. With more careful study design, osmotic stress and devitrification can be avoided, which should result in more robust procedures.

8.13.3 Aneuploidy

Due to the known disruptive effects of cold exposure on the metaphase II spindle as discussed above, several investigators have examined rates of aneuploidy and polyploidy after freezing MII stage mouse oocytes. In 1992, Sterzik and colleagues noted a relatively small increase (4 % *vs.* 0 % in controls) in the incidence of hyperploidy in 2-cell stage embryos developed from frozen and thawed oocytes using 10 % DMSO and 10 % sucrose exposure at room temperature and applying a standard equilibrium freezing procedure to -32 °C prior to liquid nitrogen storage. The total incidence of all types of aneuploidy and polyploidy were not significantly different compared to controls. Using this freezing procedure, there was also a significant reduction in the proportion of frozen and thawed oocytes that underwent cleavage after IVF compared to unfrozen oocytes (34 *vs.* 75 %). Bouquet and colleagues published two reports describing an analysis the results of freezing mouse oocytes in 1.5M DMSO to -80 °C. In one report (Bouquet *et al.*, 1993), they determined that the incidence of sister chromatid exchange was about 50% higher in all treatments (CPA exposure at either 0 °C or room temperature, cooling to 0 °C, and complete freeze and thaw cycles) compared to controls. There was no significant difference among the treatments themselves. The proportion of polyploidy embryos originating from frozen and thawed oocytes was about twice as high (15.1 % *vs.* 7.1 %) compared to unfrozen oocytes. There was no difference in the proportion of haploid embryos, suggesting that digyny was responsible for the results. The results in their second report were similar, with the frozen and thawed group showing an increase of ~ 5-fold in polyploidy compared to control oocytes. They also looked at the developmental potential of the embryos after freezing, and the proportion of oocytes undergoing cleavage after

IVF was reduced to about half the level of the controls (30 *vs.* 64 %). Their results showed a reduction in cleavage for oocytes having been exposed to DMSO for 15 min at 0 °C (49 % cleavage) and room temperature (40 % cleavage). Bos-Mikich and Whittingham showed that exposure to DMSO at 37 °C for 60 min resulted in a very high incidence of aneuploidy (76.9 %, with an approximately equal mix of hypo- and hyperploidy) after parthenognetic activation. The average incidence of aneuploidy increased by varying amounts due to varying combinations of exposure to 1.5M DMSO and cold temperature (0 °C: cooling oocytes in the presence of DMSO for 30 or 60 min resulted in a 14 and 11 % aneuploidy rate, respectively; in the absence, 19 and 17 %, respectively). The incidence of retention of the second polar body did not differ across treatments, however. These results suggest that, while aneuploidy may be a factor in loss of developmental potential after cryopreservation attempts in mouse oocytes, the probability is relatively small. This may be a result of the ability of the MII spindle to undergo normal repair after prolonged cooling, as described above.

8.13.4 Direct Comparison of Equilibrium Freezing, Rapid Cooling, and Vitrification Methods

Several investigations have tested slow-cooling methods directly against rapid cooling methods in the same study, with only some of the later methods containing vitrifiable concentrations of CPA. Surrey and Quinn compared rapid cooling using 3.5M EG plus 0.25, 0.5, or 1.0 M sucrose to slow-cooling with either 1.5 M DMSO or PG, the later containing 0.2 M sucrose (Surrey and Quinn, 1990). For the first three methods, the authors transferred the cells initially to solutions with just sucrose, and then to solutions containing both CPA prior to freezing. The average blastocyst development rates for the respective groups were 49, 59, 5, 30, and 60 percent. Untreated oocytes developed to blastocysts at a rate of 62 %. Aigner and colleagues compared slow-cooling with 1.5 M DMSO to rapid cooling with 3.5 M DMSO plus 0.5 M sucrose. They noted ice crystal formation during cooling with the rapidly-frozen treatment. Survival was higher with the rapidly-cooled oocytes compared to the slowly-cooled ones (94 *vs.* 69 %). After a 1-hour recovery in culture, 67 % of the oocytes cooled slowly had normal spindle morphology compared to 84 % of the rapidly-cooled oocytes (91 % normal spindle morphology in control oocytes). After a 3-hour recovery, the proportions were 72, 87, and 95 %, respectively). Men and colleagues showed a reduction in fertilization of oocytes cooled with an ultra-rapid method using 3.5 M DMSO and 0.5 M sucrose (59 %) compared to 72 % and 74 % fertilization for cells slowly frozen with 1.5 M PG plus 0.2 M sucrose and vitrified with VS1 solution. Lane and Gardner described superior results using cryoloop vitrification compared to slow-cooling (Lane and Gardner, 2001). The former procedure having been conducted with 2.8 M DMSO and 2.8 M EG plus 10mg/ml Ficoll and 0.65 M sucrose, and the later with 1.5 M PG in a choline chloride based solution. Fetal development was better in the vitrified oocytes (57 %) compared to those cooled slowly (26 %). Two recent reports described very poor development of oocytes having been frozen slowly with PG compared to vitrified oocytes. In the report by Valojerdi and colleagues, none of the oocytes frozen in PG developed to the morula stage, whereas nearly 60% of those vitrified in 30% Ficoll, 10.7 % acetamide, 10 % EG, and 0.5 M sucrose developed to blastocysts. Similarly, in the report by Kim et al, only 3 % of the oocytes frozen in PG survived, whereas 85 % survived vitrification on electron microscope grids in the presence of 5.5M EG plus 1.0M sucrose.

These results are very insightful in regards to the difficulty of making broad generalizations about the success of equilibrium *vs.* non-equilibrium methods. According to Aigner and colleagues, rapid cooling is superior to equilibrium freezing. According to Men and colleagues, rapid cooling is inferior to equilibrium freezing or vitrification. Lane and Gardner conclude that vitrification is superior to equilibrium, as do Valojerdi and Salehnia and Kim and colleagues. However, according to the results of Surrey and Quinn, rapid cooling can be equally effective compared to equilibrium freezing, depending upon the specific nature of the method utilized. While their conclusions are all correct in general, the more precise conclusions should be that some methods of equilibrium freezing are as good as or superior to non-equilibrium methods, and some methods of non-equilibrium freezing are superior to equilibrium methods. The most appropriate question to be asked as a result of such an analysis is why some methods are better than others. Obviously, cooling rate is only one of many factors in a cryopreservation procedure, and the particulars of the other factors may favor one method over the other. Designing experiments to understand the fundamentals behind the reasons is the mark of good scientific inquiry.

8.13.5 Bovine Oocytes

Numerous reports have appeared in the literature describing efforts to cryopreserve bovine oocytes using slow cooling methods (Table 4). Overall, this approach has been unsuccessful for MII bovine oocytes, with very low blastocyst yield *in vitro*. Investigators have tested several variables, including post-thaw culture period, different CPA (Glycerol, DMSO, or PG;) and at different concentrations (1 or 1.5 M), the use of different sugars as post-thaw osmotic buffers (sucrose, lactose, or trehalose;), different cooling rates (0.3, 0.6, or 0.9 °C;), concentrations of sucrose in the freezing solution (0, 0.1, or 0.2;), and polarization of cytoplasmic lipids. None of these variables has had a marked effect on the success of the procedure as measured by overall post-thaw developmental potential. As discussed above, bovine oocytes are highly prone to chilling injury. Long time periods of exposure to cold temperatures seems to be the common factor among the studies cited above, and is likely the major contributor to the failure of slow cooling as a methodology. As a result, researchers have turned their attention to rapid cooling methods to cryopreserve bovine oocytes. The earliest reports attempted vitrification in ¼ cc straws (Table 4b). In 1992, Fuku and colleagues directly compared a vitrification method to a slow cooling method. Surprisingly, the slow cooling method proved superior. However, the oocytes were exposed to the highly concentrated vitrification solution (2 M DMSO, 1 M acetamide, and 3 M PG; DAP213) in a single step. Thus, the osmotic shock of this procedure may have been responsible for much of the loss of developmental potential. As supporting evidence for this hypothesis, Hamano and colleagues showed that a higher developmental potential could be preserved if this CPA was diluted from the cells in a gradual fashion (12 steps; 10 % blastocyst yield) than in a more rapid fashion (2 steps; 0 % blastocyst yield). A 1-step addition procedure was also used in this investigation.

In 1996, Martino and colleagues demonstrated a significant improvement in the cryopreservation of bovine oocytes using a so-called "ultrarapid cooling" method (Table 4c). In this report, they utilized electron microscope grids as a carrier for the oocytes and vitrification solution and plunged them directly into liquid nitrogen. The blastocyst yield ranged from 10 to 15% with all of the treatments, including the CPA exposure controls, and the authors suggested that these results demonstrated that chilling injury could be overcome when such rapid cooling was employed. Since this time, several other devices/methods to achieve rapid cooling have been developed and tested with bovine oocytes, including OPS, solid surface vitrification (SSV;, gel loading pipette tips, glass capillaries, and the cryotop . While many of these reports demonstrate higher survival and blastocysts development compared to the use of slow-cooling methods, others have shown more limited success, even when using identical methods. The inventors of these ultra-rapid cooling devices/methods and other authors have argued that the success of these devices suggests that ultra-rapid cooling is a necessity to achieve effective cryopreservation of bovine oocytes. However, others have argued that the use of such devices is cumbersome, and may not be necessary if improvements of the overall vitrification approach are made. In a series of experiments, Otoi and colleagues took a more systematic approach to develop improved vitrification solutions and methods for their addition and removal to try to account for potential osmotic damage during the procedure, and demonstrated that the use of a ¼ cc straw could achieve results comparable to the so-called ultra-rapid cooling devices. Testing various stepwise CPA addition methods, EG concentrations, and CPA removal procedures, this group achieved blastocysts development up to 20 % (the average development for the best procedure was 10% compared to 41 % for untreated oocytes). They determined that a 3-step procedure was superior to a one or two-step CPA addition. Most reports on bovine oocyte vitrification use a 2-step CPA addition with no justification for such an approach. Such improvements provide strong support for utilizing a fundamental approach to improving methods to cryopreserve mammalian oocytes. Despite the success of the trials by Otoi and colleagues (Otoi *et al.*, 1998), utilization of ultra-rapid cooling devices has become the established paradigm for mammalian oocyte vitrification. This has had unfortunate side effects (in my opinion) in that the avenues of investigation have been apparently narrowed. Across the field of mammalian oocyte cryopreservation, much of the current research continues to proceed in a trial-and-error fashion with the assumption that the use of ultra-rapid devices is a prerequisite. This has the unfortunate effect of eliminating the consideration of investigations into procedures which my be at least as successful, and perhaps even better from the standpoint of cell viability and user-friendliness. This seems particularly true in the human assisted reproduction community, where a robust method for oocyte cryopreservation is increasingly needed.

8.13.6 Porcine Oocytes

To date, very few investigations focusing on cryopreserving MII porcine oocytes have appeared in the literature. Even though bovine oocyte cryopreservation has been very difficult, porcine oocytes have proven even more recalcitrant to cryopreservation (see Table 5 for an overview). In 1999, Nagashima and colleagues reported only a 6.3 % development rate (to the 8-cell or morula stage) for porcine oocytes after having been vitrified in 40 % EG and 1.0 M sucrose in a standard holding medium with ¼ cc straws, and having been fertilized by sub-zonal sperm injection. The oocytes also underwent lipid removal *via* centrifugation and micromanipulation prior to vitrification. None of these oocytes developed to blastocysts in culture. The method used to vitrify these oocytes was very unorthodox, and may account for much of the loss of developmental potential. The cells were initially incubated in 10 % glycerol for 10 min prior to transferring them to the vitrification solution.

Failure of embryo development after cleavage was also reported by Rojas and colleagues. This group used a 7-step CPA addition (0.5 M increments up to 30 %, then transfer to 40 % EG) and a 4-step EG removal process using sucrose as an osmotic buffer. The oocytes were also incubated with 7.5 μg/ml Cytochalasin B prior to vitrification. Ten percent of the vitrified and warmed oocytes (OPS method) underwent cleavage after IVF, and none progressed further. Somfai and colleagues also investigated the effects of Cytochalasin B at the same concentration on porcine oocyte vitrification using the solid surface vitrification method. They used a 2-step CPA addition procedure (4 % EG for 10 – 15 min at 37 °C, then 35 % EG with 5 % PVP and 0.3 M trehalose (≤ 30 sec exposure) as the final vitrification solution). They measured post-warming viability using several assays, including vital dye staining (both propidium iodide and fluorescein diacetate) and artificial activation followed by *in vitro* culture. In their first experiment they demonstrated that exposure to the CPA without cooling had no effect on viability as measured by the vital dye assays. They went on to show that CPA exposure also had no effect on the proportion of oocytes developing to blastocysts after activation. However, vitrification with or without Cytochalasin B caused a significant proportion of the oocytes to lose their developmental potential (2.5 to 5 % blastocyst development compared to 26 to 27 % with untreated oocytes) Gupta and colleagues used a nearly identical vitrification procedure to that used by Somfai and colleagues just described. In an experiment using GV-stage oocytes, they showed a slight improvement in the outcome when a combination of EG and DMSO was used, so they applied this combination to MII oocytes as well.

They also determined that this method caused a significant decrease in the developmental potential of the oocytes (only 3.4 % progressed to blastocysts compared to 20.1 % for the control treatment after IVF).

Fujihara et al. investigated the effects of Paclitaxel (Taxol®) on vitrification at concentrations up to 5 μM. These authors used the cryotop device for vitrification and fertilized the oocytes using ICSI. Concentrations of paclitaxel actually reduced the developmental potential in a concentration dependent manner (7.4, 4.5, 1.2, and 1.2 % blastocysts rate for 0, 0.5, 1.0 and 5.0 μM, respectively). Control oocytes developed to blastocysts at a rate of 38%. These results were similar to those described in a more recent report by Shi and colleagues (Shi *et al.*, 2006). In their study, none of the oocytes vitrified using OPS either in the presence of absence of 1 μM paclitaxel with 20 % EG, 20 % DMSO, 18 % ficoll 70, and 0.3 M sucrose developed to blastocysts. They did see an improvement in the proportion of oocytes undergoing cleavage with paclitaxel, however (24.3 % vs. 5.6 %). Wu and colleagues noted that OPS vitrification with 40 % EG and 0.6 M sucrose caused all of the MII oocytes to fail to undergo cleavage division after IVF. In an ultrastructural analysis, they showed that the cortical granules in the oocytes were not released as a result of the vitrification procedures, suggesting that zona hardening was not the reason for failure of cleavage. They also showed that the mitochondria appeared normal in their electron micrographs, but a noticeable change in the level of homogeny of the lipid vesicles, and vacuoles appeared in the cytoplasm. They also demonstrated a significant disruption in the microtubule staining patterns (13 % *vs.* 80 % for the controls) and actin staining patterns (37 % *vs.* 72 % for the controls).

As very large number of studies have been undertaken to advance our understanding of oocyte cryobiology. Some significant insights into the nature of oocyte cryoinjury have resulted from these investigations. For example, our understanding of the differential sensitivity of oocytes across taxa to cooling has provided a means to explain why slow cooling methods are relatively successful with mouse oocytes, but fail to preserve the viability of oocytes from cattle and pigs. Furthermore, cytoskeletal elements in oocytes (e.g.microtubules and microfilaments) have been shown to be easily damaged by various aspects of cryopreservation. Like all cells, oocytes are damaged by the formation of intracellular ice and high concentrations of cryoprotectants, yet damage from cryoprotectants can be mitigated to a large degree by reducing the osmotic stress associated with their addition and removal. Vacuoles are often seen in ultrastructural analyses of oocyte vitrification, yet the nature of these vacuoles is still uncertain. As was highlighted throughout this chapter, cryopreservation is a multi-step process, with each step having the potential to cause cell damage. Many of the variables in a cryopreservation procedure are interdependent, as discussed in the section on the science of cryobiology. As a result of this equivocal nature of study outcomes, it is often very difficult to ascribe cryoinjury to a specific component of a cryopreservation method which makes overcoming the injury quite challenging.

An interesting note was made in the report by Gupta and colleagues. The authors mentioned that some of the droplets of solution became opaque during cooling or warming, and they were discarded from their analysis. This suggests that the solution they used was not a stable vitrification solution. Their solution was nearly identical to the one used by Somfai and colleagues described above, except sucrose was used as the nonpermeating sugar instead of trehalose, which was used by Somfai and colleagues. Again, this observation reflects the failure of studies investigating procine oocyte cryopreservation to account for some of the fundamental principles of cryobiology. Accounting for some of these basic principles is likely to lead to improvements in the cryosurvival of oocytes .

9

CRYOBIOLOGY AND CRYOPRESERVATION OF SPERM

Low temperature has been utilized to keep living cells and tissues dormant but potentially alive for cryopreservation and biobanking with great impacts on scien- tific and biomedical applications. However, there is a critical contradiction between the purpose of the cryopreservation and experimental findings: the cryopreserved cells and tissues can be fatally damaged by the cryopreservation process itself. Contrary to popular belief, the challenge to the life of living cells and tissues during the cryopreservation is not their ability to endure storage at cryogenic temperatures (below −190°C); rather it is the lethality associated with mass and energy transport within an intermediate zone of low temperature (−15 to −130°C) that a cell must traverse twice, once during cooling and once during warming. This chapter will focus on (1) the mechanisms of cryoinjury and cryopretection of human sperm in cryopreservation, and (2) cryopreservation techniques and methods developed based on the understanding of the mechanisms.

9.1. INTRODUCTION: AN OVERVIEW

The pioneering studies in cryobiological science starting from the middle of the last century is based on the idea that whether life span and longevity could be extended by storing various frozen tissues and organs or even human body for many years. Possible cellular and tissue destruction during the freeze-thaw processes became the subject of curiosity. The male reproductive cell, spermatozoon, was first discovered in 1677 by Van Leeuwenhoek and was called "Animalacula." The early studies to obtain pregnancy by artificial insemination were done by Lazzaro Spallanzani in 1776. The studies with scientific basis and modern vital cell freezing studies began in 1949 after the discovery of glycerol as a cryoprotectant by Polge et al. and spermatozoon became the first mammalian cell to be successfully frozen . Following the success in freezing spermatozoa, Whittingham et al. successfully froze the mouse embryo by using dimethylsulfoxide

(DMSO) as a cryoprotectant, in 1972. Nowadays, millions of children have been born from cryopreserved sperms or embroys. Sperm cryopreservation has been successfully applied in various fields to benefit the human kind and animals, including assisted reproduction, rescuing the endangered species, and saving the ecosystem. However, we also know that damage can be caused to the cells during the cryopreservation process. In this review, the mechanisms of damage to the sperm cells during the process of cryopreservation will be taken under spotlight and we will try to elucidate them in a cause-effect manner.

9.2 CRYOBIOLOGY

Cryobiology is a multidisciplinary science, studying the physical and biologi- cal behaviors of living materials (e.g., cells and tissues) at low temperatures.

Cryobiology contains many disciplines such as, cellular biology, theriogenology and molecular biology, engineering and mathematics, veterinary and human medicine, intensive and extensive farming on land and in watery environments. Optimization of the cryopreservation procedure of spermatozoa needs all the above-mentioned disciplines because of the complex cellular structure, activation and capacitation mechanisms of spermatozoon.

9.2.1 Dormant or dead? Two aspects of freezing of cells

Freezing biological time occurs when the cells are cooled down in controlled manner under temperatures lower than necessary for continuing their normal physiological activities. Damaged or dying cells exhibit characteristic changes that cause structural differentiation under destruction process and are dragged into a possible death. In the present case, there are two types of scenarios, where cells may be damaged or killed: necrosis and apoptosis.

9.2.1.1 Necrosis

The cell death takes place through one of the pathways such as necrosis, necro- biosis, or apoptosis. In necrosis, a pathological cell death, the cell has been severely damaged as a result of sudden and extreme trauma, and soon death occurs. Because the cell loses control of permeability of plasmalemma, the cytosol rapidly swells by excessive inflow of water and ions, and organelle becomes excessively swollen. As a result, the cell lyses without spending energy. The cell contents and endotoxins leak out of the collapsed or inflammatory cells into the area where inflammatory changes begin. In this case, neighboring cells and tissue elements may also be exposed to damage as well. Factors causing damage by disrupting the local homeostatic bal- ance of the cells and tissues such as hypoxia, viral, and bacterial infections, toxins, radiation and changes in temperature are the main causes of necrosis. Necrosis, as an irreversible process, is also defined as a transition of cell into a definite death and loss of all of its physiological functions.

9.2.1.2 Apoptosis

Apoptosis, which is a programmed cell death, differs from necrosis with a variety of morphological, biochemical, and physical changes.

Essentially, apoptosis is a physiological cell death, a natural process that occurs during embryonic development and periodically in organs such as mammary gland and uterus, which undergo cyclic changes. In this case, the cell systematically ceases all of its functions and breaks down by incorporating its structural components into vesicles [9]. The main histological differences between apoptosis and necrosis are summarized in Table 1.

9.3 CRYOBIOLOGY AND CRYOPRESERVATION OF SPERM DOI

Apoptosis consists of three stages as, final decision, execution, and cleaning phase. The cell receives nonreversible lethal apoptotic stimulation beyond the time of decision. In the execution phase, condensation of chromatin in the cell, shrinking of the cytoplasm, formation of buds in the cell membrane, fragmentation of DNA, and formation of apoptotic bodies occur. These changes are affected by different enzymes such as proteases, lipases, and nucleases. In the final stage of the cleaning process, apoptotic bodies are phagocytosed by macrophages [tingible body macrophages (TBM)] or neighboring cells . The main difference between the consequences of apoptosis and necrosis is that the apoptotic cell never leaks the cytoplasmic contents into the extracellular space and destroys the genetic content before phagocytosis. All these events are of great importance in the removal of cytotoxic T lymphocytes containing virus-infected cells and activated harmful granules. In this respect, apoptosis is a clean way of cell death.

Table 1. Comparison of the effects of necrosis and apoptosis on cells, organelles, and tissues.

Cellular results	Necrosis	Apoptosis
General cellular	The cell is swollen, massive cell death occurs in a very large area.	Cells die in groups, apoptotic bodies are changes formed.
Organelles	Damaged	Not damaged
Mitochondria	Mitochondria, due to lack of ATP swell and break down	Mitochondria are swollen, cytochrome-C is released.
Cell membrane	There is structural deterioration, selective permeability control is lost.	It is intact, the surface takes a crater appearance, vesicles are formed.
Nucleus	Chromatin loses its normal organization and is in the form of thick chromatin yarns. Pyknosis, karyorrhexis, or karyolysis develops.	Chromatin is fragmented, it is concentrated in the form of a hat, nucleolus is dispersed
Causal factors	Anoxia, starving, physical and chemical traumas that lead to ATP deficiency.	They are physiological and pathological conditions that do not lead to ATP deficiency.

Cellular results	Necrosis	Apoptosis
Effects on the tissue	The inflammatory events develop, degenerated cell debris is found in the environment and these are engulfed by phagocytes. There is a common tissue destruction.	There is no inflammation, the resulting apoptotic vesicles are phagocytosed by neighboring cells and macrophages, a rapid involution occurs without collapse in the tissue.

Stress factors such as heat and cold shocks, oxidative stress, ultraviolet light, and ionizing radiation trigger the activation of stress-activated protein kinases (SAPKs) such as c-Jun N-terminal kinase (JNK) and p38. JNK is an important element of the signal pathway leading to apoptosis in response to stress conditions. Apoptosis related to our subject arises as cold-induced apoptosis, rather than the lethal effect of long-term storage, freezing damage or cold-struck damage occurs. Contrary to the assumption that very low temperatures are lethal to cells, the cells are more severely damaged at moderate temperatures between −15 and−60°C. All chemical or biological reactions are almostly ceased at the liquid nitrogen temperature (−196°C). The only reaction that can occur at −196°C is the one arisen from cosmic radiation. A 200–400 rad dose of radioactivity can damage 63% of a cell population. The earth is exposed to 0.1 rad cosmic radiation per year.This corresponds to the amount of radiation that the frozen mammalian cell will be exposed to for 2000–4000 years at −196°C. Cells that are stored at temperatures above −80°C, deteriorate over time and continue their cellular activities due to presence of solutes with different ion concentrations in the environment that are not fully frozen. Death can occur anytime, depending on temperature, animal breed, cell type, and freezing medium. For a cell, which can be stored theoretically for 4000 years at −196°C, it can be considered that biological time has stopped.

9.4 STRUCTURE OF BIOLOGICAL MEMBRANES

All gamete cells, oocyte, and sperm, have a liquid mosaic membrane structure that are mainly composed of various phospholipids and proteins. The cell membrane has at least three major tasks: to separate the cells from the external environment, to ensure that they have a specific shape, and to control the exchange of various solutes between the cell and the external environment. The major structural components involved in the membrane structure are phospholipids, glycolipids, transmembrane proteins, peripheral proteins, and cholesterol.

According to, membrane integrity is dependent on four important factors as follows:

- Preservation of membrane integrity by bonds between the lipids to prevent lateral displacement of the lipids;
- In order to preserve the stability of the lipids, they have to bond with the build ing blocks of the cell skeleton, with proteins forming a naturally occurring column;
- The presence of large integral membrane proteins which serve to provide diffusion between the intracellular and the external environment and to stabilize intracellular balances; and

- The presence of phospholipids and aquaporins with selective permeability in the membrane structure.

There are two layers, internal and external, formed by different types of phos- pholipids linked together in a chain form in the cell membrane. Positively charged phospholipids such as phosphatidylcholine and sphyngomyelin tend to be present in the outer leaflet of the cell membrane, whereas anionic phospholipids such as phos- phatidylethanolamine, phosphatidylserine, and phosphatidylinositol tend to be pres- ent in the inner leaflet. The interactions between these two phospholipid layers work to form the transmembrane channels by cholesterol and proteins. Phospholipids, which provide fluidic and variable membrane structure and are predominant in the docosahexaenoic acid chain, constitute 65–70% of total membrane lipids.

In order to make a more detailed explanation of the biological membrane structure, the first necessity is to identify the phospholipid and the above-mentioned sub-groups of phospholipids.

9.4.1 Phospholipids

Phospholipid consists of a phosphate group, one or two fatty acid groups attaching to this group, an alcohol group, and the glycerol or sphingosine backbone linking them. Phospholipids, which form the majority of the cell membrane, also increase the resistance of the cell to cold shock. It was also found that resistance to cold shock was higher in live sperm with high phospholipid/cholesterol ratio.

9.4.2 Phosphatidylcholine

It is a membrane lipid that consists of a phosphate group, two fatty acids, and the glycerol backbone of choline. Phosphatidylcholine, also known as lecithin, is present in high amount in the membrane and plays an active role on membrane stability. Simpson et al. [21] found that phosphatidylcholine exhibited protective properties against cold shock and that phosphatidylcholine-rich spermatozoa exhibited better motility levels than the spermatozoa deprived of phosphatidylcho- line. Sariozkan et al. obtained a higher level of motility in the group with 5% lecithine compared to the control group for bull sperm (Figure 1).

9.4.3 Phosphatidylinositol

It is a membrane lipid that consists of a phosphate group, two fatty acids, and a glycerol backbone of inositol. It has functions in the growth and division of the cell, exchange of membranous materials, participating in the cytoskeleton structure and binding to the target proteins (membrane proteins) to maintain membrane stability and fulfilling their tasks [23]. Luconi et al. found that human sperm with phospha- tidylinositol exhibited higher motility level (Figure 2).

9.4.4 Phosphatidylserine

It is a membrane lipid consisting of two fatty acids and glycerol backbone of serine. It causes the start of capacitance or apoptotic changes through displace- ment between membrane foliage.It is emphasized that phosphatidylserine is localized mainly in the head and middle region of a normal spermatozoon, emphasizing phosphatidylserine externalization in apoptotic spermatozoa, and 14% externally phosphatidylserine apoptotic spermatozoon in normal semen. Removing these externalized phosphatidylserine apoptotic spermatozoa from the environment, results in an increase in fertility. Wilhelm et al. also investigated the effect of phosphatidylserine and cholesterol supplemen- tation on sperm parameters after freezing/thawing, and consequently, samples with phosphatidylserine and cholesterol showed higher motility and viability than the control group (Figure 3).

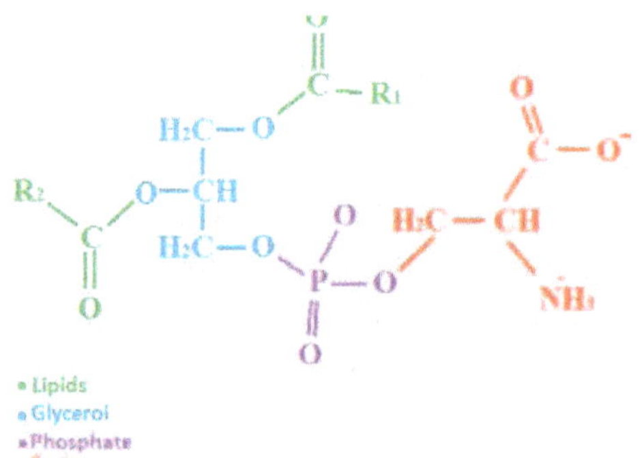

Fig. 1: Diagram of phosphatidylcholine molecule.

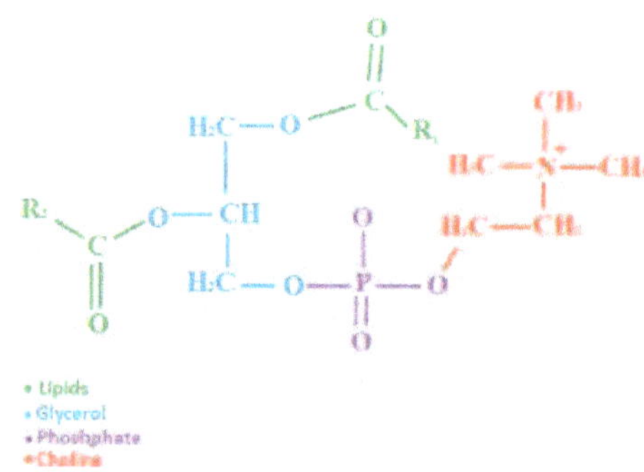

Fig. 2: Diagram of phosphatidylinositol molecule.

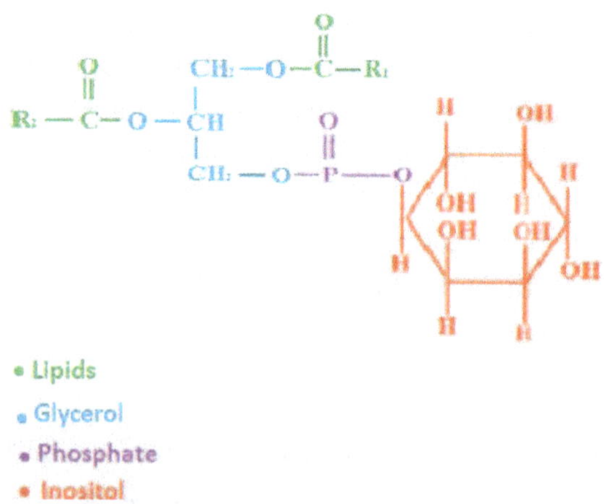

Fig. 3: Diagram of phosphatidylserine molecule.

9.4.5 Phosphatidylethanolamine

It is a class of lipids that consist of two fatty acids and a glycerol backbone of ethanolamine. It has the ability to build hydrogen bonds with other membrane proteins with its two electrons. This ability contributes to membrane integrity and plays a role in membrane reforming during phase transitions (Figure 4).

9.4.6 Sphingomyelin

Unlike the other membrane lipids mentioned above, the backbone of sphingo- myelin is formed by sphingosine (an unsaturated aminoalcohol) instead of glycerol. It is a membrane lipid consisting of a fatty acid and a sphingosine backbone of choline group (Figure 5).

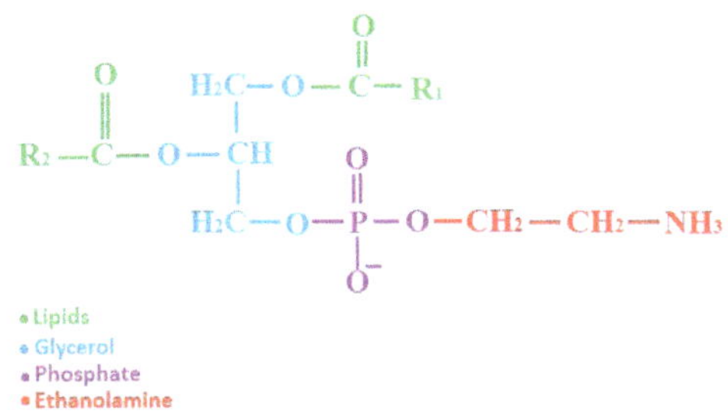

Fig. 4: Diagram of phosphatidylethanolamine molecule.

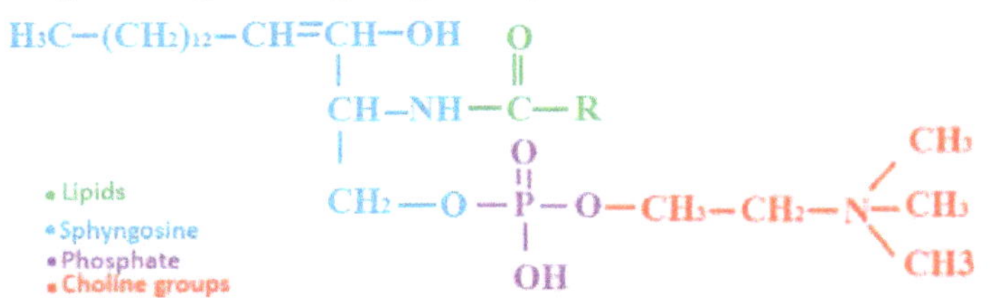

Fig. 5: Diagram of sphingomyelin molecule.

Although phospholipids constitute the great majority of membrane structure, the most effective molecule is cholesterol that provides integrity to both the phos- pholipids and other components. The effect of cooling and freezing on the sperm membrane is directly related to the cholesterol/phospholipid ratio in the membrane, saturation in the hydrocarbon chains (the chain formed by the addition of hydro- gen bonds between the hydrogen and carbon atoms of the lipids themselves), and protein/phospholipid ratio [28]. At the same time, it was proven that cholesterol in phosphatidylcholine, phosphatidylethanolamine, and cholesterol mixtures was observed more intensely in the areas of phosphatidylcholine . The cholesterol molecule is considerably smaller in size than phospholipids and therefore has the ability to move freely in the membrane. The cholesterol molecule is mostly concentrated in portions, where hydroxyl groups are rich in phospholipid ester carbonyls. If cholesterol was extruded from the spermatozoon plasma membrane, there was an increase in fluidity, permeability, and fusion capacity in the membrane structure, and the structure of the membrane was damaged. When cholesterol was added into membrane, the membrane stability and motility increased, whereas capacitance decreased.

Cholesterol plays an effective role in the phase transitions in the membrane structure. As the cell membrane passes from the liquid crystal phase to the lamellar gel phase, the membrane cholesterol densifies the structure of the hydrocarbon chains of the fluid phospholipids and causes degradation of the lipids in the gel phase. There are various proteins along with lipids on the cell membrane.

Proteins, which have the functions of contributing to the membrane integrity by forming the attachment area to the lipids, and allowing the cells to exchange materials from the outside, are divided into two membrane proteins of, namely, the peripheral membrane and the integral membrane. Peripheral membrane proteins can be found both on the exterior and interior of the cell. It has the task of forming protein-protein bonds and participating in the skeletal structure of the membrane. Integral proteins, which are another type of proteins, are opened both internally and externally of the cell and interact with both environments. It has functions, such as creating a holding place for cytoskeleton and glycoproteins, generating responses by receiving signals from various chemical substances, and providing the exchange of substances between the inner and outer environments of the cell.

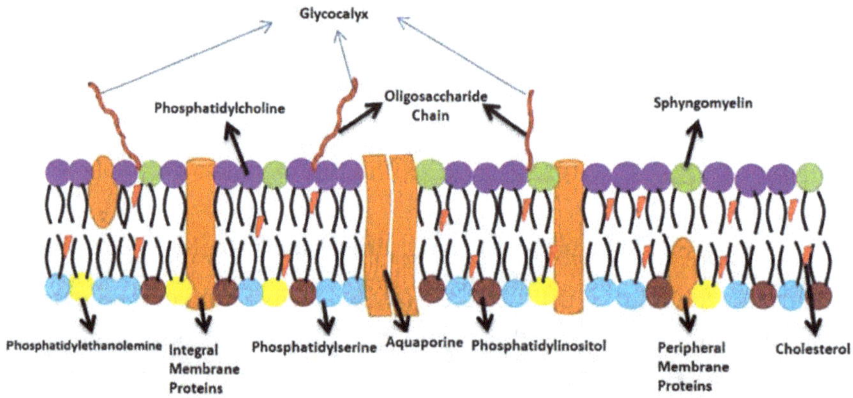

Fig. 6: A schematic representation of cell membrane components is shown.

During freezing and thawing, protein channels called aquaporins (AQP) have functions to control the water inlet and outlet to the cell. Aquaporins are composed of four different monomers, which are independent of each other, in many cell types, from bacteria to plants and to mammalian cells. These four different mono- mers combine and form a tetramer. While each monomer is only water-permeable, the fifth channel in the middle allows passage of both water and various ions. It is much faster and easier for water to pass through aquaporin channels when compared to membrane lipid. Although, there are 200 kinds of aquaporins in all living species, there are only 13 kinds of aquaporins in mammals. AQP7, AQP8, and AQP11 were identified on the spermatozoon membrane. While AQP7 is responsible for sperm glycerol metabolism, AQP8 is responsible for the water exchange of the cell. AQP11 is involved not only in the membrane structure, but also in the tail as well as in the exchange of cellular matter .

Cells use oligosaccharide chains (sugar chains formed by combining 2–20 monosaccharides) that extend out of the membrane as a tentacle or an arm to perceive various structures in the extracellular environment, interact with other cells, and initiate the necessary physicochemical reactions. These sugar chains are termed glycolipid chains when derived from lipids and glycoprotein chains when derived from proteins. These glycolipids and glycoprotein chains extend outward to bring a network-like structure around the cells. This network-like structure is called glycocalyx (Figure 6).

9.5 DAMAGE AND PHASE CHANGES IN THE MEMBRANE STRUCTURE

The cell membrane is a fluidic mosaic structure with cholesterol, integral and peripheral proteins, lipids, and many more.

There are many different types of phospholipids in the cell membrane. The changes that occur during freezing-thawing also originate from the structural differences in these phospholipids. The main factor that drives the structural differences is the dimensional changes between the head part forming the phospholipids and the acyl chains. These dimensional changes directly affect the various changes in the membrane structure at the later stage as well as lipids depending on the lipid type.

If in a phospholipid, the head and acyl chains are similar or identical in size, the phospholipid molecule is in a cylindrical form. Phosphatidylcholine is as an example for this kind of phospholipids. This kind of phospholipids is termed as bilayer lipids and are not significantly affected by phase changes during cell cooling. In the homogeneous phospholipid distribution that occurs during the phase change, these types of phospholipids in the form of bilayers are gathered together to form the bilayer membrane model; in other words, they continue to create the lamel- lar phase of the liquid crystal. If in a phospholipid the head and acyl chain are in different sizes, the phospholipid is in an inverted or flat cone-like form. This kind of phospholipids is called non-bilayer form phospholipids. Phosphatidylethanolamine is as an example for this type of lipids.

When the cell starts to cool down, the non-bilayer form of phospholipids undergoes a phase transition from the liquid crystal phase to the gel phase [36]. If the head and acyl chains are of different sizes in a phospholipid, this kind of lipids are also called non-bilayer lipids. During the phase changes, the non-bilayer lipids in the aggregate come together to bring up the different structures, which will be mentioned in the next section. Non-bilayer form phospholipids mainly consist of 2 types of phospholipids:

Type 1: Polar (+charged, hydrophilic) head in type 1 non-bilayer phospholipids is larger than apolar (–charged, nonpolar) tail. For this reason, they are in the form of an inverted cone. When the membrane is cooled, the negatively charged heads are directed outward and the neutral apolar parts are inwardly arranged to form micelle-like structures depending on temperature changes.

Type 2: For polarity of type 2 form phospholipids, the polar head is smaller than the apolar tail. For this reason, they are in the form of a flat cone, and when tem- perature changes, the negatively charged head part is arranged in the apolar neutral part outwardly and called hexagonal II (HII) (Figure 7).

Non-bilayer lipids are distributed heterogeneously on the cell membrane and carry out some vital tasks for the cell, including:

i. Providing membrane flexibility,
ii. Participating in the barrier activity against the outer environment,
iii. Protecting their stability by binding to the membrane's peripheral proteins through creating a surface to attach to the proteins, and
iv. Building the structures that integral proteins can hold, etc.

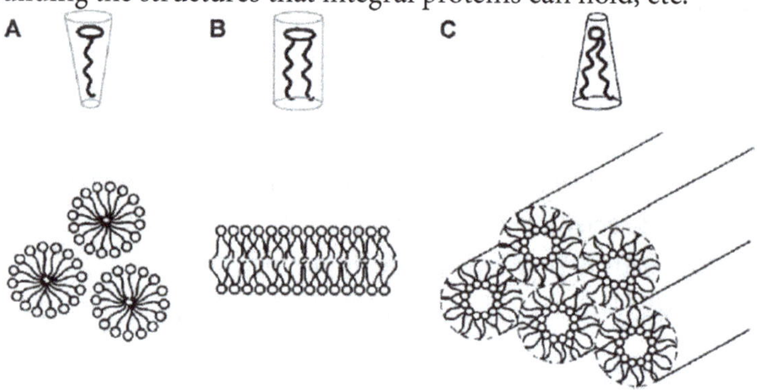

Fig. 7: This schematic representation depicts phospholipid-shaped non-bilayer type 1 lipids as indicated by the letter A and they are schematized at the bottom just below. In B, bilayer form phospholipid is schematized and bifilar lamellar form with a normally arranged phospholipid model was also depicted in the case where the cells were not subjected to heating or cooling (at a 36°C temperature). In C, non-bilayer type 2 lipids and hexagonal form II formulated due to heat exchange are schematized.

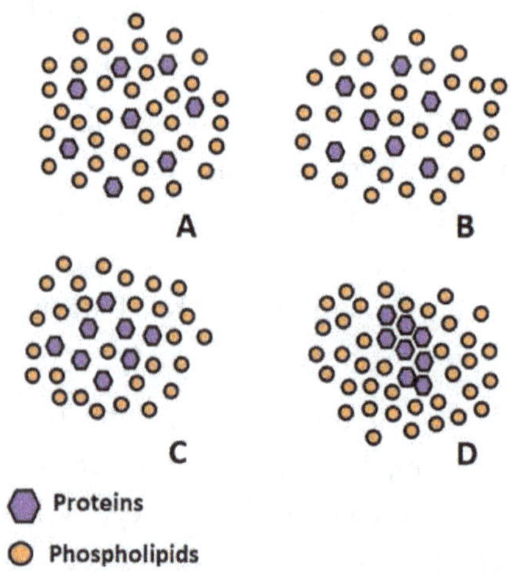

Fig. 8: In this schematic, the effect of cooling on the cell membrane structures is shown.

The phospholipid-protein components, which are in a heterogeneous structure starting from A, show clusterings due to cooling (B, C) as mentioned above, and membrane proteins appear to be tight in the center while phospholipids appear to be like honey pellets in the last stage (D).

Membrane phospholipids enter a phase transition generally between 36 and 5°C. These phase changes are species-specific and cause different responses to temper- ature changes depending on the animal's spermatozoa. When cell cooling starts, some phospholipids in the membrane structure undergo phase changes. The phospholipids in the non-bilayer form are bound to each other by hydrocarbon-chain bonds. As the hydrocarbon chains undergo crystallization and hardening, the lipids passing through the hexagonal Form II become more tightly bonded to each other, and the membrane proteins weakly bound to phospholipids cause aggregation on the membrane. They are tightly agglomerated to occupy less space than normal (Figure 8).

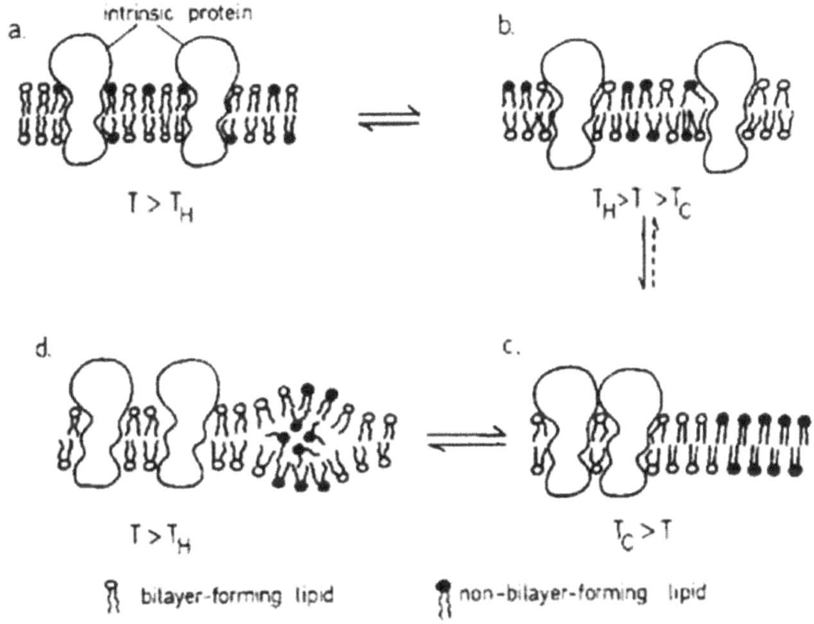

Fig. 9: Bilayer and non-bilayer phospholipid interactions during cooling and after re-thaw process. The figure shows the phase separations in biological membranes associated with cooling biological membranes from their growth temperature (a) to a temperature below the gel (b) to liquid-crystalline phase-transition temperature (c) and reheating to the growth temperature (d). Lipids that tend to form hexagonal-II phase are indicated by a solid head group.

Since proteins and various phospholipids are arranged in a heterogeneous structure in the normal arrangement of cells, exchange of substances is achieved in almost every region of the cell membrane. When the membrane structure is deteriorated in this way, substance exchange occurs especially where protein aggregation is intense. In this case, the non-bilayer phospholipids in specific interactions with proteins are separated from proteins and transformed from fluidic mosaic to homogeneous form.

The phospholipids that are interlocked with the proteins and hydrocarbon bonds that come together due to phase changes during cooling of the cells are in the nonlamellar phase at −196°C. However, during thawing, proteins and phospholipids do not return to their original situations. In this process, the membrane in the form of non-bilayer phospholipids came together and converted into the structures in the form of phospholipid micelles and hexagonal Form II. However, during thawing, these phospholipids cannot recombine to form bilayer phospholipids and turn into a fluidic mosaic membrane structure; therefore, the membrane structure has permanently been damaged (Figure 9). In this case, the non-bilayer phospholipids interacting with proteins are separated from the proteins and converted into a homogeneous form, different from the mosaic form.

Proteins are one of the basic building blocks of the cell and carry out many functions in the cell metabolism. Hydrogen bonds in the water molecules play a very important role in the denaturation of proteins. In the hydrophobic polar (HP) protein model, amino acids are reduced to polar (represented by P, charge or dipole) or nonpolar (H) point entities. Hydrophobicity is defined as the desire to reduce surface area of proteins that are interacting with water, and is related to the conversion of neighboring molecules into a similar energy-attractive structure. The hydrophobic molecules in the solution produce a more energetically favorable structure compared to the water bodies by forming ice-like cages in their surroundings at low temperatures. The cell systematically tries to reduce the energy by increasing the number of cages. This is the cold denaturation of proteins. On the opposite side, at higher temperatures, water molecules do not form cages. Irregular water molecules around the hydrophobic amino acids in proteins are energetically less advantageous than the water bodies. Thus, the proteins try to hide their hydrophobic parts in the building blocks. Biological structures, especially proteins, undergo structural changes during temperature changes. For this reason, nucleic acids and many polysaccharides need to stay in a certain temperature range with specific ionic strength and pH value to fulfill their functions. Cold protein denaturation occurs at a temperature range of 0 to −20°C, reversibly or irreversibly, but mostly irreversibly. Globular proteins often undergo partially reversible denaturation.

During denaturation, the covalent bonds between the atoms forming the peptide bonds between the amino acids of the proteins are not broken, but only hydrogen bonds change their structure. Protein denaturation during freezing or thawing leads to protease leaks from liposomes presenting in the acrosome and loss of membrane integrity.

9.5CRYOPROTECTIVE AGENTS (CPA)

Prior to the discovery of cryoprotective agents, some successful cooling proto- cols relied on extracellular ice formation suppression and sustaining cell viability by various chemical agents [7]. Mammalian cells had not been successfully cryopre- served until 1949 by Polge,

Smith, and Parkes when glycerol was accidentally found to have protective functions in freezing cells [3]. It has also been observed that if the correct dose of cryoprotectant is used and the cells are cooled down at the optimal cooling rate, the survival potency of cells is increased .

Cryoprotectants are used to avoid or decrease the cold shock damage and intracellular ice formation during freezing, recrystallization during thawing, and membrane destabilization. Cryoprotectants can reduce the freezing point and the proportion of salts and solutes in the sample by increasing the amount of liquid fraction, and suppress ice formation both outside and inside of the cells. In bio- logical structures, while the hydrogen bonds between the membrane phospholipids are connected by the oxygen atoms contained in the water molecules, the cryopro- tecting substances such as glycerol are replaced by water. These bonds have utmost importance for membrane integrity. It has been reported that supplementation of 0.2 M sucrose or trehalose into bull semen during freezing and thawing increases sperm viability. A good cryoprotectant should be water-soluble and have minimal toxic effects .

9.5.1 Division of Cryoprotectants According to Their Mechanism of Action

Cryoprotectants can be divided into two groups according to their mechanism of action, as penetrating and non-penetrating cryoprotectants.

9.5.1.1 Penetrating cryoprotectants:

Penetrating cryoprotectants have low molecular weight and therefore they have the ability to enter the cells. In this way, they can affect both intracellular and extracellular environments. When cryoprotectant is added into the cell suspension, cytosolic water moves to the exterior milieu of the cell due to the water chemical potential difference between inside and outside of the cells. The penetrating cryoprotectant penetrates into cells because of the concentration difference. This process lasts until equilibrium between intracellular and extracellular environments for both water and cryoprotectant is reached. The freezing point of the intracellular medium is decreased, and the intracellular ice formation (IIF) can be eliminated or prevented. Penetrating cryoprotectants penetrate into the cells, form new hydrogen bonds with water molecules by breaking the hydrogen bonds between them, and thereby change the structure of water. In this way, the cryoprotective function exhibits by preventing the cells from reaching high concentration of ions and avoiding extreme dehydration due to water loss during freezing.

Penetrating cryoprotectants such as glycerol, dimethyl sulfoxide (Me2SO or DMSO), ethylene glycol, formamide, 1,2-propanediol, 2,3-butanediol, and propylene glycol affect spermatozoa during cryopreservation by changing the physical properties of intracellular solution, decreasing the intracellular ice formation, increasing the resistance to old shock, regulating the protein and lipid structure of the cell mem- brane, and increasing membrane fluidity. In the following part, two most widely used penetrating cryoprotectants, glycerol and DMSO, will be discussed.

Glycerol: Many penetrating cryoprotectants have been tested, giving different results in cryopreservation of gamet cells of different species. However, the most commonly used penetrating cryoprotectant is still glycerol, which was discovered in 1949 by Polge et al. as a result of sperm vitrification. As shown in Figure 10, a glycerol molecule contains three hydroxyl groups.

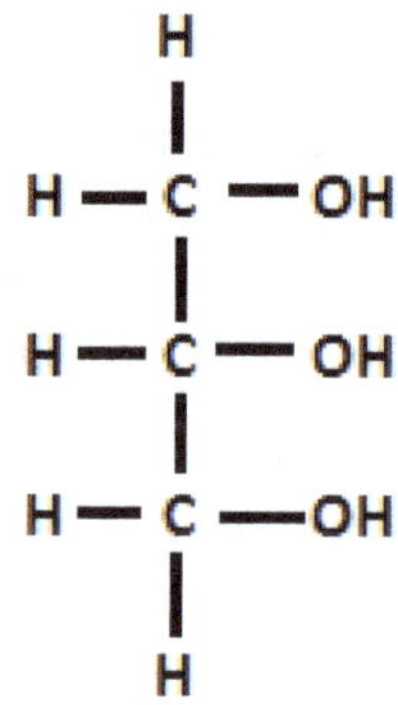

Fig. 10: Diagram of a glycerol molecule.

Each glycerol molecule is capable of binding to three water molecules. Because of the smaller molecular size, glycerol can easily pass through the membrane pores. With the decrease of temperature, the hydroxyl bonds formed between water molecules get hardened and solidified. For this reason, the frozen water expands and damages the cell. As a cryoprotectant, glycerol exerts its function by preventing water molecules from becoming large-volume ice crystals by binding to the hydrogen atom of the water molecules and interlocking during freezing.

In addition to the osmotic effect on cells, glycerol directly acts on the cell membranes. Amann and Picket have shown that glycerol exerts its primary effect in the extracellular environment, but it also intracellularly affects the cell membranes and organelle membranes. It binds to phospholipid groups in the plasma membrane, reduces the membrane fluidity, and also forms a particulate clump in the membrane by interacting with membrane proteins and glyco- proteins . At the same time, it creates structures similar to gap junctions (small connection channels between cells) in the membrane, reduces the electrical capacitance of the membrane, and causes large-scale rearrangement of membrane structure. Glycerol causes a decrease in the membrane fluidity by providing interactions between the inner and outer layers of the membrane through regulating the structure of fatty acid acyl chains. Glycerol also affects the polymerization (creating multiple monomers to form larger and differently shaped macromolecules, called polymers) and depolymerization (decomposition of large macromolecules into smaller monomers) of microtubules, which indirectly affect the plasma membrane. Changes in the microtubule structures in the tail of spermatozoon influence the interactions between microtubule-associated proteins.

While glycerol has many benefits as a cryoprotectant, its use in high dose has a detrimental effect on the cells. These harmful effects appear in at least seven cellular activities:

- It denatures proteins.
- It increases the viscosity of cytosol.
- It causes changes in the polymerization and depolymerization of α and β tubulins, the basic proteins of the microtubules found in the spermatozoon tail.
- It causes structural changes in the microtubuli.
- It acts on the bioenergetic balances.
- It acts directly on the plasma membrane and glycocalyx (a meshwork cell coat that is formed mainly by external proteins, externally-located glycoproteins and glycolipids providing various chemical interactions between the cell and the external milieu).
- It creates osmotic stress due to slow penetration through the cell membrane than other cryoprotectants.

It is measured that the motility and integrity of both sperm and acrosomal membranes that were frozen with different concentrations of glycerol (2, 5, 10, and 15%) and dimethyl sulfoxide (2, 5, 10, and 15%) in their study on Rhesus monkeys. The highest motility was measured as 45.5% in the 5% glycerol group. In terms of membrane integrity, the control group had the highest score (77.9%), while the 5% glycerol group had lower value (61.6%). Values of acrosomal integrity were 91.2 and 82.4% for the control group and 5% glycerol group, respectively. When these two parameters taken into consideration, DMSO had lower augmenting effects on Rhesus monkey semen than glycerol.

Awad and colleagues compared glycerol, ethylene glycol, and methanol as cryoprotectants and measured the sperm motility post thawing. The best result was obtained with 3% glycerol (72.4% motility). Motilities of sperm cryopreserved with 3% ethylene glycol or 3% methanol were significantly lower (56.9 and 22.6%, respectively).

Dimethyl sulfoxide (DMSO): Dimethyl sulfoxide (DMSO) is a liquid at room temperature with color scale changing from colorless to yellow. It is soluble in water, ethanol, acetone, diethyl ether, benzene, chloroform, as well as in aromatic compounds. Apart from being a cryoprotectant for semen, DMSO also has anti-inflammatory properties, bacteriostatic and tranquilizing effects, diuretic and local analgesic activity. It is a penetrant and carrier, and it can strengthen the effect of insulin (Figure 11).

When used in concentration of 5–10%, DMSO protects against damage on mouse T and B lymphocytes, human embryos, and many cellular sys- tems during freezing and thawing. Sometimes, DMSO has even higher protective potential than that of glycerol in freezing sperm of some animal species, such as elephants. Despite that DMSO has the same mechanism of action with glycerol, DMSO has lower penetrability in human and porcine spermatozoons . At the same time, DMSO has better protective capacity than glycerol and ethylene glycol in sperm freezing/thawing process and is more active in testicular tissues than glycerol and 1,2-propanediol.

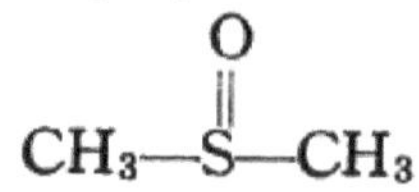

Fig. 11: Diagram of DMSO molecule.

Anchordoguy et al. used different cryoprotectants (DMSO, glycerol, proline, sucrose, and trehalose) for shrimp sperm and measured the viability. DMSO showed the highest viability (56.3%), followed by glycerol (29.3%). Proline and sucrose gave similar viability (24.5%), whereas trehalose gave relatively lower viability (16.2%). The group without cryoprotectant did not show any viability after thawing.

Two used different concentrations of DMSO and glycerol in this study of cryo-preservation of semen of black grouper fish and graded the motility scores from 0 to 4. The level 0 means no motility, 1 means 0–25% motility, 2 means 25–50% motility, 3 means 50–75% motility, and 4 means 75–100% motility. Results of the study revealed that 10% DMSO resulted in level 3 of motility, 20% DMSO resulted in level 4, and 30% DMSO resulted in level 1. However, glycerol solutions with different concentrations showed level 0 motility.

9.5.1.2 Non-penetrating cryoprotectants:

Non-penetrating cryoprotectants cannot penetrate through the sperm mem- brane and only act in the extracellular environment. They also function by incorpo- rating the membrane structure or reducing the freezing point of the medium. They are generally divided into two groups: low molecular weight cryoprotectants and high molecular weight cryoprotectants. Non-penetrating cryoprotectants with low molecular weight include three subgroups as monosaccharides (glucose and galactose), disaccharides (sucrose and trehalose), and trisaccharides (raffinose and melezitose). Cryoprotectants with high molecular weight [e.g., polyethylene glycol (PEG), ficoll 70, polyethylene oxide (PEO), polyvinyl alcohol (PVA), poly- vinyl pyrrolidone (PVPP), etc.] exert their functions by reducing the ice crystals formed externally during the freezing/thawing process to decrease cellular damage. As a special and effective cryoprotectant, trehalose will be discussed as follows.

Sugars used during sperm cryopreservation serve as an energy source and an osmotic pressure regulator, cryoprotectant, and cellular ice formation reducer. As a disaccharide formed by binding two D-glucose molecules, trehalose extends the distance between the membrane phospholipids by binding to the polar part of the phospholipids in the cell membrane. These cavities formed in the membrane prevent the formation of ice in the cell by helping the outflow of water from the cell during freezing, preventing harmful effect of cellular dehydration, and stabilizing the cell membrane.

Phospholipids accumulate due to van der Waals forces when energy is drawn from the environment by cooling sperm and transition occurs from liquid crystal phase to gel phase. During thawing, irregular voids occur in the cell membrane. This causes damage in the membrane structure, irregular ion and water leakage both into and out of the cell [68]. When trehalose is added into the medium, it forms gaps between phospholipids in the cell membrane and these gaps prevent the phospholipids from aggregation during freezing and protect the entity of mem- brane structure after thawing (Figure 12). In a study by Bucak et al. on freezing ram semen, two different doses of gluta- thione (5, 10 mM), taurine (50, 100 mM), and trehalose (50 and 100 mM) were used as cryoprotectants, and motility of the semen samples were measured at 0, 6, 24 and 30 h post thawing. The highest motility was obtained in samples containing 50 mM trehalose. Hu et al. investigated cryopreservation capacity of trehalose on five different groups of bull sperm. Group 1 was the control group without trehalose, whereas groups 2–5 contained trehalose with concentrations of 25, 50, 100 mM trehalose, and 200 mM. The group with 100 mM trehalose had the highest semen motility, mitochondrial activity, acrosome integrity than those of the other.

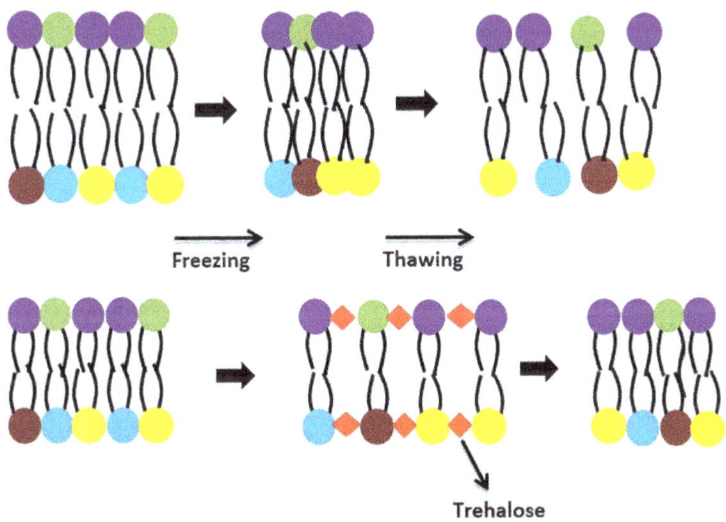

Fig. 12: Schematic drawing of cryoprotective effect of trehalose on membrane phospholipids groups.

Öztürk et al. studied the effect of arginine and trehalose on bovine sperm samples in their study. Trehalose showed higher (50.5% subjective, 64.7% CASA) motility than arginine (29% subjective, 12.2% CASA) when compared to the control group . Uysal and Bucak studied rapid (10°C/min from 5 to –25°C, 50°C/min from –25 to –130°C) and slow (0.5°C/min from 5 to –25°C, 50°C/min from –25 to –130°C) cooling rates with different doses of trehalose (0, 50, 100, 150 mM), and investigated sperm motility, survivability, membrane integrity, and morphologically abnormal sperm rates after thawing. Evaluations showed that the medium with 100 mM trehalose resulted in the best scores (sperm motility 72.0%, morphologic sperm abnormalities 25.5%, sperm viability 75.9%, and sperm membrane integrity 68.2%) . Bucak and his colleagues also used trehalose (50, 100 mM), taurine (25, 50 mM), cysteamine (5, 10 mM), and hyaluronan (0.5, 1 mM) as antioxidants in their study of ram sperm cryopreservation. Motility, acrosomal damage, viability, and total abnormality parameters were examined. The group with 50 mM trehalose, 25 mM taurine, and 5 or 10 mM cysteamine provided the best results.

9.6 MOLECULAR STRUCTURE OF WATER IN LIQUID AND SOLID PHASES

As the main source of life, water is different from most of the other molecules in the nature. Depending on the amount of energy in the environment, the water molecules undergo changes in their structure and phase transitions occurring between liquid, solid, and gas phases. Water molecules are in the liquid phase at the mammalian body temperature and perform very important tasks in almost every living cell.

The water molecule consists of two hydrogen atoms binding to a negatively charged oxygen atom at an angle of 104.52° with covalent bonds of 95.84 picometer in length (Figure 13). Hydrogen bonds are formed between water molecules (Figure 14). The fluid structure of the water at room temperature is based on the fact that these bonds are constantly breaking and re-bonding [75].

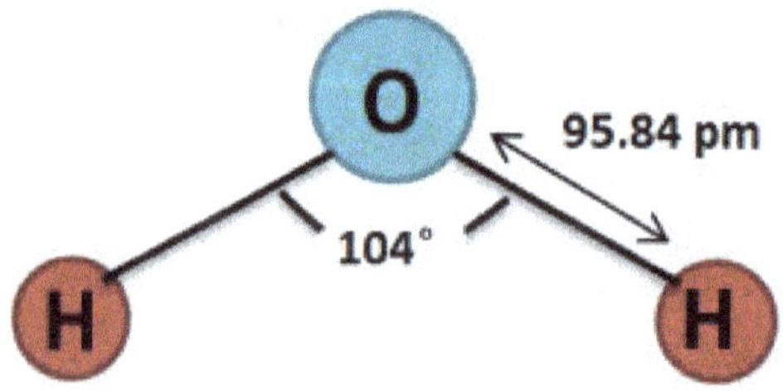

Fig. 13: Schematic drawing of the hydrogen bridge bond.

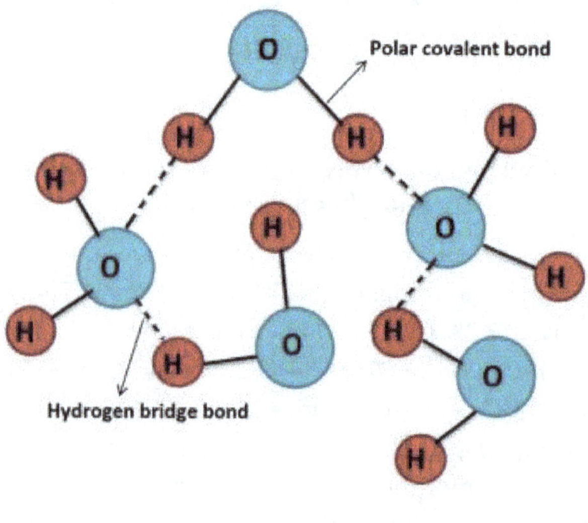

Fig. 14: Schematic drawing of water molecules.

As mentioned earlier, water molecules are in liquid form in living organisms. However, during cooling, the covalent bonds between oxygen and hydrogen are shortened and the hydrogen bonds become hardened and extended. So, they tend to line up in a symmetrical manner and to expand. Important effects on water- based biological molecules should not be overlooked. The water itself is described as a biomolecule, even as the 21st amino acid. In some cases, the ability of protein molecules to perform their normal functions depends on the presence of water molecules on their surface. The bond between water and protein surface is formed by giving the proton in the hydrogen atom of water to the protein or by taking the proton in the hydrogen of the amine group and its derivatives (NH, NH2, NH3), which forms the backbone of proteins. Water molecules also build bridges ("water bridges") to provide interactions between the atoms of different protein molecules.

9.7 CRYOBIOLOGICAL EVENTS DURING FREEZING AND THAWING

During cooling and thawing, spermatozoa undergo various changes in their chemical and physical status. The first one is the phase change of lipids in the sperm membrane, generally happening between 17 and 36°C. Different lipids have dif- ferent phase transition temperatures. Proteins are normally found in asymmetric position between the membrane lipids in the liquid crystal structure. When the lipids transform to gel phase, membrane structures are disrupted and ion metabolism is regulated. When the membrane lipids are in liquid crystal or gel phases, the leakage of liquid electrolyte through the membrane is minimal. However, during phase changes of lipids, the member permeability is increased, which causes solute leakage. Lipid phase changes occur not only during cooling but also during thawing. Fluorescent diacetate leaching was observed in the spermatozoa treated with fluorescent diacetate and re-dissolved by cooling to −5, −10 and−20°C.

The second change that takes place is the transformation of the cytosolic water in the cell into ice. Spontaneous ice formation occurs between −5 and−15°C. Generally, above −5°C, the internal and external milieu of the cell are supercooled, and no ice formation is observed. At the temperature between −5 and−10°C, ice formation starts in the extracellular environment, then cells are dehydrated and intracellular environment is supercooled. The fate of the cryopreserved cells, life or death, depends on the following cooling and thawing process.

9.7.1 Cryobiological Damages During Freezing

As mentioned in the two factor hypothesis proposed by Peter Mazur, damage arising from freezing occurs through two components. The first one can be defined as the direct damage caused by ice (ice injury), and the second one as the effect of the solution ion concentration on the cell (solution injury). However, this situation appears in different manners in male and female gametes. While intracellular ice formation is observed in mammalian oocytes and embryos at different degrees of cooling, the case is not the same for spermatozoa. The main factors include the pressure on spermatozoa caused by ice crystals formed in the external environment and the harmful effect of the high concentration of salt and mineral inside the cell.

Two-factor hypothesis of cryoinjury: The two-factor hypothesis suggested by Peter Mazur can be explained as follows: Factor 1: During cooling, extracellular ice formation happens first, which leads to the increase of solute concentration outside of cells. If the cooling rate is slow enough, the cells will have enough time to dehydrate and the cytosolic water will outflow from the cells until it is balanced with extracellular water. Dehydration of cells may include a few steps: separation of water molecules from the cytoplasmic liquid environment to the lipid part, diffusion in the lipid bilayer, and migration from lipid to external fluid, all of which are associated with activation energy (threshold energy value for a chemical reaction to takes place) (Figure 15). In this case, injury to the cells is majorly from the highly concentrated solutes in the cells [16]. Some biological constructs (e.g., eight-cell embryos) may be resistant to solution effects.

Factor 2: If the cooling rate is rather high, the cells will not have enough time to discharge cytosolic water into the extracellular environment, and water will freeze inside the cells [16](Figure 16).

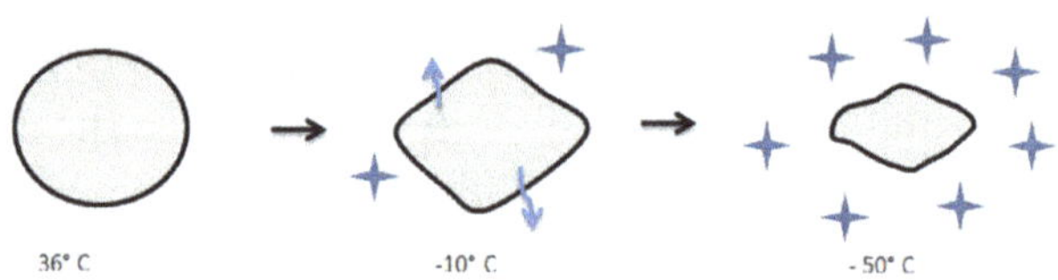

Fig. 15: chematic of factor 1 in "two-factor" hypothesis of cryoinjury. During slow cooling procedure, intracellular water has enough time to outflow from the cells.

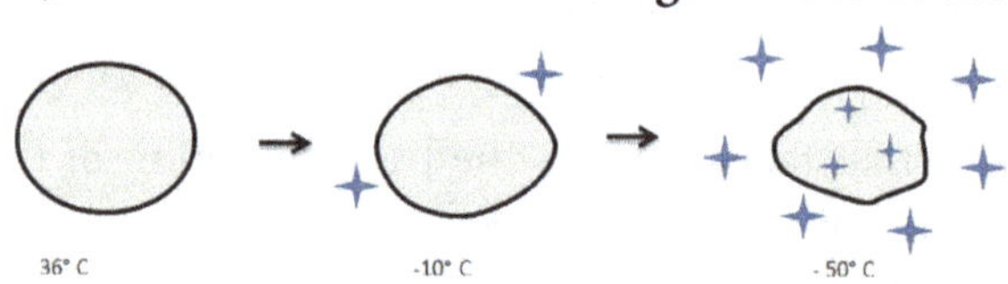

Fig. 16: Schematic of factor 2 in "two-factor" hypothesis of cryoinjury. If cells are cooled rapidly, water does not have time to outflow and freezes both inside and outside of the cells.

In more detail, when the cells are cooled slow enough, the cells dehydrate and an increase in the intracellular solute concentration is observed before reaching the freezing temperature. In long term, the cells are damaged due to severe dehydra- tion and the increased intracellular ion concentration, which allows the exchange of some critical ions (e.g., potassium) via channels in the membranes [83]. The increase in the ion concentration causes the water-insoluble proteins in the cell to interfere with these ions and transit into a soluble form. The osmotic pressure remains unchanged as these ions bound to proteins causing an artificial decrease in intracellular ion concentration. Protein concealment of intracellular potassium and chloride ions allows the entry of sodium from exterior, and the cytosolic ion concentration increases again. During defrosting, the extracellular ice first melts, dilutes the outer environment, and water enters the cells. As the amount of cyto- solic water increases, the potassium and chlorine-binding proteins liberate from the bonds and the hidden potassium and chloride reappear. As the amount of intracellular ions increases, the cells begin to rehydrate, during which process cracks in the cell membrane and cell lysis were observed.

On the other side, if the cells are cooled down too quickly, the intracellular water will not have enough time to balance with the extracellular environment, freeze inside the cells, and damage the intracellular structure and cell membrane. The inhibition of intracellular ice formation by cryoprotectants is undoubtedly helpful on the survival of the cell. However, the increase in the amount of intracellular solution leads to damages such as protein denaturation, cell lysis (cell membrane breakage), and mitochondrial and nucleus damage.

The morphological and physiological differences in the spermatozoa of differ ent animal species must be taken into consideration. Thus, the sperm of different species will have different responses to cryopreservation; therefore, the optimal cryopreservation protocol would be different for them. For example, the optimal cooling rate was 100°C/min for the bull sperm, whereas 1–10°C/min for human, 30–50°C/min for pig , and 50–60°C/min for ram.

Morris used different cryoprotecting agents (CPAs) in the sperm cryopreser vation with both rapid and slow cooling procedures (Figures 17 and 18).

In the study, sperm cryodamages were ultrastructurally demonstrated. In the Scanning Electron Microscopy (SEM), vacuole-like structure and head, neck, tail deformations were visualized (Figure 19). In the Scanning Transmission Electron Microscopy (STEM), membrane deformations, head, axoneme, acrosome, mitochondrium damages, and double tail were observed (Figure 20). Ice crystals formed in the extracellular medium also act on the cytosol and may result in intracellular ice formation. When cells are in close proximity or in contact.

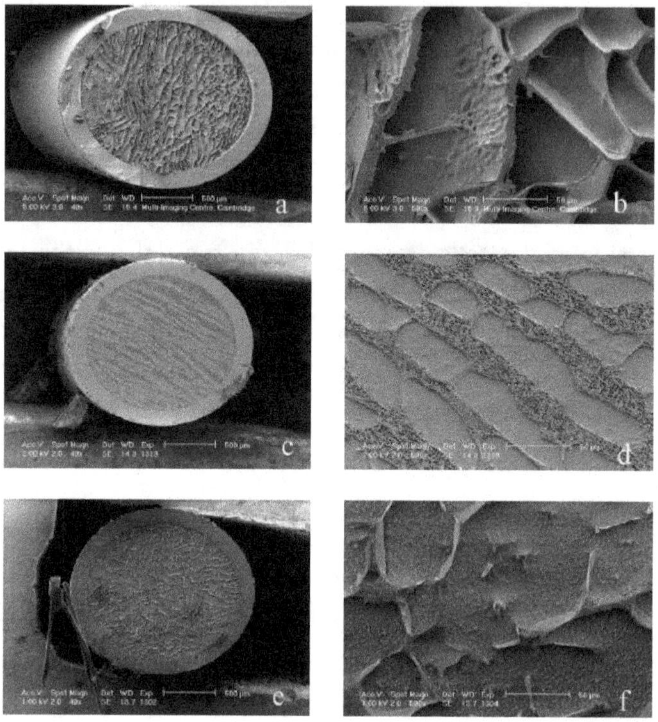

Fig. 17: In (a) and (b), samples with added glycerol (3 volumes of 0.1 M sodium citrate to 1 volume of 0.33 M fructose and 1 volume of 0.33 M glucose + 4 ml of egg yolk +3 ml of glycerol) were frozen at 10°C/min visualized with a Cryo-scanning electron microscope. In (c) and (d): Glycerol added samples were frozen at 3000°C/min. In (e) and (f), sperm were frozen at 3000°C/min without glycerol. In (a), there are channels occurs (in which the spermatozoa can be protected) under the effect of glycerol, whereas (c) shows narrow channels as opposed to (a). In (e) no cryoprotectant substance is added and it is observed that the formed ice does not leave enough space for spermatozoa. This is illustrated in more detail in Figure 18. with any of extracellular ice crystals, the aquaporin channels in the cell membrane are also affected. Small ice crystals penetrate through the aquaporins into the cytosolic milieu and may trigger intracellular ice formation .

Damage during thawing: Rapid cooling causes the formation of small intracellular ice crystals. Such small ice crystals have higher energy than large ice crystals because of their small radius curves and they are thermodynamically more irregular. Due to this free energy difference, they show melting at lower temperatures than large ice crystals. Recrystallization is a phenomenon that small ice crystals combine together into larger ones, which cause defects such as various cuts and defects on organelles and membranes in the cell by physical pressure or crystals .

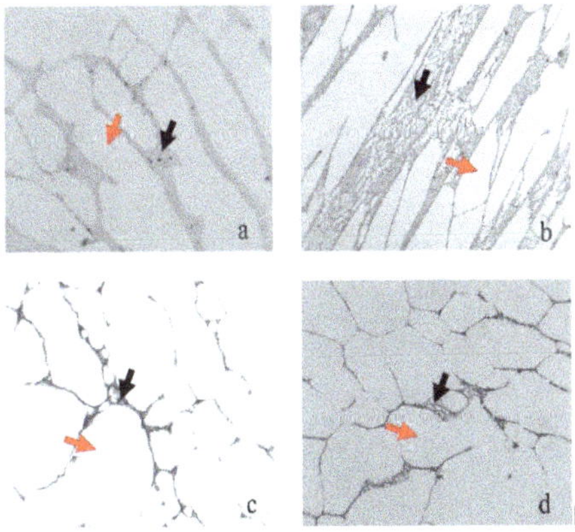

Fig. 18: In the above photographs, glycerol was added and frozen at a rate of 10°C/min (a), glycerol added and frozen at 3000°C/min (b), frozen at 10°C/min without glycerol (c), and frozen at 3000°C/min without glycerol (d) samples are illustrated. In the photo shown with the letter A, there is an unfrozen section between the ice masses (red arrow), where cells locate themselves, in the photograph b, the same frozen part is observed but in this part there are small pieces of ice in pieces and it poses a danger to the cells. While there are much smaller frozen canals on c photo, there is small ice crystals in the d picture, even in the unfrozen canals (black arrow) .

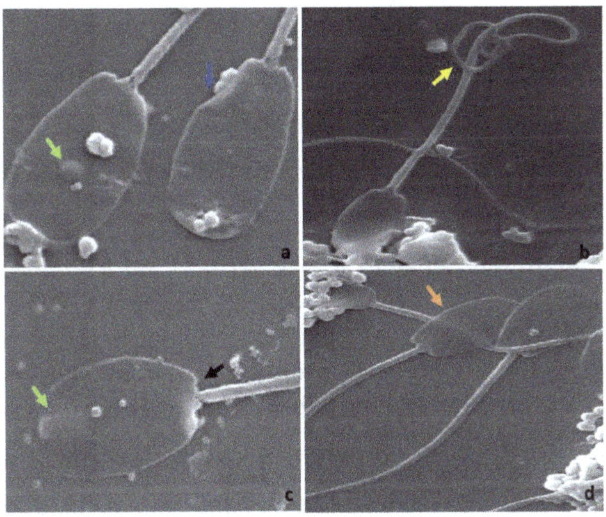

Fig. 19: SEM micrographs of frozen-thawed spermatozoa. (a) Vacuole-like structures (green arrow), head damage (blue arrow). (b) Tail damage (yellow arrow). (c) Vacuole-like structures (green arrow), neck damage (black arrow). (d) Normal sperm (orange arrow). a, c: 10,000×, b: 5000×, d: 3000×.

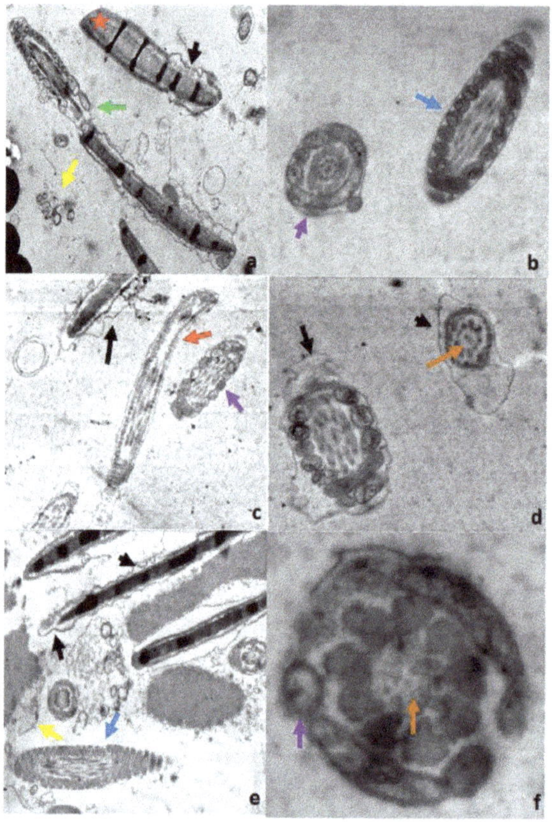

Fig. 20: STEM micrographs of frozen-thawed spermatozoa, (a) nucleus (red star), membrane damages (black arrow), neck damage (green arrow), membrane fragments (yellow arrow). (b) Mithochondrial damage (purple arrow), undamaged middle region (blue arrow). (c) Membrane damages (black arrow), mitochondrial damage (purple arrow), damaged middle region (red arrow). (d) Membrane damages (black arrow), axoneme damage (orange arrow). (e) Membrane damages (black arrow), membrane fragments (yellow arrow), undamaged middle region (blue arrow). (f) Undamaged middle region (blue arrow), axoneme damage (orange arrow).

Damage due to intracellular ice formation and recrystallization has been demonstrated in yeast, plant cells, hamster tissue culture cells, and turkey sperm. During defrosting, recrystallization generally starts at −40°C, and it becomes intense at the temperature range of −25 to −20°C. The recrystallization mechanism consists of three types, isomass, active, and migratory.

1.Isomass: Isomass is a form of recrystallization that changes the shape of ice crystal. It can also be defined as the tendency of uneven rough surfaced or pro- truding ice particles combined to form larger particles having a more oval and smooth surface (Figure 21).

Fig. 21: Schematic drawing of isomass recrystallization.

2.Active: In the active recrystallization model, two or more smaller ice crystals combine to form a larger ice crystal (Figure 22).

Fig. 22:Schematic drawing of active recrystallization

3.Migratory: In the migratory recrystallization model, small crystals around a large crystal melt into a large crystal structure (Figure 23).

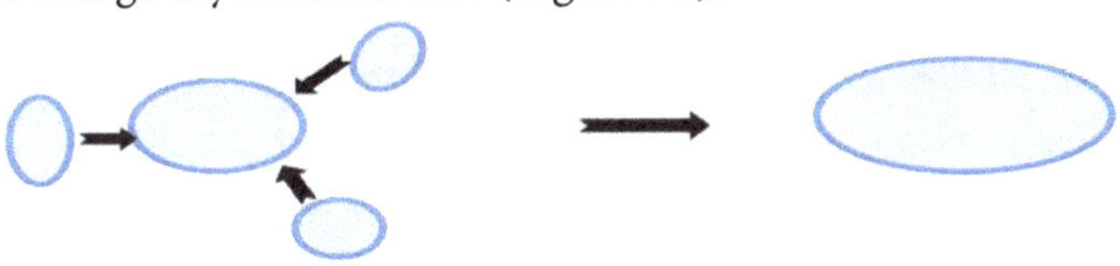

Fig. 23: Schematic drawing of migratory recrystallization.

Results of the previous studies revealed that the effect of ice formation or recrystallization depends on the crystal size [93] and on the total amount of ice in the cell [94]. In this context, rapid thawing protocol is more beneficial in terms of motility, viability, and membrane-organellar integrity, since small ice crystals do not have enough time to recrystallize when a fast warming protocol is applied. However, Mackenzie has shown that yeast cells cooled in optimal rate do not survive if they are quickly thawed, but sustain vitality when they are slowly thawed.

Another damage during thawing is due to the osmotic stress. This is especially true in frozen thawed rabbit, mouse, and bovine embryos and also freeze- thawed lymphocytes and hematopoietic cells . In addition, remaining cryoprotective agents inside the cell are incorporated into the cytosolic water, lead to cell swollen, and cause various damage in the cell membrane, possibly cell lysis. In the fertilization process, when the sperm reaches the fertilization zone, it enters between oviductal epithelial cells, infiltrates into the oocyte zona pellucida, undergoes acrosome reaction, opens a pore in the zone pellucida, and finally reaches oolemma. Disturbances in sperm membrane function due to cryopreservation directly affect this entire fertilization mechanism. In this case, fertilization cannot occur under vivo conditions. A spermatozoon with poor motility will have difficulty reaching the fertilization zone, hence has a poor ability to fertilize ovum.

It has been found that after the freeze-thaw procedure, human and bull sperma- tozoa contain more calcium than unprocessed samples, due to impairment of the selective-membrane permeability of the spermatozoa. Also, the affected spermato- zoa cannot exert normal capacitation and fertilization due to high calcium con- centration of the cytosolic environment, which also leads to premature acrosomal defects and sperm hyperactivation, causes losses in both capacitation and acrosome reaction .

The DNA damage: Spermatozoa with DNA damage have relatively limited fertilization ability. These cells have fertilization defects and potentially transfer abnormal genetic materials to the offsprings. These genetic disorders might result in abnormality in embryonic development and aneuploidies (a numerical chromosomal disorder in which the chromosome number differs from the normal, diploid (2n).

Because the spermatozoa with serious genetic damage have low in-vivo and in-vitro fertilization scores, a limited number of embryo could be harvested and embryos will have many of the developmental problems. In the light of these findings, it was concluded that sperm chromatin/DNA integrity is an important prerequisite for healthy fertilization and embryonic development. In addition to abnormal DNA/nuclear protein interactions arising from intracellular ice crystals , scientists attributed the DNA damage during freeze-thaw processes to two different mechanisms:

The first mechanism proposed by Zribi et al. is due to DNA damage in cryopreservation, and is thought to be caused by the activation of caspases and apoptosis. According to this hypothesis, apoptotic body-like structures are formed in the sperm cells during the freezing process. Apoptotic changes show themselves as nuclear fragmentation, chromatin condensation, mitochondrial expansion, and unusual changes in the cell membrane. These changes produce the enzymes from cysteine protease group, called caspases, which play an important role during apoptosis. These enzymes are found in the cytoplasm as pro-caspases in zymogen granules and their activation requires a two-step cleavage in the specific aspartic acid residues to form active holoenes. The caspases activate sequentially (water- fall-like reaction sequence), leading to a proteolytic helmet. Initiator caspases transduce death signals initiated by apoptotic stimulation to effector caspases. The effector caspases are related proteins, for example, actin or fosdrin from cytoskel- etal proteins, lamin-A, the nuclear membrane protein, and poly (ADP-ribose) polymerase (PARP) involved in DNA repair break down apoptotic cell morphology . DNA fragmentation is the latest sign of apoptosis, which occurs due to cas- pase-3 activation. The caspase-3 enzyme inhibits DNA repair by inactivating DNA repairing enzymes namely poly ADP-ribose polymerase (PARP) Table 2.

Current studies have shown that there is a positive correlation between the presence of PARP protein and sperm maturation. PARP homologs PARP-1 (75 kDa), PARP-9 (63 kDa), and PARP-2 (60 kDa) were found in injected sperm. In another study [108], oxidative stress and PARP inhibition occurred in early apoptotic spermatozoa, with water leakage, cell shrinking, and picnosis (volume reduction of the cell nucleus and chromatin condensation). The apoptosis-inducing factor released by the mitochondria (flavin), an enzyme that contains adenine dinucleotide helps oxidization during oxidative phosphorylation, causes DNA fragmentation in flavoprotein. During freezing and thawing, an increase in apoptosis markers was observed. Changes in the apoptotic membrane permeabil- ity, membrane potential, caspase activation, and phosphatidylserine externalization due to cryopreservation are also observed .

Table 2. Caspase family subgroups and their members [8].

Subgroup	Function	Member
I	Activator (initiator caspases)	Caspase 2
		Caspase 8
		Caspase 9
		Caspase 10
II	Executioner (Effector caspases)	Caspase 3
		Caspase 6
		Caspase 7
III	Inflammatory mediators	Caspase 1
		Caspase 4
		Caspase 5
		Caspase 12
		Caspase 11
		Caspase 13
		Caspase 14

The second mechanism proposed by Thomson et al. is that DNA damage is caused by caspase activation and oxidative stresses in the apoptotic cells exposed to extreme hostility. It is highly possible that freeze-thaw process results in the formation of free oxygen radicals that damage the nuclear DNA. With the weakened DNA repair mechanism, the frozen-thawed sperms are vulnerable to oxidative attacks. An increase in the mitochondrial membrane potential due to changes in its membrane fluidity during sperm cryopreservation was also observed, which leads to the formation of free oxygen radicals. Accordingly, DNA damage to the spermatozoon single and double strand breaks by the released free oxygen radicals are very common. The formation of free oxygen radicals reaches peak level in human sperm and seminal leukocytes at 4°C . Frozen semen samples containing high percentage of sperm are more prone to DNA damage.

Although the cryopreservation adversely affects antioxidant activity, interestingly, it makes spermatozoa less vulnerable to free oxygen radical damage.

The DNA damage in the cryopreserved spermatozoa mainly occurs at early stages during/after thawing. In a previous experiment, the highest DNA dam- age score was observed within the first 4 h after thawing. Therefore, freeze-thawed semen samples should be used immediately in the clinical settings . However, Isachenko et al. suggested that freezing or cryoprotectant did not make any difference in DNA integrity in their study Table 3.

Table 3. Different studies showing the effects of freezing/thawing protocols on DNA damage.

Cryopreservation Method	Results
Equilibration at 37°C, freezing in liquid nitrogen at -80oC, storage in liquid nitrogen at-196°C.	The freezing/thawing procedure causes DNA damage int the sperm.
Freezing up to-20°C, Freezing inliquid nitrogen, storage in liquic nitrogen at-196°C.	
Freezing with liquid nitrogen under computer control.	
Programmable freezer	
Freezing in liquid nitrogen vapor Equilibration at 37°C, freezing in liquid nitrogen at −80°C, storage in liquid nitrogen at −196°C Programmable slow cooler + vitrification.	
Cooling to −20°C, freezing in liquid nitrogen at −100°C, storage in liquid nitrogen at −196°C.	

The mitochondrial damage: Mitochondria and other membrane-bound organelles are found in the major- ity of eukaryotic cells and the mitochondria have a vital role in ATP synthesis in cells via oxidative phosphorylation . Mitochondrion is found in nine fibrous columns located in the body of the spermatozoon and serves as an energizing engine for the semen . Plasma membranes and mitochondrial membranes show similar sensitivity to cryopreservation. High levels of cholesterol and poly- unsaturated fatty acids cause the membrane to become more fluidic under cold conditions.

Mitochondrion is unique double membrane-bound organelle in the cell. There is an intermembraneous space between the inner and outer mitochondrial membranes. Matrix facing surface of the inner membrane have oxidative- phosphorylation enzyme system and the cytoplasmic part consisting of structural and functional subunits in which many enzymes are organized. These proteins are crucial for the metabolic processes required by the cell and for the maintenance of the cellular structure. Morphologically, the mitochondria show different charac- teristics in the sperm of different animal species. While in humans, rhesus mon- key, dogs and other mammals, the mitochondria are in oval form, whereas they are cylindrical in fish . The size and number of mitochondria in a given cell may vary according to the bioenergetic needs of species. Mitochondrion has its own DNA (mtRNA) involved in the synthesis of messenger RNA, ribosomal RNA, and transfer RNA, which are highly needed for the internal metabolism of the cell [135]. In the mammalian spermatozoa, mitochondria are located at the middle part of the cell and will form multiple disulfide bonds to form the mitochondrial capsule.

Sperm mitochondria, unlike those of the somatic cells, include cytochrome c, cytochrome c oxidase subunits, protein isoforms, and isoenzymes. Mitochondria play a very fundamental role in ATP production by oxidative phosphorylation. In the production of mitochondrial ATP, glycolysis uses carbon fuels such as pyruvate, glutamine, and amino acids. At the same time, mitochondria are involved in other processes such as production of free radicals and apoptosis. The release of free oxygen radicals such as superoxide anion, hydrogen peroxide, and nitric oxide is important. If the free oxygen radical release exceeds the capacity of the antioxidant defense system of a cell, peroxidation of the biological membranes, especially sperm plasma membrane and undesirable conditions can induce oxidative stress, DNA-related aging, and apoptosis. Most impor- tantly, the mitochondria create the activation energy required for the movement of the sperm.

ATP formed by oxidative phosphorylation in mitochondria allows motility through microtubules. For this reason, any disorder in the mitochondrion metabo- lism leads to decreased motility. Oxidative phosphorylation in the mitochondrion requires two basic components in the inner membrane of the mitochondrion, the ATP synthase, and the respiratory chain.

During cryopreservation, two types of damage occur in mitochondria. The first is the direct damage in the mitochondrial DNA, deteriorations in the inner and outer membranes of the mitochondrion. The second is the indirect damage caused by losing ability to carry out genetic coding for mitochondrial activity. According to Irvine et al., motility and velocity decrease in sperm with high DNA damage. The nuclear DNA segments damaged during freeze-thawing process may be responsible for the loss of mitochondrial functions.

A change in the mitochondrial membrane fluidity also causes changes in the membrane potential and in exchange and release of free oxygen radicals through the mitochondrial membrane. Peroxidative damage caused by an increase in the concentration of free oxygen radicals is associated with sperm plasma mem- brane damage and disruption of the aczonemal structure .

Cryobiological events that occur during freezing not only cause physical dam- age but also disrupt the chemical structures. The cooling of the cell results in the production of free radicals, and it also suppresses the natural defensive mechanism that the cell develops against the formation of free radicals.

Depending on their low activation energy, the free oxygen radicals are more difficult to be removed at low temperatures than the normal enzymatic cleansing. Increasing amounts of free oxygen radicals affect various biomolecules including membrane lipids, proteins, and nucleic acids through nonenzymatic pathways and cause the formation of new free oxygen radicals in the environment. Transition metals such as copper and iron, which are highly effective in intracellular metabolism, such as the cytochromes of the mitochondrial electron transport chain, are important catalysts for the radical chain reactions. In particular, when catalysts are present in the species capable of redox cycling, they exchange electrons mutually between the oxidation sites.

The disruption of the homeostatic balance of the metal inlet and outlet of the cell during cooling leads to the formation of free oxygen radicals, damages the cell wall by thickening. Under normal conditions, while the formation and decomposi- tion of free oxygen radicals are in equilibrium, the disruption of this equilibrium causes continuous formation and association of free oxygen radicals. Restricted free oxygen radicals cause damage to the cellular membrane by impairing lipid peroxidation. Under normal conditions, the cells have ways to repair such damage, but irreversible damage during storage and thawing can lead to cell death.

During irreversible cell damage or cell death, calcium is an important agent in the cell necrosis. Due to the cold shock, unbalanced free radical formation causes an increase in intracellular calcium. All of these events have increased the importance of antioxidants and calcium channel blockers (calcium enters and exits through the calcium channels in the membrane from the extracellular medium, and this entry- exit mechanism is controlled by the so-called calcium channel blockers) during freezing and thawing. These blockers consist of three prototype agents, dilithazem, verapamil, and nifedipine, and many sub-agents such as amlodipine, benidipine, cilnidipine, felodipine, isradipine, nilvadipine, etc.

Free oxygen radicals are separated into three groups;

- Oxygen free radicals: superoxide anions (O^2), hydroxyl radical (OH), and hyperoxyl radical (HOO).
- Non-radical species: hypochloric acid (HOCl) and hydrogen peroxide (H2O2).
- Reactive nitrogen species and free nitrogen radicals: nitroxylon, nitrogen oxide, peroxynitrite, etc..

Chemically, radicals are molecules having one or more unpaired electrons. This chemical state causes an excessive amount of reactivity with an electrically irregular structure in the predominant molecules. As in the other cells, energy is also produced aerobically in the spermatozoa. Energy is produced by oxidative phosphorylation in mitochondria and by oxidation of nicotinamide adenine dinucleotide (NADH) and is stored in the form of adenosine triphosphate (ATP). In this whole electron trans- fer chain, oxygen is reduced to free radicals in high electroactivity by taking four electrons. As a result of this reduction reaction, water becomes clear (Figure 24).

Lipid peroxidation: Reactive oxygen groups, which occur during the electron transport chain in the mitochondria, act on membrane lipids, rich in unsaturated fatty acids, causing deterioration of their structure. This is called lipid peroxidation.

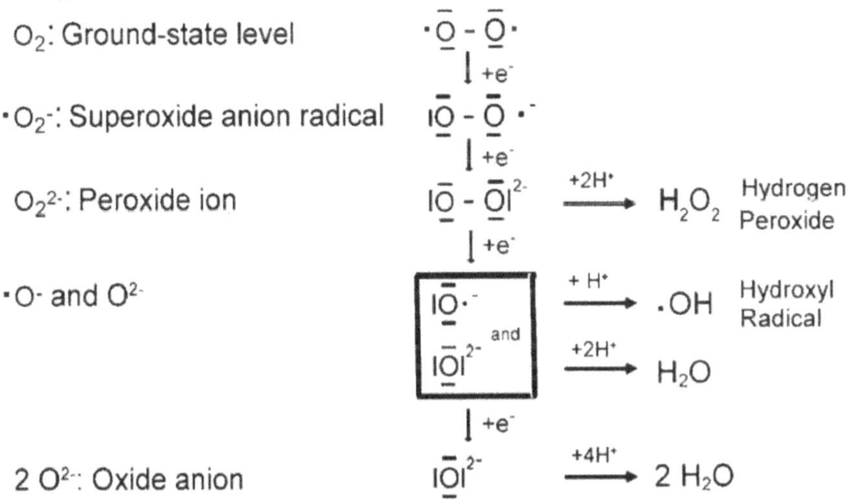

Fig. 24: Formation of reactive oxygen species [145].

Lipid peroxidation is exerted in three phases, initiation, propagation, and termination: Initiation phase: In the initial phase, the highly reactive OH– or hydroxyperoxyl radical (–HO2) removes a proton (H+) from the medium by affecting the polyun- saturated fatty acids in the lipid layer and converts their existing structure into a lipid radical and releases water. Low reactivity oxygen groups such as H2O2 cannot initiate this reaction. In the present case, lipid radicals can be in various forms with electrons in delocalized state. At this stage, lipid radicals can react with molecular oxygen and are converted to lipid peroxides.

Propagation phase: During the propagation phase, the lipid radical formed in the previous phase merges with the oxygen in the molecular state and forms the lipid peroxide radical. The lipid peroxide radical reacts with a neighboring fatty acid to form a new lipid peroxidase and becomes itself a lipid hydroperoxyl. The ongoing reactions in this way are also referred to as the "radical chain reaction." The lipid peroxidation event affects 60% of the membrane lipids [147].

Termination phase: In the last phase, two or more lipid radicals bind to each other to form hydroperoxyl radicals such as malandialdehyde (MDA), 4-hydroxy-2-alkene, and 2-alkenal, which have dangerous properties in the last stage (Figure 25).

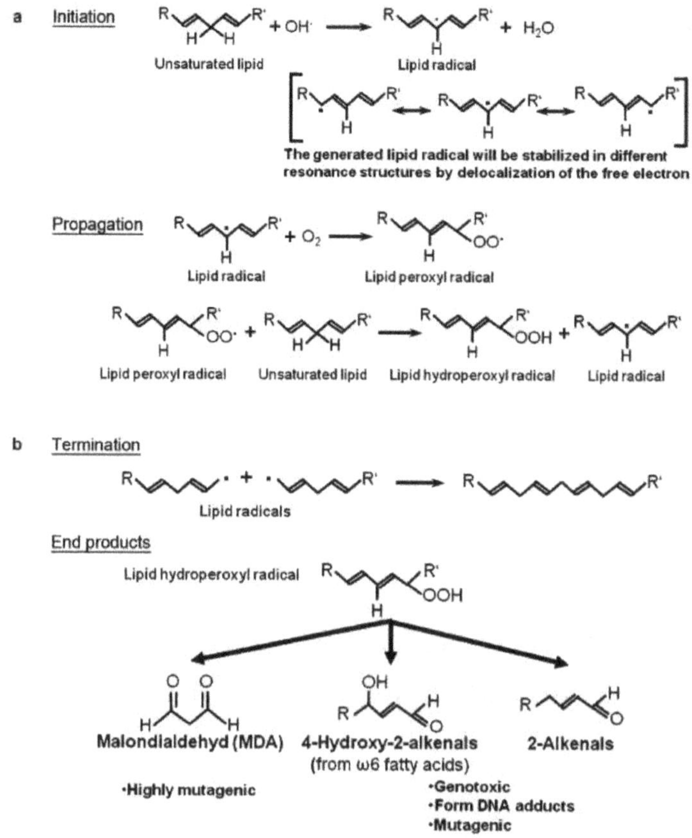

Fig. 25: Formation stages, propagation, and end products of lipid peroxidation .

9. 8 ANTIOXIDANTS AS CELLULAR DEFENSE MECHANISM

9.8.1 Antioxidants Exert Their Effects By:

- Preventing free oxygen radicals from starting chain reactions,
- Stopping chain reactions that have begun,
- Preventing the formation of free oxygen radicals, and
- Performing peroxides by breaking down and reducing local oxygen concentra tion.

Under normal conditions, spermatozoa reduce antioxidants to counteract with free oxygen radicals that are produced metabolically. These molecules are mainly enzymes. The most known antioxidant enzymes are superoxide dismutase (converts the superoxide anion to H_2O_2), glutathione peroxidase (detoxifies the organic per- oxides), and catalase (converts H_2O_2 to water). However, an abnormally high amount of free oxygen radicals released by the freeze-thaw procedure, damage the cell by overcoming its antioxidant defense capacity. In such cases, different antioxidants are supplied into the diluent solutions.

9.8.2 Antioxidants are Divided into Primary and Secondary Antioxidants

9.8.2.1 Primary antioxidants:

The antioxidants which blocks the chain reaction by cut- ting in the half enter this group. Antioxidants of this group are mainly: tocopherols (vitamin E), butyl-hydroxyl-anisole (BHA), butyl-hydroxy-toluene (BHT), and ethoxiquine.

9.8.2.2 Secondary antioxidants:

These kinds of antioxidants prevent oxidation of lipids by delaying their oxidation. They act by binding to metal ions, which catalyze the oxidation of lipids, by binding oxygen itself, and by absorbing UV rays. Some anti- oxidants such as citric acid, amino acids, ethylene diamine tetra acetic acid (EDTA) and certain phosphoric acid derivatives, ascorbic acid (Vitamin C), ascorbyl palmitate, sulfites, erythorbic acid and sodium erythorbate, and glucose oxidase are examples of secondary antioxidants. Bucak et al. obtained better spermatological parameters in the Ankara goat and bull spermatozoa frozen in the antioxidants added to the dilution solutions after thawing.

9. 9. MEMBRANE STRUCTURE IN THE SEMEN CELLS IS COMPLEX

In this chapter, we provide detailed information about cryobiology and cryo- preservation of sperm. Necrosis and apoptosis of the cell during freezing are described. The structure of the biological membranes is detailed, the damages of the membranes during the cryopreservation process are mentioned and some cryo- protectants used against membrane damages are given. The structure of water, the mechanism of freezing and ice formation, and the effect of intracellular/extracel- lular water presence on cryopreservation were explained. These data were enriched with electron microscope images. Apart from physical damage, DNA damage, mitochondrial damage, ROS formation, lipid peroxidation damage, and antioxida- tive defense mechanism are mentioned. A comprehensive section on cryobiology and freezing of spermatozoa has been tried to be prepared, and it is aimed to give detailed information about cryopreservation

The membrane structure in the semen cells is complex. Several types of phos pholipids and proteins, cholesterol and oligosaccharides are involved in cell mem- brane structure in a manner appropriate for the cell structure and are effective in providing intracellular and extracellular balance. The osmotic balance between the intracellular and extracellular environment, DNA integrity, the mitochondrial energy system, and many other systematic ovum fertilization by spermatozoa and the transfer of genetic material are all due to the complete and orderly functioning of this biological system as a whole. Although freezing and storing the semen samples at −196°C has been a groundbreak in artifi- cial insemination, despite all the precautions taken, a slight decrease in the motility and vitality caused by cold shock occurred during freezing procedure. Rapid cooling of the cells disrupts the membrane structure, and cells are sub- jected to physical pressure from both inside and outside due to the formation of intracellular and extracellular ice. Although the slow cooling of the environment removes the physical impression of the ice, the cells are trapped in the solutes with a high ion concentration due to excessive water loss. For this reason, they are exposed to chemical deterioration.

To avoid these two situations, scientists have made efforts to determine the optimum cooling rate, and finally, optimal cooling rates have been determined for the semen of different mammalian species. In addition, various cryoprotectants have been used to prevent ice formation and deterioration of membrane integrity, but these substances with various side effects have also damaged the cells when supplemented. DNA, which has a highly complex structure, is damaged even when the cool- ing rate and cryoprotectant are applied at optimal levels. The free oxygen radicals, which are formed as a result of slow metabolism, terrorize the cellular milieu, deteriorate the metabolism, and disrupt almost all functions of the cell. Against these toxic agents, cells produce protective mechanisms called antioxidants, but they are ineffective against high concentrations of free radicals, which are formed excessively in the cold environment. Artificial antioxidants added into the solu- tions do not provide the desired success so far. Sperms can also be damaged due to recrystallization during thawing.

10

IMMUNOCYTOCHEMICAL STAINING OF THE METAPHASE II SPINDLE IN MAMMALIAN OOCYTES

10.1 INTRODUCTION

Investigation into microtubule function in mature mammalian oocytes in relation to ryopreservation is one area of study for the present authors. An increasing number of couples are seeking the assistance of physicians for reproductive dysfunction, and cryopreservation of gametes and embryos plays a role in this type of therapy (http://www.cdc.gov/nccdphp/drh/ART99/index99.htm).

Unfortunately, the success of human oocyte cryopreservation is low. The metaphase II (MII) spindle in mature oocytes has been shown to be sensitive to the non-physiologic conditions imposed upon cells during cryopreservation, and it is believed that the disruption of the MII spindle is a principal cause of the loss of developmental potential in frozen-thawed human oocytes.

Using fluorescence immunochemistry with tubulin-specific antibodies allows one to visualize microtubular structures in cells. This technique has been applied in an effort to understand normal and abnormal microtubule structure and function in many of the reports cited in this chapter. Although we use this technique exclusively on oocytes, it is applicable to embryos and other cell types. In conjunction with confocal microscopy, this technique allows high resolution assessment of microtubules, which may provide a means of understanding their function, and in many cases malfunction.

10.2 MICROTUBULES

Microtubules are one of the three principal components of the cytoskeleton. They can be very static structures, as those found in cilia or flagella, or dynamic structures such as those found in the cytoplasm. They have several functions, most involved with either movement of the cell or components therein. It is the structure of microtubules that confers their ability to facilitate movement, and this ability allows such important processes as protein and organelle translocation and DNA segregation.

Microtubules are assemblies of subunits of α- and β-tubulin proteins. These subunits associate via non-covalent interactions to form tube-like structures with a diameter of approximately 24 nm. The αβ dimers can associate and disassociate rapidly causing the growth and shrinkage of microtubules, a process termed dynamic instability. This dynamic behavior allows microtubules to "search" the 3-dimensional space inside of a cell for specific targets. As an example, this phenomenon has been visualized in living newt lung cells as a mechanism to capture kinetochores to the developing mitotic spindle.

Movements of proteins, vesicles, and entire organelles within a cell are facilitated by microtubules in conjunction with microtubule motor proteins, of which kinesins and dyneins are two major families.

Microtubules have an inherent polarity, with both (+) and (-) ends dictated by the orientation of the tubulin dimers. Generally, kinesins are (+) end directed motors, moving toward the (+) end of the microtubule, while Dyneins translocate cargo toward the (-) end. There are, however, exceptions to this rule. Movements that rely upon microtubule motor proteins include the beating of cilia and flagella (dyneins), positioning of the endoplasmic reticulum and golgi apparatus and movement of vesicles between, and the migration of chromosomes during mitosis and meiosis. It is this last example that is of particular importance in developing oocytes.

10.2.1 Microtubules and the Spindle

The separation of chromosomes during mitosis by the mitotic spindle is one of the earliest studied, and most fascinating, events occurring in cell biology. During the M-phase of the cell cycle, a highly choreographed series of events occurs, eventually culminating in cytokinesis. Microtubules play an integral part in all of these events, and changes in microtubule dynamics and location occur during this period.

In mammalian cells, prior to the dismantling of the nuclear membrane in preparation for mitosis, the centrosome duplicates, and each daughter centrosome becomes located at opposite sides of the nucleus. Microtubules and kinesin motor proteins facilitate this movement, and along with the microtubule organizing centers (centrosomes) develop into the mitotic spindle. Unlike G-phase microtubules, M-phase microtubules are restricted in location to the meiotic and mitotic spindle. Not only is the location of these microtubules modified, the dynamics of M-phase microtubules also changes rapidly. The rates of polymerization and depolymerization in spindle microtubules increases markedly compared to their G-phase counterparts. The average lifetime of a mitotic microtubule is less than one minute, compared to approximately 10 minutes for G-phase microtubules.

Because of this dynamic nature, the spindle is continuously changing in size and shape during mitosis.

Upon nuclear envelope breakdown, the kinetochore microtubules of the spindle probe the cytoplasm and "search" for the chromosomes. It is believed that either the (+) end of the microtubule achieves a direct hit on the kinetochore, or it passes the kinetochore and a lateral interaction occurs whereby the microtubule is captured and the chromosome is moved to the (+) end via kinesin motors. Either way, the kinetochore eventually caps the (+) end and stabilizes it, preventing further depolymerization. A microtubule from the opposite pole eventually attaches in a similar manner to the kinetochore of the sister chromatid and the sister pair migrate to the midpoint of the spindle. Once all of the chromatids have become aligned between the spindle poles, the proteins holding the sister chromatids together are cleaved and the chromatids migrate toward the spindle poles. Once all of the chromosomes have segregated to the respective spindle poles, cytokinesis commences and 2 daughter cells are formed which contain exact copies of the genome along with one compacted centrosome. Cytoplasmic microtubules are again nucleated from this centrosome, and the cell cycle begins anew.

10.2.2 Microtubule Function in Mammalian Oogenesis and Fertilization

In the mouse, meiosis is initiated as early as embryonic day 12 in the oogonia, transforming the cells into oocytes. Shortly after birth, all of the oocytes have reached the dictyate stage of meiotic prophase, where they will remain arrested until stimulated to resume meiosis at the time of ovulation. This development of oocytes is similar in all vertebrate species, and in those with a long reproductive span, such as humans, the oocytes can remain arrested at prophase I for decades.

Oocytes are recruited to resume growth and development during the life of the organism. They proceed through the stages of meiosis during maturation, and the pattern of microtubule activity resembles that described for somatic cells in the M-phase. Investigations of microtubule organizing center (MTOC) position and microtubule morphology in oocytes from mouse, bovine, porcine, and human have been reported and show very similar patterns during the stages of oogenesis.

At the time when oocytes are arrested at the first meiosis, only very short microtubules are present, and they are nucleated from the centrosome in mouse oocytes. Upon breakdown of the germinal vesicle and resumption of meiosis, a dramatic change in the arrangement of the centrosomes occurs, with one remaining near the chromatin and another migrating into the cytoplasm. In addition, microtubules begin to nucleate around the chromosomes. In many mammalian species, the centrosome lacks the pair of centrioles in the maturing oocyte, the centrioles having been lost during oocyte maturation. It is not until further embryonic development that centrioles reappear. A bipolar spindle forms during this time as the number of microtubules increases. In addition, although remnants of the microtubule organizing centers in the cell periphery still exist, they do not nucleate microtubules. Thus upon the onset of metaphase I, microtubules are only located within the spindle.

In the anaphase I stage, the pairs of sister chromatids are separated by the microtubule-based spindle and half of the DNA is extruded from the cell in the first polar body. The cell then enters Metaphase II (MII) and another spindle complex develops. As in metaphase I, microtubules are only localized in the spindle of MII oocytes (see Figure 1). At this stage the oocyte is arrested for a second time, with activation usually commencing upon interaction with a spermatozoon during fertilization.

Microtubules play a predominant role in the events which occur during fertilization (Schatten and Schatten, 1987). Initially, activation triggers the resumption of meiosis and results in the final meiotic division. The remaining sister chromatids are separated by the spindle and half are extruded into the second polar body. This results in the creation of a haploid set of maternal chromosomes to complement the haploid set of paternal chromosomes introduced by the spermatozoon. The sperm is dismantled during the early stages of fertilization, with the chromatin undergoing decondensation culminating in the formation of the male pronucleus. A pronucleus also develops around the female chromosomes at this time.

Upon activation in mice, cytoplasmic microtubules again form, radiating form the peripheral MTOC, and these microtubules eventually transport the male pronucleus from the cortex to the center of the egg cytoplasm. During this transport, the female pronucleus is also captured by the microtubules and both pronuclei migrate toward the same location. At the time when the pronuclei are in close apposition, a sheath of microtubules forms around each. After pronuclear dissolution and intermingling of the chromosomes from the two pronuclei, a process called syngamy, the chromosomes and microtubules develop into a spindle and the first embryonic cleavage division commences. A similar process occurs in fertilization from other mammalian species with an interesting exception. The focus of microtubule organization is the centrosome which develops from the centriole that is contributed by the spermatozoon. An MTOC develops from this organelle and nucleates microtubules which eventually cause the migration of the pronuclei and syngamy as in the mouse.

10.2.3 Pathologies Associated with Microtubule Dysfunction in the Oocyte

Given the importance of microtubules for gamete interaction and subsequent embryonic development, it is not surprising that microtubule dysfunction in oocytes and zygotes has serious repercussions. Incorrect chromosome segregation during meiosis and improper union of the 2 haploid sets of chromosomes can lead to developmental failure of the embryo.

There have been several reports on the correlations between male infertility and sperm centrosome failure. It is investigated the correlation between fertility and centrosome organization and sperm aster morphology in bulls and found a strong correlation between those factors. The bull with the highest fertility demonstrated superior aster formation and morphology in comparison to bulls with lower fertility. In another study investigating 211 presumed failed to fertilize human oocytes, many of the oocytes had been penetrated by one or more sperm but had arrested development at very early stages. Although not rigorously tested, it was thought that some causes of this arrest were defects in the sperm centrosome. In a similar study , failed-fertilized and pronuclear-arrested oocytes derived from clinical cases where the males had severe oligoteratozoospermia were examined for microtubule patterns. Abnormal patterns were seen, and the conclusions were that centrosome dysfunction may have contributed to the developmental arrest. Although as these studies show that male factor incidence of infertility can result from defects in microtubule organization and function, an even more common problem with fertility related to microtubule function in humans results from maternal aneuploidy. The estimated level of aneuploidy in humans, given that it is nearly always lethal, is quite surprising. Although it is estimated that only 0.3 percent of newborns are aneuploid it is estimated that greater than 20 percent of human conceptions result in aneuploid embryos. It is believed that most aneuploidies are maternal in origin, and most of those arise from mis-segregation during the first meiotic division. Although the precise mechanism or mechanisms which result in aneuploidy are unclear, a correlation between maternal age and aneuploid frequency has been known for decades. The increase in incdence of trisomy in pregnancies in older women is striking, with the numbers of recognized pregnancies being trisomic rising from approximately 2 percent in women 25 years old to approximately 35 percent in women over 40 years old.

The trend in delaying childbearing in developed countries is well documented , as is the changes in microtubule patterns in aged mammalian oocytes. A better understanding of the causes of microtubule alterations in ageing oocytes could assist in maintaining the fertility of those women who desire to have children yet choose to delay childbearing.

Materials

- 35mm petri dishes.
- Oocyte transfer pipettes.
- Methanol with 5mM EGTA, pre-chilled in –20 degree freezer.
- Thick glass slides with a concave well, for example Fisher Catalog Number 12-565A.
- Immunochemistry Blocking Buffer: PBS, 0.2 percent (V/V) Triton X-100, 3mM sodium azide, 0.1% saponin (mass/V), 130 mM glycine, 2% bovine serum albumin (mass/V), and 5% Horse or Donkey serum (V/V) (see **note 1**).
- Microscope slides.

- Microscope slide coverslips #1.5 (see **note 2**).
- Anti-Tubulin primary antibody, preferably clone DM1-A (see **note 3**).
- Secondary Antibody labeled with fluorophore of choice. We use a donkey-antimouse antibody with a Texas Red→ conjugate, from Jackson Immunoresearch (see **note 4**).
- Sytox Green DNA stain from Molecular Probes, catalog # S7020 (see **note 5**).
- ProLong Antifade reagent from Molecular Probes, catalog # P7418 (see **note 6**).
- Waxy material for mounting coverslips to slides, to prevent the coverslips from squishing the cells upon mounting: 50% petroleum jelly and 50% paraffin wax (see **note 7**).

Methods

i. All treatments described here are performed at room temperature unless otherwise noted. However, since microtubules are cold labile, ensure that the cells are maintained at physiologic temperature prior to fixation.
ii. Fix the cells by transferring them into cold methanol (-20 degrees C) for 6 minutes; transfer as little liquid with the cells as possible (see **notes 8 & 9**).
iii. Evacuate the transfer pipette of all solution (see **note 9**).
iv. After fixing the cells, transfer them to approximately 2 milliliters of blocking buffer in a 35 mm petri dish and hold them at 4 degrees C (see **note 10**).
v. Dilute the primary antibody in blocking buffer at a ratio of 1 microliter of antibody solution to 9 milliliters of blocking buffer. Keep the solution on ice until you are ready to use it. Mix it thoroughly by carefully inverting the tube at least 3 times, and transfer the cells to a 35mm dish containing 1.5 – 2 milliliters of the diluted primary antibody.
vi. Maintain the cells in this solution for at least 1 hour at room temperature. If desired the cells may be held in this solution overnight at 4 degrees C (see **note 11**).
vii. Transfer the cells to a 35mm dish containing 1.5 – 2 milliliters of blocking buffer as a wash. Transfer the cells with as little volume as necessary. They should be held in this solution for at least one hour. They may be washed for a few hours or overnight if desired (see **note 12**).
viii. Dilute 5 microliters of the secondary antibody in 7.5 milliliters of the blocking buffer (see **Note 4**).
ix. Transfer the cells to a 35mm dish containing 1.5 – 2 milliliters of the secondary antibody diluted as described above. Hold the cells in this solution for at least 1 hour, or overnight at 4 degrees C (see **note 11**).
x. Transfer the cells to a 35mm dish containing 1.5 - 2 milliliters of blocking buffer as a wash. They should be held in this solution for at least one hour (see **note 13**).

xi. Get one each of the 2 types of tubes from the ProLong Antifade reagent box (see **note 6**). The liquid in the clear tube needs to be thawed. This can be done at room temperature or in a water bath. Carefully (see below) transfer 200 microliters of the clear solution into the amber tube. There is a crystalline substance in the bottom of the amber tube. Upon transfer, use the pipette tip to suspend the crystals in the liquid (with the action of a mortar and pestle). Once the crystals are suspended, let the solution sit for at least 15 minutes to allow the crystals to dissolve. If DNA counterstaining is desired, to the remainder of the liquid in the clear tube, add 0.25 microliters of the Sytox Green stock solution (see **note 5**). After 15 minutes, draw the solution in the amber tube into a clear pipette tip to make sure all of the crystals have dissolved. Return the 200 microliters of the solution from the amber tube to the clear tube. Mix thoroughly but carefully! This solution is very viscous and foams easily. If excessive foaming occurs, centrifuge the tube at maximum rpm in a benchtop microcentrifuge. Keep the tubes capped as often as possible.

xii. Draw up 20 microliters of antifade reagent into a pipette and set the pipette down on the bench next to you. Using a dissecting microscope, transfer the cells to a microscope slide with as little liquid as possible. Suck up as much of the liquid as you can without removing the cells. Grab the pipette containing the antifade reagent and watch the remaining solution on the slide evaporate. When it has nearly all evaporated (this will facilitate the cells sticking to the glass), quickly but carefully add the antifade reagent to the cells. Adding more just makes a mess, adding less will create air bubbles in the solution as it dries. Once the cells are covered with the drop, add a small amount of the waxy substance to the corners of a coverslip, and carefully place the coverslip onto the slide, waxy side down, avoiding the creation of air bubbles in the antifade reagent. Carefully press down on the corners of the coverslip with a pipette tip just until you see the oocytes being squished. The coverslip will rebound slightly and the distance between the coverslip and oocytes will be minimal (see **note 14**).

xiii. Give the slides at least one hour to dry prior to looking at them under the fluorescence microscope. Although it may not be necessary, we keep the slides under aluminum foil during the drying process. Seal the coverslips with clear nail polish after the drying time. We recommend imaging the cells within 2 days to maximize the signal to noise ratio.

Notes

i. Serum from a species closely related to the host of the secondary antibody is important during the blocking step. We use horse serum since we are using a donkey antibody. Donkey serum is available from Jackson Immunoresearch, and either should work well for this application. The blocking buffer we use was derived empirically. Due to the presence of sodium azide, this material is considered toxic, and should be handled and disposed of appropriately. This buffer should be stored at 4 degrees C. Periodic filtration through a 0.45 micron filter can be performed if precipitates form over time. We make 1 liter volumes if we will be using all of it within approximately 30 days. If longer storage is expected, we make up a 1 liter volume of the solution without the BSA and Serum, and then make a 200 milliliter volume of working solution by adding BSA and serum to 200 milliliters of the stock solution.

ii. Most high power microscope objectives are designed to work with microscope coverslips with a thickness of 0.17mm, which is equivalent to #1.5. Check the objectives on your microscope to verify this before mounting the cells, and use appropriate coverslips. Although this sounds trivial, it is important for getting good images. We use 18mm square coverslips. If larger coverslips are used, use a proportionately larger volume of antifade reagent when mounting the cells.

iii. We use a primary antibody from Sigma (St. Louis, MO. USA, Catalog # T9026) without further purification. We also aliquot our stock solution in 5 microliter volumes and store these in a -80 degree freezer. Monoclonal antibody clone DM1-A has proven to be an excellent antibody for general use in labeling tubulin structures. The antibody dilutions discussed in this chapter are only relevant to the antibody sources we cite. Other sources of antibodies may work just as well, but the dilutions necessary to achieve a good signal-to-noise ratio will differ and will need to be determined empirically. If a different antibody is used, ensure that a secondary antibody is chosen that will bind to it (in other words, if the primary antibody is not from a mouse, do not choose a secondary antibody that binds to a mouse antibody).

iv. The secondary antibody comes in the form of a lyophilized powder and we use the manufacturer's recommendations for resuspension. Add 400 microliters of quality room temperature water (DI or Milli-Q) to the powder and gently mix by pipetting. After about 10 minutes, draw the solution into a clear pipette to ensure complete dissolution. We either maintain this solution in a refrigerator if we expect it to be used within 2 months or we freeze aliquots diluted 50 percent in glycerol. To do this, add 400 microliters of glycerol to the stock solution and aliquot 10 microliter volumes of this dilution into 0.5 milliliter tubes and store the aliquots at -80 degrees C. To use this solution, thaw one tube and add the entire contents to 7.5 milliliters of blocking buffer. Add 1.5 to 2 milliliters of this solution to a 35mm petri dish for the final working solution. We use glycerol with these antibodies for long-term storage but not for the primary antibody, only because that is what the manufacturer recommends. We believe that either way would work for both antibodies, but assume that the manufacturers have recommended the methods that work best for their products.

v. Sytox green from Molecular Probes is a superb DNA fluorophore. We aliquot the original product in 1 microliter volumes and dilute 0.25 microliters of this in 1 milliliter of the antifade reagent. We recommend it's use if it can be obtained.

vi. The Prolong Antifade reagent has proven to be an exceptional antifade reagent. We have received positive comments from multiple core facilities on its photostable properties. The kit comes with multiple amber tubes each containing a crystalline precipitate, and two bottles of a transparent solution. We thaw these bottles and aliquot this solution in 1 milliliter volumes into transparent 1.5 milliliter microcentrifuge tubes. This way, a tube can be thawed for each procedure instead of having to repeatedly thaw and freeze the entire solution in the bottle. We use transparent tubes because it is easier to tell when the solution has been thoroughly mixed. The final solution is very viscous and foams easily, so handle it carefully. Be sure to mix the solution thoroughly prior to its use. We recommend using this reagent if it can be obtained.

vii. The "waxy material" is a homemade concoction of half petroleum jelly ("Vaseline") and half paraffin wax. To make the waxy material, take a 500 ml 178 beaker and put it on a hot plate. Add approximately an equal volume of paraffin wax (judge by the size of the blocks) and petroleum jelly. The wax and jelly can be purchased at any large grocery store. When the wax has fully melted, stir the two together to mix well, and draw the solution into 30 milliliter syringes (without needles). The material will solidify as it cools. Like many things in life, the precise ratio of the two materials is not critical. As long as it is solid but flows when pressure is put on the syringe plunger it will be fine.

viii. We find that methanol fixation provides a quick, safe, and easy alternative to fixation in formalin. In addition, it seems to provide better access to intracellular antigens during immunocytochemistry. If a few oocytes are being fixed, we recommend using thick glass slides with a concave well. Adding about 4 drops of methanol to the well and transferring as little medium with the oocytes as possible has the best outcome. The oocytes are fragile while in methanol. Therefore, be gentle with them during fixation. Have the slides and methanol equilibrated in a freezer prior to fixation, and return the slides with the fixative and oocytes to the freezer during fixation. We keep the slides in a 60mm Petri dish during this procedure. If more oocytes are being fixed, 35 mm dishes containing methanol may be used. However, the oocytes tend to move around in the methanol during fixation thus making it more challenging to find and remove them after the procedure. It can be done with practice, however.

ix. Because methanol precipitates proteins, if methanol and protein-containing medium are mixed in the transfer pipette, it can become clogged with protein precipitate, possibly leading to the loss of the oocytes. We find it essential to evacuate the liquid out of the pipette before going into the methanol to retrieve the fixed cells. Simply evacuate the pipette before moving into the methanol, then allow the pipette to draw up methanol prior to picking up the cells. Keep the oocytes near the end of the pipette during transfer from one medium to another. If the oocytes are too far up into the pipette when fixed, they can rupture during transfer into the blocking buffer.

x. Blocking buffer is used to lower the non-specific binding of antibodies in the cells. Holding the cells in this buffer will prevent this and reduce the background staining of the cells. We do not keep the dishes in a humidified chamber. Instead, we check on the volume of the liquid and add additional medium if necessary to prevent the all of the medium from evaporating. The holding time can range from days to weeks, if desired. Just be sure to check on the volume of the liquid to avoid complete evaporation. We have found that staining mouse oocytes with this procedure produces high levels of background staining when compared to oocytes from other species. This can be reduced by holding the cells in the blocking buffer for a few days, however. For oocytes from other species, a few hours of blocking prior to incubation with the antibodies usually will suffice.

xi. The times in which the cells are held in the antibody solutions have been derived empirically. We have found that overnight incubations maximize the signal, and ensure full saturation of the cellular antigen. Shorter incubation times can be performed if time is of the essence, but we recommend at least 1 hour in all of the solutions (antibody and wash).

xii. The precise time in the wash buffer is not critical, up to a point. At least one hour is recommended at room temperature. We have washed them for as long as overnight and found that the signal is still strong. Longer washes have shown to reduce the signal, thus we recommend washes no longer than overnight. We recommend the washes to be performed at room temperature, even if going overnight. This is especially true for mouse oocytes.

xiii. The time for the final wash can be as short as one hour. If control cells are used in the experiment (which we strongly recommend), they can be viewed under the microscope (briefly) to assess the level of background staining. If it is low after 1 hour, then the cells can be mounted on slides. We usually place the cells into the final wash buffer in the morning and then mount the cells in the afternoon. The signal should remain strong at least until the next morning. When the cells have been labeled with the secondary antibody, keep them protected from light until mounting. We cover the dishes with aluminum foil during the wash step.

xiv. This step requires some practice. It may be frustrating at first, but with experience it will become routine. Performing it properly, however, will result in high-quality imaging and less frustration in the long run. Since a small number of cells are usually being observed, it helps if they are all together on the slide. Thus, once the coverslip has been applied, allow the antifade solution to slowly spread out across the area under the coverslip. If the movement is too fast, the cells will detach from the glass and move with the flow. It is not a problem for imaging the cells, just more frustrating to find the cells during imaging. The cells don't always stick to the glass, so don't be surprised if they move. We have found that within 2 days the fluorescence signal begins to diminish, especially if the coverslips are not sealed with nail polish. Therefore we recommend imaging the cells within 2 days (1 day preferably). Although we do not store the slides in a freezer, reports in the literature suggest this to prevent loss of the signal. We recommend imaging with fluorescence microscopy as soon as possible. Sealing the edges of the coverslip with clear nail polish is recommended, but not necessary if the slides are handled carefully at this point. Give the samples at least 1 hour to dry before imaging. Overnight drying time is recommended. Be careful not to apply pressure to the coverslip before complete drying, which might cause the coverslip to move relative to the slide and destroy the sample.

11

FUNDAMENTAL OF GAMETE CRYOBIOLOGY

11.1 INTRODUCTION: AN OVERVIEW

The era of reproductive cryobiology began in 1949 with the discovery of glycerol as a potent cryoprotectant. In that year, Polge and colleagues first reported the successful use of glycerol for sperm cryopreservation. Since this initial fundamental technological step, much progress has been made in the cryopreservation of reproductive cells and tissues. New technologies of slow freezing, freeze-drying and vitrification have been developed, and nowadays not only can all types of reproductive germ cells be frozen, but cryopreservation of ovarian and testicular tissues has also found a wide application . Considerable advancements have thereby been achieved in the biotechnology and storage of important murine lines, in genome resource banking for endangered species, in agriculture by germplasm transfer of genetically superior animals, and in human reproductive medicine .

In human medicine, especially, the storage of reproductive cells and tissues in special cryobanks and their use in reproductive medicine and assisted biotechnology for artificial insemination and in vitro fertilization, intracytoplasmic sperm injection and embryo transfer have become an indispensable part of modern therapy: these have introduced not only new concepts and opportunities but also new ethical considerations in the life of individuals and the entire society .However, despite this crusade of reproductive cryobiology, most attempts to improve cryopreservation systems, even when successful, have been made in empirical ways. Although the use of the protocols developed thus far for slow freezing and vitrification result in adequate therapeutic outcomes in human assisted reproduction, further improvement of cryobiological techniques would be desirable, especially in the cryopreservation of oocytes and embryos . Even more critical is the situation in many animal systems, where only small improvements in basic techniques have been made since the early 1950s. Whereas in the dairy cattle breeding industry considerable successes have been achieved with respect to freezing technology, insemination and embryo transfer, successful applications of these to the breeding of pigs, wild animals or even beef cattle have been rather restricted. It has become clear that protocols which are optimal for one cell type are not transferable to another cell type, and even not applicable within this cell type from one species to another. Although differences between species in sperm physiology and biochemistry and in physiology and anatomy of the female genital tract may be major sources of this wide biological variability , a relatively poor understanding of the fundamentals of the cryobiological processes themselves may also play a large part in limiting progress.

Historically, research in the field of cryobiology has proceeded almost independently along two parallel modes: 1) via a biophysical approach which is obviously limited in its applicability to living cellular systems; 2) via a cellular approach which is to a high degree empirical and often lacks detailed quantitative analysis. The cultural gap between these two has been narrowed somewhat in recent years as cryobiology has developed as an autonomous science. However, especially in reproductive biology and medicine, these two different research approaches continue to run in parallel on their Euclidean path. A primary example of this latter problem is the evident discrepancy between theoretically predicted optimal cooling rates and the empirically obtained cooling rates that result in maximal cell survival. This review will therefore focus on basic aspects of cryobiology and cryoinjury from the point of view of fundamental physical principles, and on the phenomena and biological cell properties which appear to be mainly responsible for the heterogeneous outcomes and poor rates of improvement in conventional technologies (e. g. slow freezing) of gamete cryopreservation.

11.2 BASIC PRINCIPLES OF CRYOBIOLOGY

Cryobiology as a science describes the effects of sub-zero temperatures on live organisms. The complexity of cryobiology is deeply rooted in the complex behavior of living cellular systems and their responses to massive temperature changes. Thus understanding of the effect of cryobiological processes at the cellular and molecular level requires interaction between many fields of biological and physical science: cell physiology, pathology and biochemistry, biophysics and physical chemistry, thermodynamics and mathematics.

In a biophysical approach to cryobiology, the behavior of cells is predicted based on models of mass solute transfer at various temperatures. Most calculations used for the estimation of optimal cooling protocols are based on the so-called"ideal solution" theory. In this section, the basic terms and concepts of biophysics relevant to cryobiology will be re-iterated: solution theory, osmosis, water and solute transport. Because of space limitations, the literature referring to the basics of biophysics has not been extensively cited, and only a few crucial final quantitative biophysical equations have been addressed. The biophysical and thermodynamic principles of cryobiology have been summarized and reviewed many times, and these books or publications may be consulted for further details.

11.3 CELLULAR WATER

Water plays a crucial role in the life of the cell: not only does the living state of the cell depend entirely on intracellular water as a vehicle for the intracellular contents (e.g. hydrophilic structures such as proteins, metabolites and ions), but the extracellular environment also consists, in a generic sense, of aqueous solutions. These solutions (intracellular and extracellular) can exist in different states, and phase changes will occur in them during freezing and thawing, containing as they do electrolytes, osmolytes, macromolecules and gases. The concentration and distribution of these solutes in the extracellular and intracellular environments undergo changes during cryopreservation.

11.3.1 Water Transport and Osmotic Properties

To describe water transport in cells, usually two major assumptions are made: 1) cells are surrounded by a membrane which is impermeable for solute but permeable for water and 2) cells are normally in a state of osmotic equilibrium.

Water can be transported across these semi-permeable membranes between compartments containing different concentrations of solutes. Changes in osmotic pressure (internally as well as externally) will produce changes in cellular volume and shape, phenomena which have been seen in a variety of cell types . According to osmotic theory, as osmotic conditions change, water movement will take place until there is no osmotic gradient across the plasma membrane and the osmotic pressure of the cytoplasm matches the external osmotic pressure:

$$\pi_i = \pi_e$$

Then the cell volume V at these onditions will fit the equation:

$$\pi_e(V - V_b) = \pi_0(V_0 - V_b) \qquad (1)$$

where V0 is the isotonic volume and Vb the osmotically inactive volume (non-solvent component of the cytoplasm). This equation in its re-arranged form is widely used in cryobiology to determine the osmotic parameters at a given temperature, and is known as the Boyle Van't Hoff Equation:

$$V = \pi_0/\pi_e(V - V_b) + V_b \qquad (2)$$

It implies that at a constant temperature cell volume changes reciprocally with osmolality. Typical Boyle Van't Hoff plots show the relative volume changes of cells in environments of differing osmolality (Fig. 1). The cell volume decreases with increasing osmolality. However, only part of the cellular volume is osmotically active: the point at which the line crosses the Y-axis corresponds to the osmotically inactive cell volume.

However, the Boyle Van't Hoff equation describes only the equilibrium cell volume without giving any information about how rapidly this equilibrium will be achieved. The kinetics of cell volume as a function of time are determined by the physical structure of the membrane and depend on the hydraulic conductivity of the membrane (L_p). The flux of water across the bilayer obeys Fick's law; so, assuming membrane semi-permeability to solute and solvent, the osmotic swelling and shrinkage of cells is given by the equation:

$$\frac{dv}{dt} = LpART(\pi i - \pi e)$$

where L_p is the hydraulic conductivity (water permeability), A is the surface area of the membrane, R is the universal gas constant, and T the temperature. The hydraulic conductivity is a property that is unique to a particular cell type. For example, there is more than one order-of-magnitude difference in water permeability between human red blood cells (Lp ~9 µm/min/atm) and lymphocytes (Lp ~0.5 µm/min/atm) . Cell separation technologies are often based on differential physical properties like density, osmotic water loss, or swelling and lysis . However, even within one cell type, the physical and chemical cell properties may vary between species: in particular, the diffusional water permeability of red blood cells varies by as much as a factor of 4 across 21 mammalian species . Cells with higher Lp will reach equilibrium faster: for example, in the case of bovine, canine or human erythrocytes, osmotic equilibrium in an anisotonic solution will be achieved within less than 1 s (Lp ~9–12 µm/min/atm [27, 28]), whereas equilibrationwill take more than 10 s in bovine or human spermatozoa (Lp ~0.7–1.8 µm/min/atm).

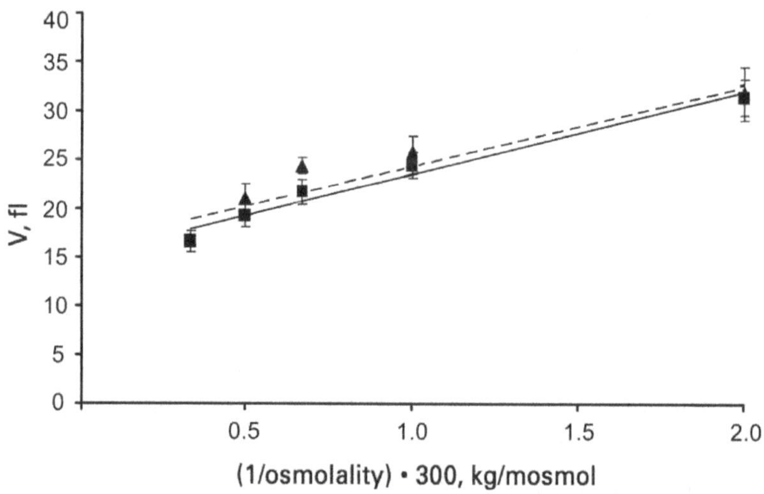

The property of hydraulic conductivity is associated with the ability of cells to survive cryopreservation stress, because it characterizes how quickly the cells can respond to a changing osmotic environment.

11.3.2 Water and Solute Transport

In most cases, the cellular environment cannot be considered as a simple aqueous solution of impermeable solute. Especially in cryopreservation, additional substances are included in the conservation media which can permeate the membrane. In this case, not only will water flow take place but also the flow of the permeable solute. The water flow will result from the water transport through the membrane due to the osmotic gradient. This osmotic gradient will be changing because the osmotic pressure of the intracellular solution in particular will depend on transport of the permeable solute across the membrane pore:

$$\frac{dv}{dt} = L_p A\left(\Delta C_i + \sigma \Delta C_p\right) \tag{4}$$

where ΔC_P and ΔC_I are the gradients for permeable and impermeable solutes across the membrane and σ is the fraction of molecules of permeable solute allowed through the pore (the so-called reflection coefficient). For a system with the solute volume V_s and the volume of osmotically active water V_w

$$Vs = Ns\,\overline{V} \tag{5}$$

$$Vw = (V - V_b - V_s) \tag{5a}$$

(N_s is the mole amount of the solute, Vs is partial molar volume of the solute), the overall solute flow

$$\frac{dv}{dt}$$

will result from the flow due to the solute concentration gradient (dependent on the permeability of solute P_s) and from the solute transport across the membrane through the pore due to the flow of water:

$$\frac{dN}{dt} = P_s A\left(\Delta Cp\right) + \left(1-\sigma\right)\overline{C}s\frac{dv}{dt} \tag{6}$$

where $_s$ is the mean solute concentration in the inner and outer compartments.

For example, if cells are more permeable to water than to a cryoprotectant, they will initially shrink because of water efflux. As the cryoprotectant subsequently enters the cell, the cell volume will increase.

In the generalized form (for k different solutes) these two thermodynamic equations are known as the coupled water-solute transport or Kedem-Katchalski model for volume flow (change in volume with time) J_v and solute flow (change in solute concentration with time) J_s.

$$Jv = \frac{1}{As}\frac{dv}{dt} = L_p \sum_k \sigma k\left(C_k^e - C_k^i\right)RT \tag{7}$$

$$J_s = \frac{1}{A_s}\frac{dN_k}{dt} = \overline{C}k\left(1-\sigma_k\right)J_{v+Ps}\left(C_k^e - C_k^i\right) \tag{7a}$$

where Ck is the concentration of the relevant solute. Two boundary cases of the Kedem-Katchalski model are especially interesting. If the solution contains only impermeable solutes, the swelling rate is given by equation (3); if the solution contains only impermeable solutes plus one cryoprotectant, the simplified equations (4) and (6) can be used for model calculations.

Theoretical Prediction of Osmotic Response From the mass transport model of Kedem-Katchalski it is apparent that the theoretical osmotic response of a cell can be characterized completely if the following parameters are known:

- Cell size (especially the surface area to volume ratio a/v)
- Fractional osmotically inactive cell volume v_b
- Water permeability l_p (hydraulic conductivity) of the plasma membrane
- Permeability coefficients (p_s) of relevant solutes
- Reflection coefficient σ of these solutes
- Activation energy e_a (to allow cellular response to be extrapolated to different temperature conditions) initial concentration of solutes.

The parameters L_p, P_s and σ can be determined by fitting experimental data to the transport equations. The isotonic cell volume can be measured directly using cell counter techniques, and osmotically inactive cell volume can be determined by using Boyle Van't Hoff plots. Activation energy can be calculated using the Arrhenius equation and an Arrhenius plot for data collected at different temperatures. The kinetics of hydration of cells during freezing can be modelled mathematically. Differences in these parameters results in differing responses to freezing/thawing conditions, whence the high degree of cryobiological variability between different cell types can be explained.

It has shown that in the majority of cases a two-parameter model for volumetric changes due to water flux V and solute flux V_s (based on equations 4 and 6) is an adequate for predicting osmotic response. For calculation of the permeability to cryoprotective agents, the measured volumes are fitted to a simplified model defined by the following equations:

$$\frac{1}{A_s}\frac{dV}{dt} = -L_p\left(C^e - C^i\right) \tag{8}$$

$$\frac{1}{A_s}\frac{dVs}{dt} = -\overline{V}sPs\left(C_s^e - C_s^i\right) \tag{8a}$$

This model has been expanded to multi-solute systems with special attention to the presence of impermeable solutes. Applications of this model in cryobiology are very valuable because the maximal volume expansion or contraction ("excursion") can be calculated in a easy and practical way according to relatively simple mathematical formulae.

11.4 CRYOINJURY AND ITS PREVENTION

The process of cryopreservation leads to morphological and functional damage of cells because it is associated with considerable osmotic changes: during freezing, cells become dehydrated, and during thawing, as water re-equilibration takes place, they are subjected to hypotonic challenge. The phase transition from water to ice is the major critical event with respect to cell survival: there is a critical region of temperature within which the cells are irreversibly damaged if they remain within it longer than a few seconds. Therefore, this section will discuss the concepts of ice nucleation and crystal growth, increasing salt concentrations, tolerable osmotic excursion of cell volume, post-hypertonic lysis, and mechanisms of cryoprotectant action.

Colligative Properties of Extracellular Solutions: Freezing Point Depression and Osmolality The cellular state and water balance in cells during freezing and thawing is determined by the colligative properties of the external aqueous solution, *i. e.* by the properties which depend on concentrations of solutes rather than on behavior of individual chemical compounds. The freezing point of a solution is primarily dependent on the concentration of solute(s) which it contains. The freezing point depression is determined by the minimum radius of an ice crystal that is stable at a given temperature, and is given by the mathematical equation

$$\Delta T = cK \qquad (9)$$

where K = 1.86 °C/mol for aqueous solutions and c the molar concentration of the solute. During freezing, the pure water precipitates out of solution in the form of ice, whence the extracellular salt concentration increases. In simple systems consisting only of two compounds, e. g. NaCl and water, binary phase diagrams can be used to describe the changes of osmolality in which the cell must exist. The osmolality of the unfrozen solution at any particular temperature is described by a simple quadratic function of temperature:

$$\pi = -0.52823T - 0.00492T^2 \qquad (10)$$

Freezing injury will take place if the cell is exposed to a critical osmolality (i.e. that at which irreversible damage or lysis occurs) [. The eutectic temperature also plays an important role in cell response to cryopreservation. At the eutectic temperature, the system is in thermodynamic balance, at which point the unfrozen fraction (solution and solute) becomes solid after spontaneous crystallization ("eutectic formation").

11.4.1 Mechanisms of Cryoinjury

The milestones in understanding mechanisms of cryoinjury were set about a third of century ago as the "Twofactor Hypothesis of Freezing Injury" by Mazur, Leibo and Chu [36] has appeared. In this work, the kinetics of cellular water loss were investigated, and both too slow and too rapid freezing was found to be associated with lethal cryoinjury. Rapid freezing will lead to intracellular ice formation because higher cooling rates lead to supercooling of cells below their nucleation temperature. In this case water will be kept in the cell, whence eventually either large lethal, or small less damaging ice crystals will be formed within. Rapid freezing may also lead to large gradients in osmotic pressure across the plasma membrane, which will become damaged at a critical pressure gradient, and intracellular freezing will also occur as a result of this damage. The building of large intracellular ice crystals is associated with cell death. Although more moderate freezing rates are associated with exposure to smaller ice crystals, these latter can re-crystallize upon re-warming and become lethal (as demonstrated with respect to embryos).

However, if freezing progresses at very slow rates, dehydration will take place over a longer period of time. During this slow freezing, the osmolality of the external medium increases as water-ice is precipitated. The cells will be exposed to high salt concentrations in the external medium and dehydrate to a degree of shrinking associated with serious cellular disruption. Cryoinjury due to the cell's sensitivity to high solute concentrations is referred as to a "solution effect".

Thus, freezing rates should be slow enough to allow the cells to minimize chemical potential and osmolality gradients across the plasma membrane and to dehydrate without being exposed to lethal salt concentrations. This "equilibrium" freezing leads to tolerable cell shrinkage, to an increase in the intracellular chemical potential, and to depression of the freezing temperature. Due to the decreased amount of intracellular water and the freezing point depression, the potential risk of intracellular ice formation is reduced, so that only extracellular ice will be formed. Thus the characteristic shape of the curve plotting cell survival against freezing rate is an inverted "U", with typically a critical cooling rate for maximum survival.

The thawing of cells also has to be performed correctlybecause the warming rate influences the survival of cells. Thawing rates comparable with the freezing rates that induce intracellular ice formation could lead to re-crystallization of small ice crystals resulting in lethal injury. Too slow warming may be lethal depending on cell type, e.g. through over-long exposure to the hypotonic stress that occurs at thawing. At this stage, possible defects in cell volume regulation may be critical. Thus, either slow or rapid warming may be required, depending on biophysical cell properties and on the cooling rates previously used.

11.4.2 Osmotic Tolerance

The shrinkage and swelling of cells is limited by their physical capacity to withstand such changes. Damage will occur beyond certain levels of shrinkage or swelling (the overall range between these levels is known as the maximal cell volume excursion). The optimal equilibrium freezing rate found for a given cell type must result in a tolerable volume excursion, i.e. a range of swelling and shrinkage that does not lead to the loss of membrane integrity and impaired cell function (e.g. motility for spermatozoa).

Tolerable volume excursions are cell type- and species-specific. Cryoinjury by Eutectic Formation Eutectic formation, i.e. the process of crystallization of the unfrozen fraction of the intracellular or extracellular medium, has thus far been relatively poorly studied in biological systems.

Eutectic crystallization is not usually observable within the normal freezing range if both inorganic salts and organic substances (e.g. cryoprotectants) are components of the system (see [40]). Due to a low concentration of electrolyte and the presence of relatively high concentrations of cryoprotectant, the impact of eutectic crystallization on cryoinjury has therefore not been seen as significant.

Hence, despite the theoretical indication that eutectic crystallization may have detrimental consequences on the cell membrane, its effect on cellular systems has been overlooked. Recently, however, quantitative information on direct cell injury due to eutectic formation has been reported. The post-thaw viability of rat prostate tumor cells decreased dramatically if eutectic crystallization took place. Moreover, cell injury was enhanced by introducing other solutes which had a higher eutectic temperature than NaCl. The authors propose two major mechanisms of cell cryoinjury by eutectic formation: (1) mechanical cell damage to the cell membrane due to extracellular eutectic formation and (2) disruption of cytoplasmic structures by intracellular eutectic formation.

11.4.3 Mechanisms of Cryoprotectant Action

A more complicated system with three or more compounds, e.g. water, sodium chloride and at least one cryoprotective additive, is used in most cryopreservation protocols. The calculation of quilibrium conditions is more complex in this case, because the cryoprotectant serves concurrently as a solvent for sodium chloride and as a solute in water. Such a system can be described by a ternary phase diagram. Melting and eutectic temperatures have been described for the three-compound systems of DMSO-NaCl-H_2O and Glycerol-NaCl-H_2O. These equations are more complex than the analogous equation (equation 10) for a binary system but they allow one to calculate the solution composition at any given temperature. In general, lowering the freezing point is associated with a higher unfrozen fraction and lower salt concentration in the extracellular space. The negative "solution effect" is thereby reduced, and the cells' exposure to hyper-osmolality diminished. Ternary phase diagrams illustrate the protective action of cryoprotectants via their effects on the initial freezing temperature and the eutectic temperature, and it has been shown recently experimentally that addition of various types of cryoprotectant leads to the suppression of eutectic crystallization, both in a simple NaCl-H2O system and in a more complex phosphate-buffered saline system. One may conclude that one way in which cryoprotective agents protect cells during freezing and thawing is by lowering the initial freezing point and by diminishing injury from eutectic crystallization.

Addition of permeable cryoprotectants will also influence the entire system state because trans-membrane movement of both solute and water will take place, thus changing the transport kinetics, cell dehydration rates and volume excursion.

Cryoprotective agents that diminish cell damage are usually divided into two general classes related to their ability to move across the cell membrane: permeable and impermeable. Their mechanisms of action are different. Cryoprotection by permeable agents can be explained by their colligative properties: they lower the temperature at which cells are exposed to a critical salt concentration. Impermeable agents, on the other hand, are thought to dehydrate cells, allowing them to be cooled rapidly before lethal injury due to solution effects takes place.

However, the effects of cryoprotectants are more complex than can be explained from thermodynamics and solution theory. In high concentrations, they can produce lethal osmotic effects during their addition and removal (hyper-osmolality) and they can be toxic [3]. Initial cellular shrinking (due to dehydration) caused by molar concentrations of cryoprotectant and subsequent swelling during its removal can both result in damage if the critical osmotic excursion is exceeded. Whether the effects of cryoprotective agents are negative or positive is greatly dependent on the kinetics of exposure. Moreover, it has been shown that the presence of these agents leads to changes in hydraulic conductivity. Various cryoprotectant species affect Lp in different ways, either increasing or decreasing it (see below), an action which leads to changes in the osmotic response of cells according to the mass-transport model. Hence, for optimization of a cryopreservation system, the optimal selection of cryoprotectant, both according to its colligative and permeability properties as well with respect to its effects on other membrane parameters and cell viability, is crucial.

11.5 DISCREPANCIES BETWEEN THEORY AND PRACTICE IN THE CRYOBIOLOGY OF GAMETES

As outlined above, several quantitative formalisms exist which allow one to predict the movements of water and solutes across cellular membranes during the freezing process. If the major membrane parameters such as hydraulic conductivity and solute permeability as well as the geometrical cell characteristics are known, the optimal equilibrium rates of cell dehydration can be calculated from the mass-transport model equations.

However, several discrepancies have already been found between theoretical approximations and experimental observations with respect to maximum cell volume response to aniso-osmolality. Moreover, empirically determined optimal cooling rates by no means agree with those calculated according to two-parameter models, especially in the case of mammalian spermatozoa. Water permeabilities extrapolated to sub-zero temperatures based on data obtained at supra-zero temperatures in the absence of extracellular ice and cryoprotectants lead to dramatic discrepancies between theoretical and experimental optimal cooling rates. Indeed, whereas observed experimental cooling rates lie in the area of 30– 50 °C/min, the mathematical calculations for an L_p value of 0.5 μm/min/atm and an activation energy of 3 kcal/mol imply that very high rates (up to 1000 °C/ min) are appropriate for the dehydration of mammalian spermatozoa. In fact, such high rates will be lethal: a major decrease in viability has been observed in a range of sperm species after cooling at no more than 100–300 °C/min. One way to account for this discrepancy between theory and practice is to consider the accuracy of mathematical approximations. For example, in the two-parameter-model, when the concentration of impermeable salts is low, the concentration of permeable solute is high ($C_{cryoprotectant} >> C_{salt}$), the Boyle Van't Hoff approximation of maximal cell volume excursion V_m (equation 1) is true, and the quotient P_{cryo}/P_{salt} approaches infinity (characteristic for freezing in the presence of any applicable cryoprotectant), the product

$$(P_{cryo}/P_{salt} - 1)(\pi 0V0 - \pi mVm)$$

is indeterminate. However, the discrepancy due to this indeterminacy should be not greater than 14 % for maximal volume excursion and therefore cannot essentially account for the discrepancy between theoretically predicted and observed optimal cooling rates. Another explanation lies in the fact that the presence of permeable substances can change the water permeability. Indeed, it has been shown in a large number of species that addition of cryoprotectants such as DMSO, EG and glycerol affects the hydraulic conductivity of many mammalian sperm species (Tab. 1). And, finally, it has been shown recently that the values of membrane parameters of mammalian spermatozoa at sub-zero emperatures in the presence of extracellular ice differ considerably from those measured at supra-zero temperature. Indeed, it has been reported that L_p at sub-zero temperatures is lower by at least one order of magnitude than L_p at suprazero temperature and that E_a at sub-zero temperatures is higher by a factor of 2 than E_a at supra-zero temperature (Tab. 2). Such divergence between membrane permeabilities at different temperatures may be related to coolinginduced changes in the sperm plasma membrane structure such as lipid phase transition, chilling injury (temperature-ependent alterations in macromolecular structures) or cold-shock. Taking into account these divergences between supra- and sub-zero permeabilities and activation energies, theoretically alculated and experimentally determined optimal rates can be brought into agreement. It is clear that further progress in avoiding the lethal consequences of cryopreservation will depend on more exact experimental determination of membrane water permeability parameters in the presence of cryoprotectant at sub-zero temperatures using differential scanning calorimetry. Unfortunately, this method is limited in that it cannot measure the permeability of cryoprotectants.

There have been attempts to improve the theoretical prediction of cooling rates based on exact solutions for the two-parameter flux model or to introduce a different approach to prevention of intracellular ice formation via restriction of intracellular super cooling. Both these approaches are likely to have crucial biophysical applications in reproductive cryobiology as they will increase the validity of volume change simulations, assuming the membrane parameters can be determined with sufficient accuracy.

Table 1: Supra-zero water and solute permeability in spermatozoa of several mammalian species (at 20–22 °C).

Species	Cryo protectant	L (µm/mn/atm)	Pcryo (x 10-3cm/min)	Method
Mouse	None	0.84	-	EPC
Mouse percoll washed	None	0.56	-	
Mouse ICR	EG	0.38	3.4	EPC
B6C3F1		0.68	1.9	EPC
Mouse ICR	Gly	0.38	2.2	EPC
B6C3F1		0.57	2.4	EPC
Mouse ICR	PG	0.25	1.2	EPC
B6C3F1		0.22	1.0	EPC
Boar	None	1.03	-	EPC
Boar	None	0.84	-	SF
Boar	DMSO	0.12	0.93	EPC
Boar	Gly	0.14	0.48	EPC
Boar	EG	0.20	1.98	EPC
Bull	None	10.8	-	TL
Bull	None	0.69	-	SF
Bull	DMSO	0.91	1.72	SF
Bull	Gly	0.29	1.75	SF
Bull	PG	0.42	2.47	SF
Bull	EG	0.39	1.49	SF
Ram	None	2.79	-	SF
Rabbit	None	0.28	-	SF
Rhesus Monkey	DMSO	1.09	1.39	EPC
Rhesus Monkey	Gly	0.91	2.21	EPC
Rhesus Monkey	PG	1.53	3.38	EPC
Rhesus Monkey	EG	1.94	6.07	EPC
Chimpanzee	EG	4.09	2.18	EPC
Chimpanzee	PG	3.91	1.75	EPC
Chimpanzee	DMSO	3.49	0.82	EPC
Chimpanzee	Gly	2.83	1.42	EPC
Man	None	1.84	-	EPC
Man	Gly	0.77	2.07	EPC
Man	DMSO	0.84	0.8	EPC
Man	PG	1.23	2.3	EPC
Man	EG	0.74	7.94	EPC

Gly: glycerol ; DMSO: dimethylsulfxide:\; propylene glycol (propane-1,2-diol); EG: ethylene glycol EPC: electronic particulrSF: stopped-fow etechique;

Table 2: Combined best-fit parameter for water transport in mammalian spermatozoa of several mammalian species in the presence of extracellular ice

Species	Cryo Protectant	Sub-zero l-p (μm/min/atm)	Sbu-zero Ea (Kcal/mol)	Spra-zeoro Ea (Kcal/mol)
Mouse	None*	0.01	22.5	
Mouse	Glycrol	0.004	15.2	8.5
Boar	None*	0.02	29.3	
Boar	Glycerol	0.005	18.1	7.8
Bull	None*	0.036	42.1	
Bull	Glycero	0.025	30.9	
	1			
Dog	None*	0.0029	15.3	
Dog	Glycerol	0.005	29.0	
Dog	DMSO	0.003	12.3	
Stallion	None*	0.02	32.7	
Stallion	Gllycerol	0.008	12.1	3.5
Man	None*	0.14	85.5	
Man	Glycerol			
	Egg Yolk	0.4	33.2	

However, improvements in the accurate determination of membrane permeability parameters and the choice of appropriate mathematical approximations will not suffice in determining cell response to cryopreservation. One of the main reasons for the discrepancy between the theoretical and experimental behavior of cells during the freezing/thawing process is the biological complexity of the cell. Biological membranes are too complex to justify some of assumptions made from the biophysical point of view. Although it is convenient to consider the cel as a "bag with salt water", there is a great deal of evidence that intracellular water and ions behave in a different way from those in ideal aqueous solutions. The cell cytoplasm does not behave like water with ions dissolved in it; there is much molecular interaction between different cell compartments where different signal transduction mechanisms are involved. Exposure to deep temperatures can affect this interaction and thereby signalling mechanisms; there is already evidence that some sperm intracellular signalling pathways can be affected during cooling and cryopreservation. Also, the cell membrane is not a simple semi-permeable barrier.

The physical and chemical composition of the membrane is complex and affected by temperature: changes in molecular structure at deeper temperatures will alter solute and water transport through pores, and changes in mobility of lipids will affect diffusion through the lipid bilayer. In particular, in cryopreserved spermatozoa, lipid diffusion in the plasma membrane has been reported to be significantly reduced relative to fresh spermatozoa.

These interactions would affect greatly biophysical models based on simplified assumptions. Analogously, the apparently relatively symmetrical external geometry of the oocyte cannot completely justify the use of current biophysical models to optimize cryopreservation protocols on the basis of the mass-transport model. The biochemical and physiological complexity of cellular organization such as cytoskeletal elements and cortical granulae must be taken into consideration. Finally, cell osmotic response is much more complex than described by the Boyle Van't Hoff relationship. There is a body of evidence that cells are able to moderate or reverse osmotically-induced swelling or shrinking by activation of ion channels and/or transport mechanisms. In spermatozoa particularly, osmotically induced shrinking or swelling is known to activate ion channels, presumably by phosphorylation or dephosphorylation, whence major intracellular ions are allowed to leave or to enter the cell and subsequent water transport takes place. Such control of cell volume involves finite processes which activate within a time range of a few seconds to several minutes in mammalian spermatozoa (and references therein). Volume regulation has also been reported for oocytes. The impact of these processes on cryopreservation response remains to be elucidated.

11.6 CRYOBIOLOGICAL HETEROGENEITY OF REPRODUCTIVE CELLS

Because of their central importance in animal breeding and human reproductive medicine, the cryobiology of reproductive cells, especially spermatozoa, oocytes and embryos has attracted a great deal of attention. Over the years major empirical advances have been made to develop successful protocols for their cryopreservation.

Similarly, more refined methods for the mathematical prediction of optimal conditions for the freezing of these cells have been established. In this section the cryobiological peculiarities of spermatozoa and oocytes will be considered.

- Transport mechanisms involved. Swelling-activated potassium and chloride channels are involved in osmotically induced volume regulation in mammalian spermatozoa. Inhibition of chloride and potassium channels leads to restriction or loss of volume regulatory function. In contrast, inhibition of K+/Cl- co-transport or Na+/K+-ATPase has no negative effect on osmotically induced volume regulation though these systems may be involved in isoosmotic volume regulation, together with other mechanisms related to sodium, potassium, chloride and osmolyte transport. NPPB: 5-nitro-2-(3-phenylpropylamino) benzoic acid; DIDS: 4,4'-diisothiocyanatostilbene-2,2'-disulfonic acid. (These and tamoxifen are all inhibitors of chloride channels.)

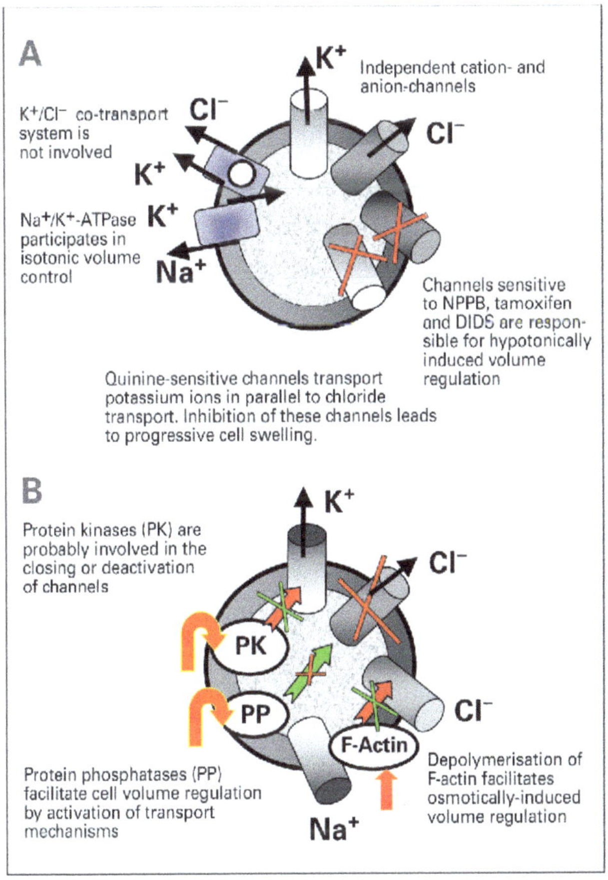

Fig. 12: Mechanisms of volume regulation in mammalian spermatozoa (modified after).

- Putative modulation and signalling pathways of volume control. The activation, regulation or deactivation of transport mechanisms can be mediated by modulation of F-actin and cytoskeletal integrity and/or by changes in phosphorylation/dephosphorylation balance (activity of protein kinases and phosphatases). Depolymerisation of F-actin facilitates volume regulation; phosphorylation appears to be involved in the closing or deactivation of channels involved in volume regulation while dephosphorylation leads to activation of ion transport mechanisms (probably chloride channels). Red arrows and lines: negative (inhibiting) action on transport and signalling mechanisms; green arrows and lines: positive (accelerating) action.

Cryobiological Properties of Spermatozoa Various studies suggest that spermatozoa of different species have specific cryobiological properties and varied degrees of sensitivity to experimental manipulation, cold shock (lipid phase transitions), freezing and osmotic tolerance. Moreover, high degrees of variability with respect to different individuals within a given species have been observed. Whereas species-specific differences in the molecular structure of the plasma membrane could explain differences in osmotic tolerance limits, hydraulic conductivity, cryoprotectant permeability and effects of different extenders on these biophysical properties, the causes of differences between individuals and especially of the heterogeneity within individual ejaculates from a particular male remain unclear.

Marked differences in sperm cell size and shape, as well in the fraction of osmotically inactive cell volume, exist between 10 mammalian species (Tab. 3). The iso-osmotic cell volume varies over a relatively narrow range (between 20–28 μm3) in human, primate and large domestic animal species like boar and bull (Tab. 3). Wider variability was observed for the sperm volume of wildlife species; murine spermatozoa, in particular, have an exceptionally high volume, reported to be about 54 μm3 as measured by the Coulter technique or about 73 μm3 when measured by magnetic resonance (Tab. 3). Mammalian spermatozoa appear to act as perfect osmometers according to the Boyle Van't Hoff relationship over a wide range of osmolalities (Tab. 3). Osmotic tolerance varies between species: porcine and bovine spermatozoa are rather sensitive to osmotic changes (maximal cell volume excursion Ve,max for maintaining motility is about 0.97–1.03), whereas murine spermatozoa have broad tolerance limits (Ve,max ~0.76–1.24). It is worth noting that there are considerable differences in sperm shape, cell volume and osmotic tolerance between mouse and man, which raises the reflection as to whether the mouse is the best model for studies on cryobiology and physiological cell function regulation in the field of human andrology.

Since freezing and thawing cause very large changes in the osmotic environment of cells, Lp appears to be the most important parameter in determining the response to exposure to deep temperatures in a given cell type. It is apparent that there is a large variation in Lp in mammalian species (Tabs. 1, 2). Estimates of Lp vary, depending upon the methodology used: e. g. time-to-lysis, stoppedflow or cell-sizing. The most accurate estimations for supra-zero permeability are given by cell sizing (particle counter technique) whereas differential scanning calorimetry is adequate for measuring sub-zero permeability.

The use of different cryoprotectants as components of semen extenders has led to considerable improvements in the cryopreservation of spermatozoa. The selection of a suitable cryoprotectant depends on (a) its membrane permeability in the given species and (b) its possible toxic effects. Not only do different cryoprotectants vary in their permeability but also a given cryoprotectant's permeability varies considerably between species (Tab. 1). Moreover, different cryoprotectants produce effects of different degrees on membrane hydraulic conductivity. In man, boar and mouse, the addition of cryoprotectants reduce the hydraulic permeability of sperm membranes; the degree of change is species-dependent. Moreover, even sperm from different mouse genotypes appear to have different water and cryoprotectant permeabilities and therefore require specific cryopreservation protocols (Tab. 1). Hence the development of new protocols or their modification by transfer to other species should be based on knowledge of specific cell osmotic and thermodynamic characteristics and the effects of different solutes and their colligative properties on these characteristics.

Sperm species	V_bV_{iso}	V_{iso} (μm^3)	Osmotic range (mosmol/kg)
Mouse	0.13*	73.4	25-900
Mouse	0.61**	54.5	75-1200
Boar (ejaculated)	0.23*	22-25.4	210-1500
Boar (ejaculated)	0.67	26.3	185-900
Boar (ejaculated)	0.63	24.1	150-900
Boar (epididymal)	0.64	25.0	150-900
Bull	0.61	23.5	150-1200
Warthog (ejaculated)	0.59	16.5	58-870
Warthog (epididyma)	0.36	21.9	58-870
Impala (ejaculated)	0.45	27.8	58-870
Impala (epididymal)	0.36	28.9	58-870
Elephant (ejaculated)	0.16	30.8	58-870
Elephant (epididymal)	0.35	27.3	58-870
Lion (ejaculated)	0.25	24.9	58-870
Rhesus monkey	0.77	36.8	75-900
Rhesus monkey	0.51	27.7	160-860
Chimpanzee	0.69	21.7	160-860
Man	0.50	28.2	145-900

11.7 CRYOBIOLOGICAL PROPERTIES OF OOCYTES

The successful freezing of oocytes will play an important role in both animal and human assisted reproduction. Use of oocyte cryobanks will provide the necessary basis for availability of large number of frozen oocytes in different state of maturation, and will allow, for example, female cancer patients to retain their fertility. The use of frozen oocytes in assisted reproductive technologies will represent a major breakthrough as an alternative to frozen embryos, as it will avoid the ethical and legal objections related to embryo cryopreservation. Generally, oocytes behave as perfect osmometers with an inactive cell volume fraction in the range of about 20 %. This parameter appears to vary relatively little between mammalian species, although a high degree of individual variation has been reported . In fish oocytes, in contrast, higher osmotic inactive volumes have been reported than in mammalian species. The oocyte is one of the largest mammalian cell types: the mean isotonic volume of human oocytes is about 2.65 × 106 μm3 . Being essentially spherical in shape, their area-to-volume ratio A/V is very low, < 0.01; their surface area is in the range 10–100 × 103 μm2, e.g. 16 × 103 μm2 for murine oocytes . They have a relatively low hydraulic conductivity and low membrane permeability towards cryoprotectants (Tab. 4), thus opportunities for improvements of cryopreservation protocols are limited. Transport of water and permeable cryoprotectants, critical for dehydration of the intracellular environment and minimisation of intracellular ice formation and osmotic stress, is less efficient than in embryos (whose surface-to-volume ratio is much higher due to their multi-cellularity and in spermatozoa (where water and solute permeabilities are higher): for comparison see Tables 1 and 4. One of the main focuses of cryobiological research on oocytes is the understanding of changes in permeability characteristics of the membrane. The reported hydraulic permeability Lp of oocytes in the absence of cryoprotectants lies in the range of about 0.4–0.5 μm/atm/min at 20 °C and shows little variation between species. Cryoprotectants affect oocyte Lp as they do in spermatozoa. However, in contrast to their effects in the latter cell type, cryoprotectants seem to increase Lp in several oocyte species, thereby accelerating and facilitating water flux; this effect was especially marked in human and bovine oocytes when DMSO was used as cryoprotectant. The maturation state of the oocyte also seems to have a crucial bearing on Lp (Tab. 4). In rat oocytes water permeability decreases during maturation whereas in bovine oocytes it increases. Such changes could be associated with disappearance or occurrence of transcripts for the broadly selective channels responsible for solute and water transfer in cells, e. g. aquaporin-9 transcript in rat oocytes.

The optimal choice of cryoprotectants for oocyte cryopreservation varies between oocytes of different species, dependant on protectant permeability characteristics (Tab. 4). The type of cryoprotectant, its concentration, and the temperature of exposure have been shown to affect greatly changes in cell volume. Even small changes in duration of exposure to cryoprotectant prior to cooling can result in drastic changes in cellular hydration. Studies have been performed on both mouse and human oocytes, using the protectants PG (propylene glycol), EG (ethylene glycol) and DMSO commonly used for cryopreservation of these cells . It has been demonstrated in particular that DMSO has a lower permeability than PG and thus would probably not be a suitable alternative. Similar permeability studies like those on spermatozoa are required to establish procedures for oocytes that will yield acceptable rates of post-thaw survival.

11.7.1 Cooling and Warming Rates

The survival of spermatozoa and oocytes and the maintenance of their fertilization potential require the use of optimal cooling and warming rates. Not surprisingly, it has been found that these rates vary between species and between cell type.

Oocyte Species	Cryo- Protectant	Stage	L_p(µm/ min/atm)	Pcryo(x 10^3 cm/min	Temp
Murine	None	Mature*	0.36	-	21oC
Murine	None	Mature*	0.40	-	22oC
Murine	DMSO	MII	0.41	0.96	23oC
Murine	PG	MII	.36	1.44	23oC
Murine	EG	MII	0.51	0.54	19oC
Murine	DMSO	MII	0.4	1.07	20oC
Murine	Gly	MII	0.98	0.02	24oC
Rat	DMSO	GV	0.68	1.29	20oC
Rat	DMSO	MII	0.45	0.56	20oC
Hamster	None	Mature*	0.45	-	22oC
Bovine	None	Immature*	0.45	-	20oC
Bovine	None	In vitro matured*	0.84	-	20oC
Bovine	EG	GV	0.5	1.32	22oC
Bovine	EG	MII	0.83	2.22	22oC
Bovine	DMSO	GV	0.73	1.6	20oC
Bovine	DMSO	MII	1.16	2.6	20oC
Goat	PG	GV	0.97	0.87	20oC
Goat	PG	MII	0.60	1.2	20oC
Rhesus Monkey	EG	Immature*	1.08	1.08	30oC
Rhesus Monkey	EG	MII	1.14	0.84	30oC
Rhesus Monkey	DMSO	Immature*		3.60	30oC
Human	None	Mature*	0.65	-	21oC
Human	None	Ff	0.84	-	21oC
Human	None	Mature*	0.40	-	20oC
Human	None	Ff	0.64	-	20oC
Human	DMSO	MII+GVBD	0.70	1.5	24oC
Human	PG	Mature*	0.53	1.68	24oC

The greatest advances in sperm cryopreservation have been achieved for bull and man. This success can be explained partly by the relatively favourable fundamental cryobiological properties of these cell species (see Tables), and partly by the general application of empirically determined slow freezing rates (1–5 °C/min) during initial cooling, followed by more rapid cooling rates after initial ice formation (50 °C/min).

Human spermatozoa in particular show a very broad survival curve in response to different cooling rates, with negligible differences in post-thaw viability between 1–100 °C/min (optimum about 10 °C/min). Despite a relative lack of detailed cryobiological investigation, the use of sperm processed in this way for artificial insemination or in vitro fertilization has been successful, even though maintenance of viability is sub-optimal (~50 % or less). However, less progress has been achieved in other species, as the simple transfer of protocols has not proven appropriate. It is clear that the various physical changes that take place in sperm cells undergoing freezing must be considered in detail in order to improve the methodology in each particular case. In general, optimal cooling rates for successful freezing of spermatozoa from many large and domestic animal species range narrowly between 30 and 50 °C/min; careful control of these rates is essential for maximum recovery of viable and functional cells. Slow equilibrium freezing of oocytes is crucial to prevent the lethal consequences of intracellular ice formation. However, the exposure to high solute concentration that necessarily accompanies this slow freezing needs to be optimized with the help of a theoretical cryobiological approach so as to avoid lysis during thawing . Currently, optimal cooling rates lie in the range 0.3–0.5 °C/min in order to promote extracellular ice nucleation at higher sub-zero temperatures (not far below zero). After the cells became dehydrated, they are plunged into liquid nitrogen with almost no risk of intracellular ice formation. Successful freezing protocols for oocytes, derived empirically as well as on the basis of known effects of medium components on membrane permeability parameters, nevertheless result in survival rates of only 40–50 %, which for practical purposes is low.

11.7.2 Particular Aspects of Gamete Cryobiology Cellular Cryoinjury of Gametes

Freezing has several severe consequences for gamete function. The overall consequence is cell death and/or limited fertilization ability. In spermatozoa, extensive damage occurs to the plasma membrane and the crosome, and to the tail configuration, with the result that motility is impaired or destroyed . The degree of damage is related to the cold-shock sensitivity of the particular species resulting from differing biochemical and biophysical membrane characteristics: while porcine spermatozoa are very sensitive, bovine and equine are less sensitive, and human are least sensitive.

There are reports about the lethal effects of intracellular ice formation during freezing on the viability of both spermatozoa and oocytes. These effects are amplified at high cooling rates. Cell injury during cryopreservation is also due to the toxic effects of cryoprotectants. Cryoprotectants in high concentration have detrimental effects on sperm motility in a number of mammalian species. This may be due in part to poor volume regulation, resulting in post-thaw lysis and reduced sperm viability after cryopreservation. Single-step addition of cryoprotectants at suprazero temperatures may result in critical volume excursions close to the limits beyond which lethal damage can occur. These limits may be only 2–7 % of isotonic volume in the spermatozoa of some species and about 30 % of isotonic volume in oocytes. If they suffer critical volume excursion, the cells are more vulnerable to subsequent freezing and thawing, thus this situation should be avoided. In addition to the detrimental physical and colligative effects of cryoprotectants, these compounds can have direct effects on intracellular structures and mechanisms. In murine oocytes the cortical actin network was disrupted by exposure to molar concentrations of DMSO. In zebrafish oocytes, the viability decreased and germinal vesicle breakdown was observed with increasing molar concentrations of PG, DMSO, EG, sucrose and glucose. From these studies, it is clear that cryoprotective agents should be present at minimal possible concentrations, and a multi-step approach should be used for their addition and removal.

For oocytes in particular, reduced fertilization ability and cell death are likely to be due very largely to the disruption of cytoskeletal elements and cortical granules at low temperatures. Cooling of oocytes resulted in a progressive disassembly of the spindle and dispersal of the chromosomes. The extent of the changes observed was dependent upon temperature and time of exposure. It should be noted that lipid phase transition in human oocytes in different maturation states appears to take place at higher temperatures than in zygotes, which explains the increased sensitivity of oocytes to cooling and cryoinjury.

11.8 SPERM PHYSIOLOGY AND THEIR RELEVANCE FOR CRYOPRESERVATION

Apart from lethal changes produced by cryoinjury in a general way, e.g. via intracellular ice formation or solution effects, there may be sub-lethal changes in spermatozoa that take place at a molecular level, impacting on their cellular function and impairing fertilizing capacity.

Spermatozoa are not able to fertilize an oocyte immediately after deposition in the female genital tract. They must first undergo changes collectively known as capacitation, during which they acquire fertilizing ability; capacitation is a prerequisite for the acrosome reaction and fusion with the oocyte. In murine spermatozoa, these changes have been shown to involve massive BSA-mediated cholesterol efflux from the plasma membrane.

This efflux takes place under the action of bicarbonate, which is a component of artificial capacitation media. It has been found that in boar spermatozoa the bicarbonate acts to stimulate a form of adenylyl cyclase and initiate cAMP-dependent pathways, one of which brings about remodelling of the phospholipid distribution in the plasma membrane, whence the cholesterol can be extracted by BSA. Further slower changes occur during capacitation, such as remodelling of the sperm surface, protein tyrosine hosphorylation, increased internal pH and Ca2+ levels, and membrane polarization. An understanding of the capacitation process can be very important in explaining the detrimental effects of cryopreservation on sperm fertilizing capacity, and will therefore have a bearing when improvements are sought. Although sperm are subjected to cryopreservation soon after ejaculation or collection, i.e. before the onset of capacitation, the physiological characteristics of thawed sperm have been found to be similar in certain respects with those of sperm that are in advanced stages of capacitation, although they do not correspond completely at the molecular level. This phenomenon is known as cryocapacitation. In particular, frozen-thawed sperm have decreased survival ability, increased acrosomal changes, shortened time interval to undergo further membrane destabilization and develop oocyte penetrating ability, and a decreased ability to bind to the oviductal epithelium.

Knowledge regarding the physiology of capacitation is therefore highly relevant to understanding the echanisms by which cryopreservation alters sperm function. Although millions of sperm are deposited in the female reproductive tract, only a small number are trapped in the distal portion of the oviductal isthmus serves as a protective storage region. Numerous studies using in vitro oviduct-sperm binding assays in several species have shown that viable uncapacitated sperm are selectively bound by the oviductal epithelium and that there is suppression of capacitation-related events such as the influx of calcium ions into the cell and tyrosine phosphorylation of sperm proteins during such binding.

This sperm storage site in the mammalian oviduct is designated a functional sperm reservoir since it fulfils the following functions: 1) selection of the fertilization-competent sperm population, 2) modulation of sperm capacitation, and 3) regulation of sperm transport to minimize polyspermic fertilization. Since the capacitation process is understood as a controlled on-going destabilization process which reduces the life span of sperm, the maintenance of sperm viability within a certain window of time and the control of capacitation are mutually associated events. To provide sufficient competent spermatozoa at the time of ovulation, capacitation should not be completed until after ovulation has occurred. At this time, conditions within the oviduct seem to reverse the inhibitory influence of the oviduct on sperm function, initiating capacitation and hyperactivation in order for the sperm cells to release and to progress toward the site of fertilization. It should be noted that sperm populations appear to undergo capacitation at different rates, thereby providing a continuous supply of "primed" cells ready to fertilize the ovulated eggs.

11.9 ASSESSMENT OF FROZEN SEMEN

The morphological and functional changes of sperm that result from cryoinjury require assessment. Whereas gross aspects such as changes in morphology and motility can be readily assessed using phase-contrast microscopy, there is a need to establish more sophisticated spermatological methods that will allow rapid and sensitive assessment of sperm function in frozen-thawed semen. Given the similarity between frozen-thawed and capacitated spermatozoa, and the fact that ejaculated spermatozoa are not able to fertilize an oocyte, it is logical to conclude that assessment of those sperm function parameters that are modified during the capacitation process must be performed under fertilizing conditions. It is known that animal fertility varies not only between males but also between ejaculates within individual males, therefore there is a need to design assays sensitive enough to detect intra- as well as inter-individual differences and even the subtle heterogeneity within one ejaculate of a given individual.

A basic understanding of the cellular processes which accompany cryopreservation is necessary for successful development of its applications in reproductive biomedicine. Physico-chemical conditions inside and outside the cell are controlled by three main parameters: extracellular solute concentration, the presence of permeable and impermeable solutes, and the cooling and warming rates. Optimization of a cryopreservation protocol requires a three-step approach: optimization for a particular cell type, for that cell type within species, and for that cell type among individuals in the given species. For reproductive medicine in particular, the heterogeneity between cells of the same type within a single sample (ejaculate) from an indivdual male must be taken into consideration. Perhaps the next decade will reveal the development and evolution of profound cryobiological approaches which will highlight the cell response produced by this subtle but crucial kind of heterogeneity. Better means of assessing the level of damage produced by cryopreservation are needed. From a technological point of view, future research should be focussed on improving techniques for the evaluation of biophysical and osmotic parameters and on establishing or developing alternative freezing technologies e.g. isochoric freezing or cryoprotectant-free vitrification.

Assays which have been shown to be sufficiently sensitive to assess the influence of cooling and/or cryopreservation on semen quality (and thereby to predict its freezability) include:

- Inducability of the acrosome reaction by calcium ionophore
- Measurement of changes in intracellular calcium content
- Measurement of volume regulatory ability
- Assessment of plasma membrane lipid disorder (merocyanine assay)
- Computer-assisted motility analysis
- Assessment of sperm chromatin stability
- Assessment of ability to bind to oviductal explants.

These and other assays have been reviewed in recent reviews of clinical and laboratory assessment of sperm function.

12

CRYOPRESERVATION AN EMERGING PARADIGM CHANGE

12.1 INTRODUCTION

In 1949 Polge, Parks and Smith1 reported on the "chance" discovery of glycerol's cryoprotective function during their efforts to preserve avian spermatozoa in the frozen state. In the following year,

Smith extended these observations by successfully cryopreserving human red blood cells (RBCs) in glycerol. These two reports identified key elements that would play a crucial role in the evolution of the field of biopreservation including the need for a cryoprotective agent (CPA), the process by which cells could successfully be exposed to penetrating CPA and the manner of freezing and thawing. In 1959 Lovelock and Bishop3 first described the use of dimethyl sulfoxide as a CPA with its advantage of enhanced permeability versus glycerol for many cell types. In the following decades incremental advances were made focusing on changes in and study of the carrier media containing the CPAs as well as the mechanisms of cell cryoinjury and cryopreservation. Most notable was the Mazur et al. report4 in 1972 which put forth the "Two-factor Hypothesis" to describe the interrelationships between cooling rates and survival as influenced by either toxic "solution effects" experienced at sub-optimal slow cooling rates or lethal intracellular ice present at high cooling rates. In effect, these studies established a biophysical foundation upon which cryopreservation experimentation rested for nearly four decades. Other notable developments were discoveries by Fahy et al. and Rall and Fahy in the mid-1980s. This group reported on the novel vitrification strategy of cell preservation in which high concentrations (approaching 8 molar) of a cryoprotectant mixture could be titrated over a concentration gradient to create a medium that when "frozen" was devoid of ice even at liquid nitrogen storage temperatures. Beginning in 1998 a series of studies revealed that perturbations in the cell's proteome and genome during and following the cryopreservation process would significantly impact survival. This effect was observed regardless of the cryopreservation protocol utilized, "optimized" or other. With the continued development of cellular-based technologies (e.g., bioreactors with stem cells, tissue engineering, etc.), there is a need for improved methods of preservation that meet the requirement for rapid return to normophysiological function. These methodologies must protect the genome and proteome thereby avoiding stress-related genetically selectivity, a criterion not met by most preservation processes, even today. In addition, the use of animal-derived products, proteins, serums, etc., raises concern over contamination with non-native components (i.e., prions, etc.,). Efforts to maintain mammalian cells in a *dormant* state capable of "on demand" restoration have centered primarily on either hypothermic (refrigerated) storage or preservation in the frozen state. Hypothermic storage often includes processes that maintain cells at temperatures above 0°C but below a normothermic temperature range (32–37°C). Advances in the hypothermic storage research have depended on improvements in organ preservation media in support of cell and tissue storage and organ transplantation including those factors related to ion balance, buffering capacity, free radical scavenging, oncotic support and nutrients.

Cryopreservation may be defined as the maintenance of biologics at sub-freezing temperatures, below -80°C and typically below -140°C. As stated above, Mazur's "Two-Factor Hypothesis" established the principles and a priori evidence of lethal conditions that need to be avoided if successful structural preservation is to be accomplished, a supposition influenced by Lovelock's classic demonstration of the relationship between survival and sodium chloride content in human RBCs. Throughout much of its development, cryopreservation research has focused less on integration of the cellular fundamentals that emerged from organ-based, hypothermic storage research and more so on physical factors, events and engineering princioples.

Cryopreservation research has had as its central focus on the physical parameters associated with the water-to-ice phase change (i.e., water flux) during the preservation process with less attention focused on the impact of the hypothermic continuum experienced by cells during freezing until a glassy storage state is achieved. Cells that are structurally preserved (avoid intracellular ice formation) remain in a state of deepening hypothermia until attaining the vitrification state of the preservation medium. Importantly, during this thermal excursion, solute levels continue to elevate due to freeze concentration. Cell function, while suppressed, does not cease until the intracellular glass transition (Tg) temperature is reached.

12.2 A LONG COLD JOURNEY

Biopreservation is now recognized as an integrative specialty defined to include processes that suppress biological aging while supporting post-preservation restoration of function. Successful cryopreservation incorporates relevant engineering principles with developments in cellular and molecular biology. Biopreservation represents the simultaneous application and management of numerous, often poorly defined (and or fully recognized), lethal conditions with the expectation of normal recovery. Biopreservation begins with a reduction in temperature typically from 37°C to the 0–10°C range. A hold temperature of 4°C is common for a brief interval (nominally 10 minutes) to allow for cryoprotectant equilibration. The selection of the hold temperature is less related to a cellular rationale and more to the fact that liquid water reaches its maximum density at 4°C. With cooling, a change in the energy balance of the cell occurs due to heat flow from the biologic. This loss of kinetic energy of molecules results in the uncoupling and re-coupling (shunting) of biochemical reactions.

The metabolome experiences imbalances that cause the failure of aerobic production of ATP, disruption of membranemediated transport (i.e., rapid gains in calcium, the loss of intracellular potassium and gain of sodium) and intracellular acidosis with pH approaching While select mammalian species have evolved mechanism supportive of whole-body hypothermia (i.e., hibernators) or seasonally adaptive heterothermy, few human cells in vivo tolerate low temperature exposure beyond a few hours. Hypothermia without manipulative intervention yields progressive cell injury during each of its three phases (cooling, maintenance in the cold and rewarming). In addition to metabolic imbalances measurable changes in cell and organelle membrane lipid domains occur in a cell. These structural characteristics (transitions) result in a change in membrane fluidity from the liquid-crystalline state to the solid gel state yielding a "leaky" membranous state. A cascade of damaging events may follow including: activation and leakage of lysosomal and lipoprotein hydrolases,35,36 activation of calciumdependent phospholipases and the release of free fatty acids, activation of the apoptotic cascade and disruption of the cytoskeletal matrix. Along with the generation of free radicals, the oxidative stressors attendant to hypothermia may result in the onset of apoptosis or gene regulated cell death.

12.3 CRYOPRESERVATION OF CELLS

Cryopreservation protocols begin with hypothermic exposures which persist through the period of active extracellular ice growth until equilibrium is reached in the glassy-state (vitrified). This journey of deepening hypothermic stress experienced by a cell has been termed the hypothermic continuum. CPA exposure represents the second step in the preservation process introducing a diversity of penetrating (membrane permeable) and non-penetrating agents contained within a carrier media to the hypothermic cell. Incubation in the cryoprotective cocktail lasts between 10–30 minutes at 4oC followed by cooling at a nominal ("optimal") cooling rate (ranging from 1 to 10°C·min-1 is common for many mammalian cells). Seeding (ice nucleation) at a temperature close to the equilibrium freezing point of the cryoprotective medium (-2 to -6°C) supports the gradual growth of extracellular ice and limits "supercooling" of the system. This supports the osmotic efflux of water from the cell and the equilibration of cryoprotective agents across the cell membrane. As extracellular ice formation continues, freezing reduces the availability of freezable water in the cell while freeze concentration of solutes increases the intracellular viscosity. When cooling rates are too rapid, cellular dehydration is inadequate increasing the probability of lethal intracellular ice formation. Non-optimal freezing effects are recognized post thaw by increased cell rupture and early stage necrosis occurring over the first few hours post-thaw. If cooling rates are too slow, prolonged exposure to multimolar levels of the freeze concentrated solutes results in cell toxicity (solution effects).4,57 An indication of "solution effect" toxicity is the appearance of delayed necrosis peaking 6–12 hours post-thaw as well as apoptosis 12–36 hours post-thaw (cell type dependent)58-60 followed by a secondary bout of necrosis related to the interplay between the progression of the modes of cell death. Cooling rate control (CRC) is accomplished by devices which support microprocessor controlled injection of liquid nitrogen to achieve active controlled rate cooling, or by passive methods often using insulated alcohol baths placed in a -80°C freezer. Active CRC devices monitor a representative sample vial, straw or bag and follow a pre-established program to achieve a desired cooling profile. Profiles are typically set to maintain a standard rate of cooling (e.g., -1°C/min for a specific cell type) over a prescribed temperature range and include a "seeding" event, nucleation through a thermal shock administered by a surge in cryogen to "flatten" the temperature rebound resulting from the latent heat of fusion of ice formation. Active CRC devices also provide records of the cooling profile. Passive CRC devices contain the sample surrounded by, but isolated from, an alcohol bath or a thermal-insulation material, and when placed in a freezer (-80°C), a curvilinear rate (e.g., approximately -1°C/min for a given cell type) is achieved in the samples. In passive CRC process seeding is often accomplished via mechanical agitation to create a nucleation event at a prescribed time during the cooling period.

Upon completion of freezing, samples are placed into longterm storage. During storage, the cell surrounded by a thin layer of vitrified cryoprotectant is maintained in a vitreous state encased in the mass of extracellular ice as long as the temperature remains above the glass transition temperature (Tg) of the solution. In addition to controlled rate cooling, high molar concentrations of cryoprotectant mixtures introduced in a stepwise manner can also be used to vitrify samples creating an ice free state both in the cell and extracellular matrix. This approach eliminates most ice crystal structure formation and has been reported to be of benefit when attempting to cryopreserve complex tissues and organs.

Retrieving samples from storage requires rapid thawing often accomplished by placing the sample in a 37–40°C stirred water bath until most of the ice melts. Once the ice has dispersed, elution of the cryoprotectant cocktail with cell culture media in a single-step or a step-wise (for high CPA concentrations) dilution process is used. Step-wise elution minimizes the volume excursions of the cell thereby preventing mechanical damage to the cell membrane and rupture. The effect of physicochemical changes that occur during cryopreservation has been extensively reported by Mazur. Despite intensive research focused on improving cell preservation, not all mammalian cells cryopreserve "equally." To highlight this issue, Lane 57 states that "Few scientific problems have proved as intractable as cryopreservation" and "...cryobiology has been straitjacketed by its need to conform to the intractable laws of biophysics. For all its successes, cryobiology has been stuck in a rut." Further, Mazur21 has stated that "The problem today (with cryopreservation) is that applying basic principles of biophysics simply cannot solve many of the remaining challenges in cryobiology."

As traditional approaches to cell storage are applied to non-terminally differentiated mammalian cells, many of these native and engineered cell types prove refractory to cryopreservation. Even in "successfully preserved" cell systems, significant death (30–70%) is often observed within 24–48 hours post-thaw. Structural protection is afforded to these cells, but mitigation of the preservation- induced stress response resulting in biomolecular-based cell death many hours post-thaw remains a critical issue. As such, it is often the case where the cryopreservation sciences have provided effective strategies for structural preservation of most mammalian cell types but, until recently, have lacked to the molecular-based tools necessary to understand and mitigate much of the post-thaw damage. Recent studies have linked numerous stress factors associated with cryopreservation to known initiators of molecular-based apoptotic cell death processes (**Table 1**).

The combination of partial physical damage to a cell coupled with cell stresses experienced during the freeze-thaw cycle can result in necrosis. Further, "cross talk" between the apoptotic and necrotic cascades may also yield secondary necrosis. This complex series of events and factors demonstrates the critical involvement that a cell's "biology" plays in cryopreservation outcome.

12.4 MOLECULAR-BASED CELL DEATH ASSOCIATED WITH CRYOPRESERVATION

Apoptotic activation in response to low temperature exposure has been documented in a variety of systems including renal cells, fibroblasts, hepatocytes, peripheral blood mononuclear cells, cord blood, spermatozoa, oocytes, ovarian-tissue, vascular tissue, et cetera. Comparison of the stressors associated with cryopreservation and those known to activate apoptosis reveals substantial overlap. An analysis of these stresses and the cryopreservation literature demonstrates the well documented involvement of apoptosis in cryopreservation failure and the benefits of cell stress response modulation to improve outcome. Although apoptosis was described in association with these reports, it was not until 1998 that apoptosis and cryopreservation failure were directly linked.8 Over the last decade, there has been the emergence of numerous studies focused on understanding the role apoptosis plays in cryopreservation failure.

The involvement of apoptotic cell death in cryopreservation has now been reported in renal cells, fibroblasts, blood cells, cornea, stem cells, cord blood, lymphocytes, sperm, ovarian tissue98 and oocytes to name a few. One aspect of this molecular involvement is the temporal component of post-cryopreservation cell death. Molecular-based cell death often takes hours to days following thawing to manifest following thawing as various pathways are activated. This results in a delay in necrotic (6 hours) and apoptotic (12-hours) activity and ultimately observable cell death. This temporal component continues to elude many investigators attempting to characterize the extent of molecular cell death following preservation in a variety of settings.

While much research has been focused on identifying and quantifying apoptosis following cryopreservation, few have detailed the initiating stresses. While this area remains in the early stage, studies have, none-the-less, begun to provide insight into the pathways associated with cryopreservation-induced molecular cell death.

These studies have implicated a host of initiation sites including the cell membrane, nucleus and mitochondria. These reports continue to solidify of the universal role molecular-based cell death exerts on cryopreservation.

12.5 MAJOR PROBLEMS IN CRYOPRESERVATION OF TISSUES AND ORGANS

Tissue/organ cryopreservation is much more complicated and difficult than cryopreservation of individual cells. First, in addition to the cell cryoinjury caused by the intracellular ice formation (IIF) and "solution effects" as described above, there are several following major problems associated with organ cryopreservation: (a) vascular damage/rupture caused by ice formation/expansion in blood vessels (water moves into blood vessels from dehydrated cells). To minimize this cryo-destructive effect, increased understanding and prediction of the fundamental mechanisms of ice formation and cell dehydration in tissues/organs are required.

Although these biophysical events have been extensively studied in single cells with cryomicroscopy techniques, experimental data in whole tissues are very limited. (b) Fracture of frozen tissue/ organ caused by the thermal stress during the warming process.

Thermal stress is one type of mechanical stress caused by nonuniform warming in a solid/frozen body (e.g., the fracture of glass when a surface is rapidly heated or cooled). To reduce the thermal stress, uniform heating is needed. Unfortunately, biological tissues have relatively low thermal conductivity and high specific heat. Therefore, the conventional heating method (e.g., heating in a stirred water bath) causes a large temperature gradient within the tissue, resulting in high thermal stress and tissue fracture. Pegg et al. and Cui and Gao, et al. have developed a slow-cooling technique for relatively uniformly warming frozen rabbit carotid arteries to prevent the fracture. However, generally speaking, the slow warming may cause intracellular ice re-crystallization killing cells. A rapid and uniform heating technology is desired to prevent both potential lethal ice re-crystallization and thermalstress- induced fracture. Apparently, rapid and uniform heating cannot be achieved by conventional heating methods. Recently, scientists have been developing single mode microwave resonance technology to achieve very rapid and uniform heating of frozen biomaterials. Problems associated with vitrification: a dramatically different approach to cryopreservation is to use either high concentrations of certain CPAs or ultra-rapid cooling rates (>106°C/min) to induce the cell cytoplasm to form a glass (i.e., to vitrify cells/tissues) rather than to crystallize. Indeed, vitrification is an ideal approach for organ cryopreservation. However, ultrarapid cooling rates are technically difficult to achieve for tissues/ organs. Several of the CPAs that are effective in ameliorating slow freeze injury also act to promote glass formation, but the required concentrations are so high, e.g., 4–8 M, that they can be very toxic to the cells/tissues. In addition, a vitrified organ is very brittle and can be easily fractured by thermal stress if not uniformly heated. (d) The lack of a single optimal cryoperservtion condition for all cell types in a tissue/organ. As we know, the optimal cryopreservation condition (e.g., cooling rate) is cell-type dependent due to the cell-type dependence of membrane permeability to water and CPAs, intracellular ice formation, osmotic tolerance limits, cryo-sensitivity and other physical/biological factors for cell cryoinjury. Because of many different cell types in a tissue or organ, it is difficult to define one single condition which is optimal for cryopreservation of all cell types. Up to now, cryopreservation (in either frozen or vitrified conditions) of large tissues and organs is not successful, in general.

12.6 IMPROVING CRYOPRESERVATION OUTCOME

Differences in the sensitivity of various cell types to cryopreservationp rocesses are well known. In an article by Van Buskirk et al. it was suggested that the basis for differing cellular survival is linked to individual cell stress response and the resultant differential activation of cell death processes. The discovery of molecular responses in cells to the preservation process has therefore resulted in a variety of attempts to control these events in an effort to improve outcome. These attempts have included alteration in solution design

Cell-based applications in cell therapy, regenerative and reparative medicine, biobanking and tissue engineering are now focusing on normal, predictable and timely return to function of the cells after cryopreservation. This is often not chieved with today's technologies and approaches. In order to address this issue, continued improvement in cryopreservation outcome will rely on the integration of cellular biology, molecular biology, biophysics, engineering and cryobiology. Furthermore, with the growing body of evidence suggesting that CPAs, such as DMSO, affect the cellular, proteome, genome and structures such as the mitochondria, the cell membrane and nucleus, it is obvious that successful preservation requires new strategies for the new definitions of success. Traditionally, cryopreservation developments focused on structural preservation of cells through the inclusion of penetrating cryoprotectants and the management of ice and chemo-osmotic perturbations. New strategies improved preservation outcome through alteration of preservation solution to mitigate some of the detrimental effects of stress that contribute to the post-thaw launch of apoptotic and necrotic cell death cascades. The literature base utilizing the integrated approach to understanding and developing new approaches for preservation grows slowly. Current studies are now focused on linking the "management" of gene regulated stress dependent effects on a cell with the traditional cryopreservation approaches. In combination cell cryopreservation outcome will doubtlessly improve to meet the increased needs in biomedical applications.

(cryoprotectant carrier media), addition of cryoprotective agent cocktails, and the incorporation of select compounds for the Targeted Control of Apoptosis (TAC) during the cryopreservation process.

The mitigation of the molecular-based stress responses to low temperature exposure and storage has been shown to be attainable with cryopreservation solution formulation that addresses both physical and cellular related events.44 The concept of specialty preservation media has evolved out of the organ preservation specialties. The Belzer and Southard team15,16 first developed ViaSpan® (the University of Wisconsin solution) to support the transport of organs (pancreas, kidney and liver). ViaSpan®, formulated for hypothermic storage, was the first solution designed to manage select putative stress factors (**Table 1**) and became the first "intracellular-like" preservation medium. In the decade that followed additional preservation solutions were developed (i.e., Celsior, HTK—Custodiol, HypoThermosol, Unisol and others). More recently, cryopreservation solution formulation has moved beyond the addition of a penetrating cryoprotective agent such as DMSO (5–15%) to cell culture media, buffered saline or these media plus serum or a protein component. Now recognized as essential to optimization of the cryopreservation process is the maintenance of proper cold-dependent ion ratios, control of pH at lowered temperature, prevention of the formation of free radicals, oncontic balance, the supply of energy substitutes, etc. Traditional media fall short in addressing changes in solution pH, free radical production, energy deprivation, etc. Accordingly, the basal properties of these historical preservation media often do not provide for protection at the cellular level. In attempt to address this issue, the cryopreservation sciences have taken lead from the organ preservation and molecular biology arenas combining these knowledge bases to increasing cell survival. Media including Viaspan, CryoStor, Unisol, Adesta, Celsior and others, to name a few, when combined with CPAs for have been reported to improve cell survival to varying degrees. Improvements have been observed in systems including hepatocytes, cord blood stem cells, PBMC's, fibroblasts, keratinocytes, blood vessels and engineered tissues. In these studies, evaluation of the cryopreservation media was conducted and correlated with improvements in cell survival, function and growth. The improvement was not noted immediately post-thaw but not until following manifestation of the molecular-based evets was the effect observed. It is now recognized that the integration of an intracellular-type solution with a penetrating cryoprotectant along with an understanding of the molecular responses of the cell at low temperature, provides for improved cryopreservation outcome. The success of these solutions is linked to an in depth knowledge and understanding of the cell death pathways activated as a result of cryopreservation-induced cell stresses. To this end, studies have suggested that the improvement in cell survival and function was due to a reduction of both apoptosis and necrosis during post-thaw recovery although the mechanism of which remains unknown.

13

DOLLY AND POLLY

13.1 INTRODUCTION

- During the winter of 1995–96, Wilmut was involved in three pivotal cloning experiments conducted at Roslin. In the first, Wilmut and his team of scientists performed embryonic cell nuclear transfer by using cultured embryonic cells that were nine days old. This was similar to the experiment that led to the creation of Megan and Morag. However, the new experiment involved a different sheep breed; the cells used for nuclear transfer came from a Poll Dorset sheep.
- This first experiment resulted in the birth in 1996 of four Poll Dorset clones: Cedric, Cecil, Cyril, and Tuppence.
- The second experiment, the team used fetal fibroblasts isolated from sheep fetuses after 26 days of development; these cells served as nucleus donors for transfer into an enucleated egg.
- This experiment resulted in the birth of two clones, taffy and tweed. In the third experiment, the scientists isolated adult cells (in this case, mammary gland cells) from a six-year-old ewe and used these cells as nucleus donors for transfer into egg cells; this technique inspired the later development of a procedure called somatic cell nuclear transfer (scnt).
- And his team constructed 277 embryos containing adult cell nuclei that were implanted into 13 surrogate mothers, only one of which became pregnant. This pregnancy was carried to term successfully. The Finn Dorset lamb, born on July 5, 1996, was Dolly.

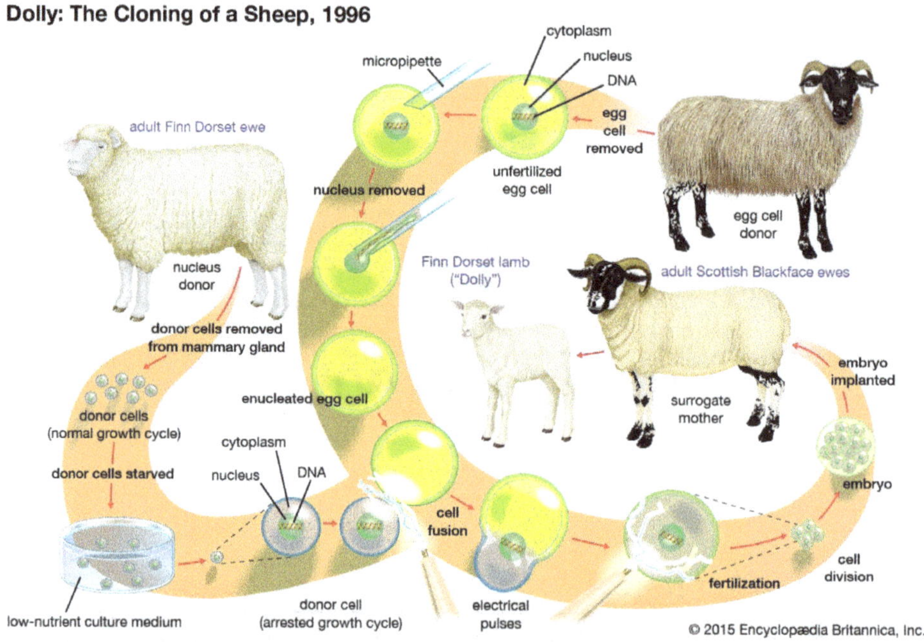

Dolly the sheep; cloning: Dolly the sheep was successfully cloned in 1996 by fusing the nucleus from a mammary-gland cell of a Finn Dorset ewe into an enucleated egg cell taken from a Scottish Blackface ewe. Carried to term in the womb of another Scottish Blackface ewe, Dolly was a genetic copy of the Finn Dorset ewe.

In 1997, following the publication in the journal *Nature* of a summary of their work leading to Dolly, Wilmut, Campbell, and the Roslin Institute instantly became known for having opened the door to a new era of controversial cloning research. The cloning of Dolly generated speculation in the media and in the scientific community about the possibility of cloning humans. Wilmut considered human cloning impractical for both ethical and scientific reasons. From his work with sheep, he knew the dangers of cloning; many embryos died following implantation, and those embryos that survived and developed to term as full-grown fetuses sometimes died immediately following birth or were born with birth defects.

Dolly standing in her pen at the Roslin Institute, near Edinburgh.

Wilmut was not interested in cloning simply for the sake of producing cloned animals, and neither was his team of scientists at Roslin. They still had problems to solve concerning their work on pharming. In 1997 Wilmut and his colleagues generated Polly, a Poll Dorset clone made from nuclear transfer using a fetal fibroblast nucleus genetically engineered to express a human gene known as *FIX*. This gene encodes a substance called human factor IX, a clotting factor that occurs naturally in most people but is absent in people with hemophilia, who require replacement therapy with a therapeutic form of the substance. Polly—along with two other sheep engineered to produce human factor IX that also were born in 1997—represented a major advance in pharming. The successful birth of Polly marked Wilmut's last major cloning experiment.

Later career

Throughout Wilmut's career at Roslin, he had been slowly moving away from research relying on embryonic stem cells, primarily because culturing embryonic stem cells from sheep embryos was extraordinarily difficult and impractical in terms of cost and time. In 2000 Wilmut was promoted to head of the department of gene expression and development at the Roslin Institute, and his research interests shifted from animals to humans. He was particularly interested in uncovering the genetic mechanisms that control embryonic development and the role that these mechanisms play in human disease. In 2005 he accepted a position as chair of reproductive science at the University of Edinburgh. He maintained a relationship with the Roslin Institute, acting as a visiting scientist. Wilmut also directed the Medical Research Council's Centre for Regenerative Medicine, located in Edinburgh, and led research efforts into cellular reprogramming.

Wilmut received several awards during his career, including the Ernst Schering prize in 2002 and the Paul Ehrlich and Ludwig Darmstaedter prize in 2005. Wilmut also was made fellow of the Royal Society of Edinburgh in 2000 and of the Royal Society of London in 2002; he was knighted in 2007. In addition to papers published in high-ranking journals such as *Nature* and *Science*, Wilmut also published several books, including *The Second Creation: Dolly and the Age of Biological Control* (2000; with Keith Campbell and Colin Tudge) and *After Dolly: The Uses and Misuses of Human Cloning* (2006; with Roger Highfield).

13.2 GENETIC ENGINEERING

Genetic engineering, the artificial manipulation, modification, and recombination of DNA or other nucleic acid molecules in order to modify an organism or population of organisms. The term genetic engineering is generally used to refer to methods of recombinant DNA technology, which emerged from basic research in microbial genetics. The techniques employed in genetic engineering have led to the production of medically important products, including human insulin, human growth hormone, and hepatitis B vaccine, as well as to the development of genetically modified organisms such as disease-resistant plants.

13.2.1 Historical Developments

The term genetic engineering initially referred to various techniques used for the modification or manipulation of organisms through the processes of heredity and reproduction. As such, the term embraced both artificial selection and all the interventions of biomedical techniques, among them artificial insemination, in vitro fertilization (e.g., "test-tube" babies), cloning, and gene manipulation. In the latter part of the 20th century, however, the term came to refer more specifically to methods of recombinant DNA technology (or gene cloning), in which DNA molecules from two or more sources are combined either within cells or in vitro and are then inserted into host organisms in which they are able to propagate.

The possibility for recombinant DNA technology emerged with the discovery of restriction enzymes in 1968 by Swiss microbiologist Werner Arber. The following year American microbiologist Hamilton O. Smith purified so-called type II restriction enzymes, which were found to be essential to genetic engineering for their ability to cleave a specific site within the DNA (as opposed to type I restriction enzymes, which cleave DNA at random sites). Drawing on Smith's work, American molecular biologist Daniel Nathans helped advance the technique of DNA recombination in 1970–71 and demonstrated that type II enzymes could be useful in genetic studies. Genetic engineering based on recombination was pioneered in 1973 by American biochemists Stanley N. Cohen and Herbert W. Boyer, who were among the first to cut DNA into fragments, rejoin different fragments, and insert the new genes into E. coli bacteria, which then reproduced.

13.2.2 Process and Techniques

Most recombinant DNA technology involves the insertion of foreign genes into the plasmids of common laboratory strains of bacteria. Plasmids are small rings of DNA; they are not part of the bacterium's chromosome (the main repository of the organism's genetic information). Nonetheless, they are capable of directing protein synthesis, and, like chromosomal DNA, they are reproduced and passed on to the bacterium's progeny. Thus, by incorporating foreign DNA (for example, a mammalian gene) into a bacterium, researchers can obtain an almost limitless number of copies of the inserted gene. Furthermore, if the inserted gene is operative (i.e., if it directs protein synthesis), the modified bacterium will produce the protein specified by the foreign DNA.

A subsequent generation of genetic engineering techniques that emerged in the early 21st century centred on gene editing. Gene editing, based on a technology known as CRISPR-Cas9, allows researchers to customize a living organism's genetic sequence by making very specific changes to its DNA. Gene editing has a wide array of applications, being used for the genetic modification of crop plants and livestock and of laboratory model organisms (e.g., mice).

The correction of genetic errors associated with disease in animals suggests that gene editing has potential applications in gene therapy for humans. Gene therapy is the introduction of a normal gene into an individual's genome in order to repair a mutation that causes a genetic disease. When a normal gene is inserted into a mutant nucleus, it most likely will integrate into a chromosomal site different from the defective allele; although this may repair the mutation, a new mutation may result if the normal gene integrates into another functional gene. If the normal gene replaces the mutant allele, there is a chance that the transformed cells will proliferate and produce enough normal gene products for the entire body to be restored to the undiseased phenotype.

13.2.3 Applications of Genetic Engineering

Genetic engineering has advanced the understanding of many theoretical and practical aspects of gene function and organization. Through recombinant DNA techniques, bacteria have been created that are capable of synthesizing human insulin, human growth hormone, alpha interferon, a hepatitis B vaccine, and other medically useful substances. Plants may be genetically adjusted to enable them to fix nitrogen, and genetic diseases can possibly be corrected by replacing dysfunctional genes with normally functioning genes.

Genes for toxins that kill insects have been introduced in several species of plants, including corn and cotton. Bacterial genes that confer resistance to herbicides also have been introduced into crop plants. Other attempts at the genetic engineering of plants have aimed at improving the nutritional value of the plant.

13.2.4 Controversy and Ethical Issues

In 1980 the "new" microorganisms created by recombinant DNA research were deemed patentable, and in 1986 the U.S. Department of Agriculture approved the sale of the first living genetically altered organism—a virus, used as a pseudorabies vaccine, from which a single gene had been cut. Since then several hundred patents have been awarded for genetically altered bacteria and plants. Patents on genetically engineered and genetically modified organisms, particularly crops and other foods, however, were a contentious issue, and they remained so into the first part of the 21st century.

Special concern has been focused on genetic engineering for fear that it might result in the introduction of unfavourable and possibly dangerous traits into microorganisms that were previously free of them—e.g., resistance to antibiotics, production of toxins, or a tendency to cause disease. Indeed, possibilities for misuse of genetic engineering were vast. In particular, there was significant concern about genetically modified organisms, especially modified crops, and their impacts on human and environmental health. For example, genetic manipulation may potentially alter the allergenic properties of crops. In addition, whether some genetically modified crops, such as golden rice, deliver on the promise of improved health benefits was also unclear. The release of genetically modified mosquitoes and other modified organisms into the environment also raised concerns.

In the 21st century, significant progress in the development of gene-editing tools brought new urgency to long-standing discussions about the ethical and social implications surrounding the genetic engineering of humans. The application of gene editing in humans raised significant ethical concerns, particularly regarding its potential use to alter traits such as intelligence and beauty. More practically, some researchers attempted to use gene editing to alter genes in human sperm, which would enable the edited genes to be passed on to subsequent generations, while others sought to alter genes that increase the risk of certain types of cancer, with the aim of reducing cancer risk in offspring. The impacts of gene editing on human genetics, however, were unknown, and regulations to guide its use were largely lacking.

13.3 CRYOPRESERVATION

Cryopreservation is based on the ability of certain small molecules to enter cells and prevent dehydration and formation of intracellular ice crystals, which can cause cell death and destruction of cell organelles during the freezing process. Two common cryoprotective agents are dimethyl sulfoxide (DMSO) and glycerol. Glycerol is used primarily for cryoprotection of red blood cells, and DMSO is used for protection of most other cells and tissues. A sugar called trehalose, which occurs in organisms capable of surviving extreme dehydration, is used for freeze-drying methods of cryopreservation. Trehalose stabilizes cell membranes, and it is particularly useful for the preservation of sperm, stem cells, and blood cells.

Most systems of cellular cryopreservation use a controlled-rate freezer. This freezing system delivers liquid nitrogen into a closed chamber into which the cell suspension is placed. Careful monitoring of the rate of freezing helps to prevent rapid cellular dehydration and ice-crystal formation. In general, the cells are taken from room temperature to approximately −90 °C (−130 °F) in a controlled-rate freezer. The frozen cell suspension is then transferred into a liquid-nitrogen freezer maintained at extremely cold temperatures with nitrogen in either the vapour or the liquid phase. Cryopreservation based on freeze-drying does not require use of liquid-nitrogen freezers.

An important application of cryopreservation is in the freezing and storage of hematopoietic stem cells, which are found in the bone marrow and peripheral blood. In autologous bone-marrow rescue, hematopoietic stem cells are collected from a patient's bone marrow prior to treatment with high-dose chemotherapy. Following treatment, the patient's cryopreserved cells are thawed and infused back into the body. This procedure is necessary, since high-dose chemotherapy is extremely toxic to the bone marrow. The ability to cryopreserve hematopoietic stem cells has greatly enhanced the outcome for the treatment of certain lymphomas and solid tumour malignancies. In the case of patients with leukemia, their blood cells are cancerous and cannot be used for autologous bone-marrow rescue. As a result, these patients rely on cryopreserved blood collected from the umbilical cords of newborn infants or on cryopreserved hematopoietic stem cells obtained from donors. Since the late 1990s it has been recognized that hematopoietic stem cells and mesenchymal stem cells (derived from embryonic connective tissue) are capable of differentiating into skeletal and cardiac muscle tissues, nerve tissue, and bone. Today there is intense interest in the growth of these cells in tissue culture systems, as well as in the cryopreservation of these cells for future therapy for a wide variety of disorders, including disorders of the nervous and muscle systems and diseases of the liver and heart.

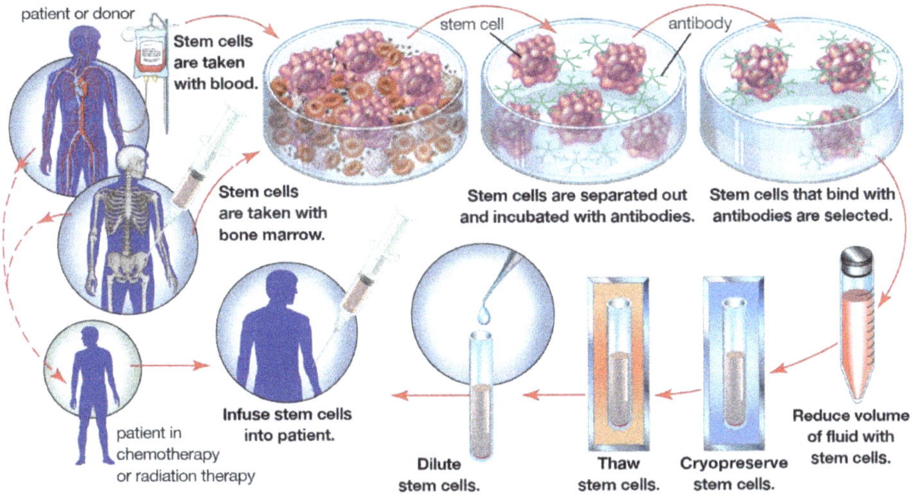

13.4 BONE MARROW TRANSPLANTATION

High doses of chemotherapy or radiation destroy not only cancer cells but also bone marrow, which is rich in blood-forming stem cells. In order to replace damaged marrow, stem cells are harvested from either the blood or the bone marrow of the cancer patient before therapy; cells also may be taken from a genetically compatible donor. In order to remove unwanted cells, such as tumour cells, from the sample, it is incubated with antibodies that bind only to stem cells. The fluid that contains the selected cells is reduced in volume and frozen until needed. The fluid is then thawed, diluted, and reinfused into the patient's body. Once in the bloodstream, the stem cells travel to the bone marrow, where they implant themselves and begin producing healthy cells.

Cryopreservation is also used to freeze and store human embryos and sperm. It is especially valuable for the freezing of extra embryos that are generated by in vitro fertilization (IVF). A couple can choose to use cyropreserved embryos for later pregnancies or in the event that IVF fails with fresh embryos. In the process of frozen embryo transfer, the embryos are thawed and implanted into the woman's uterus. Frozen embryo transfer is associated with a small but significant increase in the risk of childhood cancer among children born from such embryos.

Profound hypothermia, a form of mild cryopreservation used in human patients, has significant applications. A common use of induction of profound hypothermia is for complex cardiovascular surgical procedures. After the patient has been placed on complete cardiopulmonary bypass, using a heart-lung machine, the blood passes through a cooling chamber. Controlled cooling of the patient may reach extremely low temperatures of around 10–14 °C (50–57 °F). This amount of cooling effectively stops all cerebral activity and provides protection for all the vital organs. When this extreme cooling has been achieved, the heart-lung machine can be stopped, and the surgeon can correct very complex aortic and cardiac defects during circulatory arrest. During this time, no blood is circulating within the patient. After the surgery has been completed, the blood is gradually warmed in the same heat exchanger used for cooling. Gradual warming back to normal body temperatures results in resumption of normal brain and organ functions. This profound hypothermia, however, is far removed from freezing and long-term cryopreservation.

Cells can live more than a decade if properly frozen. In addition, certain tissues, such as parathyroid glands, veins, cardiac valves, and aortic tissue, can be successfully cryopreserved. Freezing is also used to store and maintain long-term viability of early human embryos, ova (eggs), and sperm. The freezing procedures used for these tissues are well established, and, in the presence of cryoprotective agents, the tissues can be stored over long periods of time at temperatures of −14 °C (6.8 °F). Research has shown that whole animals frozen in the absence of cryoprotective agents can yield viable cells containing intact DNA upon thawing. For example, nuclei of brain cells from whole mice stored at −20 °C (−4 °F) for more than 15 years have been used to generate lines of embryonic stem cells. These cells were subsequently used to produce mouse clones.

13.5 EDUCATION AND CRYOPRESERVATION RESEARCH

Wilmut was raised in Coventry, a town in the historic English county of Warwickshire, and he attended the Agricultural College at the University of Nottingham. In his undergraduate studies, Wilmut initially pursued his lifelong interest in farming, particularly in raising animals such as sheep. However, he soon turned his attention to animal science and basic research. In 1966, his final year at Nottingham, he received a scholarship to conduct research for a summer under English biologist Ernest John Christopher Polge in the Unit of Reproductive Physiology and Biochemistry, then a division of the Agricultural Research Council at the University of Cambridge. During this time, Wilmut performed basic experiments on animal embryos. Following his graduation from Nottingham in 1967, he returned to Cambridge, where he pursued a doctorate under the guidance of Polge, whose research was focused on improving methods of embryo cryopreservation. In 1971 Wilmut was awarded a doctorate by Darwin College, Cambridge; the title of his thesis was "Deep Freeze Preservation of Boar Semen." Wilmut remained at Cambridge and conducted extensive research on the cryopreservation of embryos. In 1973 he successfully implanted into a surrogate cow a calf embryo that had been cryopreserved. The embryo was carried to term, and Wilmut named the first-ever "frozen calf" Frostie.

13.6 GENETIC ENGINEERING AND CLONING RESEARCH

13.6.1 Pharming

In 1973 Wilmut was appointed senior scientific officer at the Animal Breeding Research Organisation (ABRO; renamed Edinburgh Research Station of the Institute of Animal Physiology and Genetics Research in 1985 and finally Roslin Institute in 1993), a government-supported research facility located in Roslin, Scot., just south of Edinburgh. At the ABRO facility, Wilmut studied embryo development and became interested in the underlying causes of embryo death in mammals. However, in the early 1980s, changes in ABRO leadership and a shift in the focus of government research projects forced Wilmut into the realm of genetic engineering. The new goal of ABRO was to generate sheep genetically engineered to produce large quantities of human proteins that would be suitable for therapeutic uses, a pursuit that came to be known as "pharming." Although Wilmut had little experience with genetic engineering and had limited enthusiasm for the project, he used his knowledge of developmental biology to obtain zygotes (one-celled embryos) from sheep and developed techniques to inject DNA into the zygote pronucleus (a haploid nucleus occurring in embryos prior to fertilization). This work eventually led to the generation of a sheep named Tracy. Tracy was created from a zygote genetically engineered through DNA injection to produce milk containing large quantities of the human enzyme alpha-1 antitrypsin, a substance used to treat cystic fibrosis and emphysema.

13.6.2 Nuclear Transfer

Wilmut's initial forays into cloning began in the late 1980s with embryonic stem cells. Wilmut and his colleagues were interested primarily in nuclear transfer, a technique first conceived in 1928 by German embryologist Hans Spemann. Nuclear transfer involves the introduction of the nucleus from a cell into an enucleated egg cell (an egg cell that has had its own nucleus removed). This can be accomplished through fusion of the cell to the egg (the technique that Wilmut used in all his later cloning experiments) or through the removal of the nucleus from the cell and the subsequent transplantation of that nucleus into the enucleated egg cell (a technique refined in the early 2000s). In 1989 Wilmut and Lawrence Smith, a graduate student conducting his thesis research at Roslin, generated four cloned lambs by using embryonic cell nuclear transfer, in which the nucleus from an embryonic stem cell was inserted into an enucleated egg. This research led Wilmut and Smith to an important discovery—namely, that the stage of the cell cycle (the sequence through which each cell progresses from one cell division to the next) at the time of nuclear transfer determined the success or failure of the experiment. They realized that the four clones they had generated happened by chance.

In 1991 Wilmut hired English biologist Keith Campbell (Smith had left the research centre in 1990), whose knowledge of the cell cycle proved instrumental in advancing the technique of nuclear transfer developed at Roslin. Wilmut and Campbell's first major success came in 1995, with the generation of two cloned Welsh mountain sheep, Megan and Morag. The following year Wilmut, Campbell, and their team of scientists decided to test a new theory based on the idea that the age or the stage of differentiation of a donor cell did not matter in nuclear transfer. Prior to this theory, nuclear transfer was believed to work only if the nucleus used as the donor for nuclear transfer came from a cell that was totipotent—i.e., having the ability to differentiate into any type of cell in the body and therefore possessing no characteristics of differentiation itself. However, observations from laboratory experiments and from Megan and Morag, who were produced using nine-day-old embryonic cells, which are presumably less totipotent than younger embryonic cells, indicated that an enucleated host egg could somehow reverse the differentiation of the donor cell nucleus, converting it back to a state of totipotency or pluripotency (slightly more differentiated than a totipotent cell). This led to the idea of using the nucleus from an already differentiated adult cell as a donor nucleus.

13.7 BIOENGINEERING

The application of engineering knowledge to the fields of medicine and biology. The bioengineer must be well grounded in biology and have engineering knowledge that is broad, drawing upon electrical, chemical, mechanical, and other engineering disciplines. The bioengineer may work in any of a large range of areas. One of these is the provision of artificial means to assist defective body functions—such as hearing aids, artificial limbs, and supportive or substitute organs. In another direction, the bioengineer may use engineering methods to achieve biosynthesis of animal or plant products—such as for fermentation processes.

13.7.1 History

Before World War II the field of bioengineering was essentially unknown, and little communication or interaction existed between the engineer and the life scientist. A few exceptions, however, should be noted. The agricultural engineer and the chemical engineer, involved in fermentation processes, have always been bioengineers in the broadest sense of the definition since they deal with biological systems and work with biologists. The civil engineer, specializing in sanitation, has applied biological principles in the work. Mechanical engineers have worked with the medical profession for many years in the development of artificial limbs. Another area of mechanical engineering that falls in the field of bioengineering is the air-conditioning field. In the early 1920s engineers and physiologists were employed by the American Society of Heating and Ventilating Engineers to study the effects of temperature and humidity on humans and to provide design criteria for heating and air-conditioning systems.

Today there are many more examples of interaction between biology and engineering, particularly in the medical and life-support fields. In addition to an increased awareness of the need for communication between the engineer and the associate in the life sciences, there is an increasing recognition of the role the engineer can play in several of the biological fields, including human medicine, and, likewise, an awareness of the contributions biological science can make toward the solution of engineering problems.

Much of the increase in bioengineering activity can be credited to electrical engineers. In the 1950s bioengineering meetings were dominated by sessions devoted to medical electronics. Medical instrumentation and medical electronics continue to be major areas of interest, but biological modeling, blood-flow dynamics, prosthetics, biomechanics (dynamics of body motion and strength of materials), biological heat transfer, biomaterials, and other areas are now included in conference programs.

Bioengineering developed out of specific desires or needs: the desire of surgeons to bypass the heart, the need for replacement organs, the requirement for life support in space, and many more. In most cases the early interaction and education were a result of personal contacts between physician, or physiologist, and engineer. Communication between the engineer and the life scientist was immediately recognized as a problem. Most engineers who wandered into the field in its early days probably had an exposure to biology through a high-school course and no further work. To overcome this problem, engineers began to study not only the subject matter but also the methods and techniques of their counterparts in medicine, physiology, psychology, and biology. Much of the information was self-taught or obtained through personal association and discussions. Finally, recognizing a need to assist in overcoming the communication barrier as well as to prepare engineers for the future, engineering schools developed courses and curricula in bioengineering.

13.7.2 Branches of Bioengineering

- Medical engineering. Medical engineering concerns the application of engineering principles to medical problems, including the replacement of damaged organs, instrumentation, and the systems of health care, including diagnostic applications of computers.
- Agricultural engineering. This includes the application of engineering principles to the problems of biological production and to the external operations and environment that influence this production.
- Bionics. Bionics is the study of living systems so that the knowledge gained can be applied to the design of physical systems.
- Biochemical engineering. Biochemical engineering includes fermentation engineering, application of engineering principles to microscopic biological systems that are used to create new products by synthesis, including the production of protein from suitable raw materials.

- Human-factors engineering. This concerns the application of engineering, physiology, and psychology to the optimization of the human–machine relationship.
- Environmental health engineering. Also called bioenvironmental engineering, this field concerns the application of engineering principles to the control of the environment for the health, comfort, and safety of human beings. It includes the field of life-support systems for the exploration of outer space and the ocean.
- Genetic engineering. Genetic engineering is concerned with the artificial manipulation, modification, and recombination of deoxyribonucleic acid (DNA) or other nucleic acid molecules in order to modify an organism. The techniques employed in this field have led to the production of medically important products, including human insulin, human growth hormone, and hepatitis B vaccine.

13.7.3 Antioxidant

- Any of various chemical compounds added to certain foods, natural and synthetic rubbers, gasolines, and other substances to retard autoxidation, the process by which these substances combine with oxygen in the air at room temperature. Retarding autoxidation delays the appearance of such undesirable qualities as rancidity in foods, loss of elasticity in rubbers, and formation of gums in gasolines. Antioxidants most commonly used are such organic compounds as aromatic amines, phenols, and aminophenols.
- Autoxidation has been found to proceed by a chain reaction; that is, a reaction consisting of a series of successive steps occurring in repetitive cycles, in each of which intermediate products called chain carriers are regenerated. Such a reaction will continue as long as the chain carriers persist. In autoxidation the chain carriers are free radicals, electrically neutral molecular fragments containing unpaired electrons. The chain can be initiated by thermally excited molecules, free radicals, metal catalysts, or light. Antioxidants, by reacting with chain carriers, terminate the oxidative chain reaction.
- An example of autoxidation that is of great commercial concern is the one leading to the rancidity of fats, oils, and fatty foods. Rancidity is caused by the degradation of the fat molecule, by reaction with oxygen, to a mixture of volatile aldehydes, ketones, and acids. The reaction can be initiated by exposure to light or by the presence of trace amounts of metals that serve as catalysts. To retard the development of rancidity, organic antioxidants, commonly tocopherol, propyl gallate, butylated hydroxytoluene (BHT), or butylated hydroxyanisole (BHA), are used. These compounds react with chain carriers by donating hydrogen atoms. The use of antioxidants for food is closely regulated in most countries. Specific limitations are normally imposed on the type and quantity of antioxidants that may be used.

13.7.4 Frostie calf

Frostie, a Hereford-Friesian calf, the first calf produced from an embryo that was frozen, thawed, and implanted into a surrogate cow. Frostie, born in 1973 and popularly called the "frozen calf," was the product of cryopreservation research conducted by British developmental biologist Ian Wilmut.

The successful birth of Frostie accelerated the development of commercial in vitro fertilization and embryo-transfer techniques for the production of livestock, particularly cattle and sheep. The freezing, thawing, and implantation techniques used to generate Frostie also led to improvements in methods used to freeze and thaw human embryos.

13.8 FOOD PRESERVATION

Any of a number of methods by which food is kept from spoilage after harvest or slaughter. Such practices date to prehistoric times. Among the oldest methods of preservation are drying, refrigeration, and fermentation. Modern methods include canning, pasteurization, freezing, irradiation, and the addition of chemicals. Advances in packaging materials have played an important role in modern food preservation.

13.8.1 Spoilage Mechanisms

Food spoilage may be defined as any change that renders food unfit for human consumption. These changes may be caused by various factors, including contamination by microorganisms, infestation by insects, or degradation by endogenous enzymes (those present naturally in the food). In addition, physical and chemical changes, such as the tearing of plant or animal tissues or the oxidation of certain constituents of food, may promote food spoilage. Foods obtained from plant or animal sources begin to spoil soon after harvest or slaughter. The enzymes contained in the cells of plant and animal tissues may be released as a result of any mechanical damage inflicted during postharvest handling. These enzymes begin to break down the cellular material. The chemical reactions catalyzed by the enzymes result in the degradation of food quality, such as the development of off-flavours, the deterioration of texture, and the loss of nutrients. The typical microorganisms that cause food spoilage are bacteria (e.g., *Lactobacillus*), yeasts (e.g., *Saccharomyces*), and molds (e.g., *Rhizopus*).

13.8.2 Microbial Contamination

Bacteria and fungi (yeasts and molds) are the principal types of microorganisms that cause food spoilage and food-borne illnesses. Foods may be contaminated by microorganisms at any time during harvest, storage, processing, distribution, handling, or preparation. The primary sources of microbial contamination are soil, air, animal feed, animal hides and intestines, plant surfaces, sewage, and food processing machinery or utensils.

13.8.3 Bacteria

Bacteria are unicellular organisms that have a simple internal structure compared with the cells of other organisms. The increase in the number of bacteria in a population is commonly referred to as bacterial growth by microbiologists. This growth is the result of the division of one bacterial cell into two identical bacterial cells, a process called binary fission. Under optimal growth conditions, a bacterial cell may divide approximately every 20 minutes. Thus, a single cell can produce almost 70 billion cells in 12 hours. The factors that influence the growth of bacteria include nutrient availability, moisture, pH, oxygen levels, and the presence or absence of inhibiting substances (e.g., antibiotics).

The nutritional requirements of most bacteria are chemical elements such as carbon, hydrogen, oxygen, nitrogen, phosphorus, sulfur, magnesium, potassium, sodium, calcium, and iron. The bacteria obtain these elements by utilizing gases in the atmosphere and by metabolizing certain food constituents such as carbohydrates and proteins.

Temperature and pH play a significant role in controlling the growth rates of bacteria. Bacteria may be classified into three groups based on their temperature requirement for optimal growth: thermophiles (55–75 °C, or 130–170 °F), mesophiles (20–45 °C, or 70–115 °F), or psychrotrophs (10–20 °C, or 50–70 °F). In addition, most bacteria grow best in a neutral environment (pH equal to 7).

Bacteria also require a certain amount of available water for their growth. The availability of water is expressed as water activity and is defined by the ratio of the vapour pressure of water in the food to the vapour pressure of pure water at a specific temperature. Therefore, the water activity of any food product is always a value between 0 and 1, with 0 representing an absence of water and 1 representing pure water. Most bacteria do not grow in foods with a water activity below 0.91, although some halophilic bacteria (those able to tolerate high salt concentrations) can grow in foods with a water activity lower than 0.75. Growth may be controlled by lowering the water activity—either by adding solutes such as sugar, glycerol, and salt or by removing water through dehydration.

The oxygen requirements for optimal growth vary considerably for different bacteria. Some bacteria require the presence of free oxygen for growth and are called obligate aerobes, whereas other bacteria are poisoned by the presence of oxygen and are called obligate anaerobes. Facultative anaerobes are bacteria that can grow in both the presence or absence of oxygen. In addition to oxygen concentration, the oxygen reduction potential of the growth medium influences bacterial growth. The oxygen reduction potential is a relative measure of the oxidizing or reducing capacity of the growth medium.

When bacteria contaminate a food substrate, it takes some time before they start growing. This lag phase is the period when the bacteria are adjusting to the environment. Following the lag phase is the log phase, in which population grows in a logarithmic fashion. As the population grows, the bacteria consume available nutrients and produce waste products. When the nutrient supply is depleted, the growth rate enters a stationary phase in which the number of viable bacteria cells remains the same. During the stationary phase, the rate of bacterial cell growth is equal to the rate of bacterial cell death. When the rate of cell death becomes greater than the rate of cell growth, the population enters the decline phase.

A bacterial population is expressed either per gram or per square centimetre of surface area. Rarely does the total bacterial population exceed 1010 cells per gram. A population of less than 106 cells per gram does not cause any noticeable spoilage except in raw milk. Populations of between 106 and 107 cells per gram cause spoilage in some foods; for example, they can generate off-odours in vacuum-packaged meats. Populations of between 107 and 108 cells per gram produce off-odours in meats and some vegetables. At levels above 5 × 107 cells per gram, most foods exhibit some form of spoilage.

When the conditions for bacterial cell growth are unfavourable (e.g., low or high temperatures or low moisture content), several species of bacteria can produce resistant cells called endospores. Endospores are highly resistant to heat, chemicals, desiccation (drying out), and ultraviolet light. The endospores may remain dormant for long periods of time. When conditions become favourable for growth (e.g., thawing of meats), the endospores germinate and produce viable cells that can begin exponential growth.

13.8.4 Fungi

The two types of fungi that are important in food spoilage are yeasts and molds. Molds are multicellular fungi that reproduce by the formation of spores (single cells that can grow into a mature fungus). Spores are formed in large numbers and are easily dispersed through the air. Once these spores land on a food substrate, they can grow and reproduce if conditions are favourable. Yeasts are unicellular fungi that are much larger than bacterial cells. They reproduce by cell division (binary fission) or budding.

The conditions affecting the growth of fungi are similar to those affecting bacteria. Both yeasts and molds are able to grow in an acidic environment (pH less than 7). The pH range for yeast growth is 3.5 to 4.5 and for molds is 3.5 to 8.0. The low pH of fruits is generally unfavourable for the growth of bacteria, but yeasts and molds can grow and cause spoilage in fruits. For example, species of the fungal genus Colletotrichum cause crown rot in bananas. Yeasts promote fermentation in fruits by breaking down sugars into alcohol and carbon dioxide. The amount of available water in a food product is also critical for the growth of fungi. Yeasts are unable to grow at a water activity of less than 0.9, and molds are unable to grow at a water activity below 0.8.

13.8.5 Control of microbial contamination

The most common methods used either to kill or to reduce the growth of microorganisms are the application of heat, the removal of water, the lowering of temperature during storage, the reduction of pH, the control of oxygen and carbon dioxide concentrations, and the removal of the nutrients needed for growth. The use of chemicals as preservatives is strictly regulated by governmental agencies such as the Food and Drug Administration (FDA) in the United States. Although a chemical may have preservative functions, its safety must be proved before it may be used in food products. To suppress yeast and mold growth in foods, a number of chemical preservatives are permitted. In the United States, the list of such chemicals, known as GRAS (Generally Recognized as Safe), includes compounds such as benzoic acid, sodium benzoate, propionic acid, sorbic acid, and sodium diacetate.

13.9 CHEMICAL DETERIORATION

13.9.1 Enzymatic Reactions

Enzymes are large protein molecules that act as biological catalysts, accelerating chemical reactions without being consumed to any appreciable extent themselves. The activity of enzymes is specific for a certain set of chemical substrates, and it is dependent on both pH and temperature. The living tissues of plants and animals maintain a balance of enzymatic activity. This balance is disrupted upon harvest or slaughter. In some cases, enzymes that play a useful role in living tissues may catalyze spoilage reactions following harvest or slaughter. For example, the enzyme pepsin is found in the stomach of all animals and is involved in the breakdown of proteins during the normal digestion process. However, soon after the slaughter of an animal, pepsin begins to break down the proteins of the organs, weakening the tissues and making them more susceptible to microbial contamination. After the harvesting of fruits, certain enzymes remain active within the cells of the plant tissues. These enzymes continue to catalyze the biochemical processes of ripening and may eventually lead to rotting, as can be observed in bananas. In addition, oxidative enzymes in fruits continue to carry out cellular respiration (the process of using oxygen to metabolize glucose for energy). This continued respiration decreases the shelf life of fresh fruits and may lead to spoilage. Respiration may be controlled by refrigerated storage or modified-atmosphere packaging. Table 1 lists a number of enzymes involved in the degradation of food quality.

Enzymes that cause food spoilage

enzyme	**food**	**spoilage action**
ascorbic acid oxidase	vegetables	destruction of vitamin C
lipase	cereals	discoloration
	milk	hydrolytic rancidity
	oils	hydrolytic rancidity
lipoxygenase	vegetables	destruction of vitamin A, off-flavour
pectic enzyme	citrus juices	destruction of pectic substances
	fruits	excessive softening
peroxidase	fruits	browning
polyphenoloxidase	fruits, vegetables	browning, off-flavour, vitamin loss

Enzymes that cause food spoilage		
enzyme	**food**	**spoilage action**
protease	eggs	reduction of shelf life of fresh and dried whole eggs
	crab, lobster	overtenderization
	flour	reduction of gluten formation
thiaminase	meats, fish	destruction of thiamine

13.9.2 Autoxidation

The unsaturated fatty acids present in the lipids of many foods are susceptible to chemical breakdown when exposed to oxygen. The oxidation of unsaturated fatty acids is autocatalytic; that is, it proceeds by a free-radical chain reaction. Free radicals contain an unpaired electron (represented by a dot in the molecular formula) and, therefore, are highly reactive chemical molecules. The basic mechanisms in a free-radical chain reaction involve initiation, propagation, and termination steps (Figure 1). Under certain conditions, in initiation a free-radical molecule (X ·) present in the food removes a hydrogen (H) atom from a lipid molecule, producing a lipid radical (L ·). This lipid radical reacts with molecular oxygen (O_2) to form a peroxy radical (LOO ·). The peroxy radical removes a hydrogen atom from another lipid molecule and the reaction starts over again (propagation). During the propagation steps, hydroperoxide molecules (LOOH) are formed that may break down into alkoxy (LO ·) and peroxy radicals plus water (H2O). The lipid, alkoxy, and peroxy radicals may combine with one another (or other radicals) to form stable, nonpropagating products (termination). These products result in the development of rancid off-flavours. In addition to promoting rancidity, the free radicals and peroxides produced in these reactions may have other negative effects, such as the bleaching of food colour and the destruction of vitamins A, C, and E. This type of deterioration is prevalent in fried snacks, nuts, cooking oils, and margarine.

13.9.3 Maillard Reaction

Another chemical reaction that causes major food spoilage is nonenzymatic browning, also known as the Maillard reaction. This reaction takes place between reducing sugars (simple monosaccharides capable of carrying out reduction reactions) and the amino group of proteins or amino acids present in foods. The products of the Maillard reaction lead to a darkening of colour, reduced solubility of proteins, development of bitter flavours, and reduced nutritional availability of certain amino acids such as lysine. The rate of this reaction is influenced by the water activity, temperature, and pH of the food product. Nonenzymatic browning causes spoilage during the storage of dry milk, dry whole eggs, and breakfast cereals.

13.9.4 Light-induced Reactions

Light influences a number of chemical reactions that lead to spoilage of foods. These light-induced reactions include the destruction of chlorophyll (the photosynthetic pigment that gives plants their green colour), resulting in the bleaching of certain vegetables; the discoloration of fresh meats; the destruction of riboflavin in milk; and the oxidation of vitamin C and carotenoid pigments (a process called photosensitized oxidation). The use of packaging material that prevents exposure to light is one of the most effective means of preventing light-induced chemical spoilage.

13.9.5 Low-Temperature Preservation

Storage at low temperatures prolongs the shelf life of many foods. In general, low temperatures reduce the growth rates of microorganisms and slow many of the physical and chemical reactions that occur in foods.

13.10 REFRIGERATION

The life of many foods may be increased by storage at temperatures below 4 °C (40 °F). Commonly refrigerated foods include fresh fruits and vegetables, eggs, dairy products, and meats. Some foods, such as tropical fruits (e.g., bananas), are damaged if exposed to low temperatures. Also, refrigeration cannot improve the quality of decayed food; it can only retard deterioration. One problem of modern mechanical refrigeration—that of dehydration of foods due to moisture condensation—has been overcome through humidity control mechanisms within the storage chamber and by appropriate packaging techniques.

13.10.1 Freezing

Freezing and frozen storage provide an excellent means of preserving the nutritional quality of foods. At subfreezing temperatures the nutrient loss is extremely slow for the typical storage period used in commercial trade.

History

Early freezing methods were based on the principle that mixing salt with ice results in temperatures well below 0 °C (32 °F). By the end of the 19th century, this method was being used commercially in the United States to freeze fish and poultry. By the 1920s Clarence Birdseye had developed two processes for freezing fish based on his quick freezing theory. His first patent, describing a method for preserving piscatorial products, involved placing food between two metal plates that were chilled by a calcium chloride solution to approximately −40 °C (−40 °F). The second process utilized two hollow metal plates that were cooled to −25 °C (−13 °F) by vaporization of ammonia. This freezing apparatus was the forerunner of the multiple plate freezer that is widely used in the modern food industry.

The freezing process

The freezing of food involves lowering its temperature below 0 °C, resulting in the gradual conversion of water, present in the food, into ice. Freezing is a crystallization process that begins with a nucleus or a seed derived from either a nonaqueous particle or a cluster of water molecules (formed when the temperature is reduced below 0 °C). This seed must be of a certain size to provide an adequate site for the crystal to begin to grow. If physical conditions are conducive to the presence of numerous seeds for crystallization, then a large number of small ice crystals will form. However, if only a few seeds are initially available, then a few ice crystals will form and each will grow to a large size. The size and the number of ice crystals influence the final quality of many frozen foods; for example, the smooth texture of ice cream indicates the presence of a large number of small ice crystals.

In pure water, the freezing process is initiated by lowering the temperature to slightly below 0 °C, called supercooling. As ice crystals begin to grow, the temperature returns to the freezing point. During the conversion of liquid water to ice, the temperature of the system does not change. The heat removed during this step is called the latent heat of fusion (equivalent to 333 joules per gram of water). Once all the water is converted to ice, any additional removal of heat will result in a decrease in the temperature below 0 °C.

The freezing of foods exhibits a number of important differences from the freezing of pure water. Foods do not freeze at 0 °C. Instead, owing to the presence of different soluble particulates (solutes) in the water present in foods, most foods begin to freeze at a temperature between 0 and −5 °C (32 and 23 °F). In addition, the removal of latent heat in foods during freezing does not occur at a fixed temperature. As the water present in the food freezes into ice, the remaining water becomes more concentrated with solutes. As a result, the freezing point is further depressed. Therefore, foods have a zone of maximum ice crystal formation that typically extends from −1 to −4 °C (30 to 25 °F). Damage to food quality during freezing can be minimized if the temperature of the product is brought below this temperature range as quickly as possible.

13.10.2 Industrial Freezers

The rate at which heat is removed from a food during freezing depends on how fast heat can travel within the food and how efficiently it can be liberated from the surface of the food into the surrounding atmosphere. Industrial freezers remove heat from the surface of a food as rapidly as possible. There are several types of industrial freezers, including air-blast tunnel freezers, belt freezers, fluidized-bed freezers, plate freezers, and cryogenic freezers.

In air-blast tunnel freezers and belt freezers, precooled air at approximately −40 °C is blown over the food products. Packaged foods, such as fruits, vegetables, bakery goods, poultry, meats, and prepared meals, are usually frozen in air-blast tunnels. The packages are placed onto dollies or hand trucks and then rolled into the freezer tunnels. In a belt freezer, food is placed on a conveyor belt that moves through a freezing zone. Bakery goods, chicken parts, and meat patties are frozen using a belt freezer.

Fluidized-bed freezers are used to freeze particulate foods such as peas, cut corn, diced carrots, and strawberries. The foods are placed on a mesh conveyor belt and moved through a freezing zone in which cold air is directed upward through the mesh belt and the food particulates begin to tumble and float. This tumbling exposes all sides of the food to the cold air and minimizes the resistance to heat transfer at the surface of the food.

Plate freezers are used to freeze flat products, such as pastries, fish fillets, and beef patties, as well as irregular-shaped vegetables that are packaged in brick-shaped containers, such as asparagus, cauliflower, spinach, and broccoli. The food is firmly pressed between metal plates that are cooled to subfreezing temperatures by internally circulating refrigerants.

Cryogenic freezing is used to freeze food at an extremely fast rate. The food is moved through a spray of liquid nitrogen or directly immersed in liquid nitrogen. The liquid nitrogen boils around the food at a temperature of −196 °C (−321 °F) and extracts a large amount of heat.

13.10.3 Quality of Frozen Foods

Improper freezing or storage of foods may result in detrimental quality changes. When foods with high amounts of water are frozen slowly, they may experience a loss of fluid, called drip, upon thawing. This fluid loss causes dehydration and nutrient loss in frozen food products.

During frozen storage, the ice crystals present in foods may enlarge in size, producing undesirable changes in texture. This phenomenon is commonly observed when the storage temperature is allowed to fluctuate. For example, ice cream stored in an automatic defrosting domestic freezer becomes sandy in texture because the ice crystals increase in size as the temperature of the system fluctuates.

Improperly packaged frozen foods lose small amounts of moisture during storage, resulting in surface dehydration (commonly called freezer burn). Frozen meats with freezer burn have the appearance of brown paper and quickly become rancid. Freezer burn can be minimized by the use of tightly wrapped packages and the elimination of fluctuating temperatures during storage.

13.10.4 Thermal processing

Thermal processing is defined as the combination of temperature and time required to eliminate a desired number of microorganisms from a food product.

13.11 CANNING

Nicolas Appert, a Parisian confectioner by trade, is credited with establishing the heat processing of foods as an industry. In 1810 he received official recognition for his process of enclosing food in bottles, corking the bottles, and placing the bottles in boiling water for various periods of time. In the same year Peter Durand received a British patent for the use of containers made of glass, pottery, tin, or other metals for the heat preservation of foods. In 1822 Ezra Daggett and Thomas Kensett announced the availability of preserved foods in tin cans in the United States. Tin-coated steel containers, made from 98.5 percent sheet steel with a thin coating of tin, soon became common. These cans had a double seamed top and bottom to provided an airtight seal and could be manufactured at high speeds.

The establishment of the canning process on a more scientific basis did not occur until 1896, when the microorganism Clostridium botulinum, with its lethal toxin causing botulism, was discovered by Émile van Ermengem.

13.11.1 Presterilization Procedures

Selected crop varieties are grown specially for canning purposes. The harvesting schedules of the crops are carefully selected to conform to the cannery operations. A typical canning operation involves cleaning, filling, exhausting, can sealing, heat processing, cooking, labeling, casing, and storage. Most of these operations are performed using high-speed, automatic machines.

Cleaning involves the use of shakers, rotary reel cleaners, air blasters, water sprayers (as shown in Figure 2), or immersion washers. Any inedible or extraneous material is removed before washing, and only potable water is used in the cleaning systems.

Automatic filling machines are used to place the cleaned food into cans or other containers, such as glass jars or plastic pouches. When foods containing trapped air, such as leafy vegetables, are canned, the air must be removed from the cans prior to closing and sealing the lids by a process called exhausting. Exhausting is accomplished using steam exhaust hoods or by creation of a vacuum.

Immediately after exhausting, the lids are placed on the cans and the cans are sealed. An airtight seal is achieved between the lid and the rim of the can using a thin layer of gasket or compound. The anaerobic conditions prevent the growth of oxygen-requiring microorganisms. In addition, many of the spores of anaerobic microorganisms are less resistant to heat and are easily destroyed during the heat treatment.

13.11.2 Sterilization

The time and temperature required for the sterilization of foods are influenced by several factors, including the type of microorganisms found on the food, the size of the container, the acidity or pH of the food, and the method of heating.

The thermal processes of canning are generally designed to destroy the spores of the bacterium C. botulinum. This microorganism can easily grow under anaerobic conditions, producing the deadly toxin that causes botulism. Sterilization requires heating to temperatures greater than 100 °C (212 °F). However, C. botulinum is not viable in acidic foods that have a pH less than 4.6. These foods can be adequately processed by immersion in water at temperatures just below 100 °C.

The sterilization of low-acid foods (pH greater than 4.6) is generally carried out in steam vessels called retorts at temperatures ranging from 116 to 129 °C (240 to 265 °F). The retorts are controlled by automatic devices, and detailed records are kept of the time and temperature treatments for each lot of processed cans. At the end of the heating cycle, the cans are cooled under water sprays or in water baths to approximately 38 °C (100 °F) and dried to prevent any surface rusting. The cans are then labeled, placed in fibreboard cases either by hand or machine, and stored in cool, dry warehouses.

13.11.3 Quality of Canned Foods

The sterilization process is designed to provide the required heat treatment to the slowest heating location inside the can, called the cold spot. The areas of food farthest from the cold spot get a more severe heat treatment that may result in overprocessing and impairment of the overall quality of the product. Flat, laminated pouches can reduce the heat damage caused by overprocessing.

A significant loss of nutrients, especially heat-labile vitamins, may occur during the canning process. In general, canning has no major effect on the carbohydrate, protein, or fat content of foods. Vitamins A and D and beta-carotene are resistant to the effects of heat. However, vitamin B1 is sensitive to thermal treatment and the pH of the food. Although the anaerobic conditions of canned foods have a protective effect on the stability of vitamin C, it is destroyed during long heat treatments.

The ends of processed cans are slightly concave because of the internal vacuum created during sealing. Any bulging of the ends of a can may indicate a deterioration in quality due to mechanical, chemical, or physical factors. This bulging may lead to swelling and possible explosion of the can.

13.12 PASTEURIZATION

Pasteurization is the application of heat to a food product in order to destroy pathogenic (disease-producing) microorganisms, to inactivate spoilage-causing enzymes, and to reduce or destroy spoilage microorganisms. The relatively mild heat treatment used in the pasteurization process causes minimal changes in the sensory and nutritional characteristics of foods compared to the severe heat treatments used in the sterilization process.

The temperature and time requirements of the pasteurization process are influenced by the pH of the food. When the pH is below 4.5, spoilage microorganisms and enzymes are the main targets of pasteurization. For example, the pasteurization process for fruit juices is aimed at inactivating certain enzymes such as pectinesterase and polygalacturonase. The typical processing conditions for the pasteurization of fruit juices include heating to 77 °C (171 °F) and holding for 1 minute, followed by rapid cooling to 7 °C (45 °F). In addition to inactivating enzymes, these conditions destroy any yeasts or molds that may lead to spoilage. Equivalent conditions capable of reducing spoilage microorganisms involve heating to 65 °C (149 °F) and holding for 30 minutes or heating to 88 °C (190 °F) and holding for 15 seconds.

When the pH of a food is greater than 4.5, the heat treatment must be severe enough to destroy pathogenic bacteria. In the pasteurization of milk, the time and temperature conditions target the pathogenic bacteria Mycobacterium tuberculosis, Coxiella burnetti, and Brucella abortus. The typical heat treatment used for pasteurizing milk is 72 °C (162 °F) for 15 seconds, followed by rapid cooling to 7 °C. Other equivalent heat treatments include heating to 63 °C (145 °F) for 30 minutes, 90 °C (194 °F) for 0.5 second, and 94 °C (201 °F) for 0.1 second. The high-temperature–short-time (HTST) treatments cause less damage to the nutrient composition and sensory characteristics of foods and therefore are preferred over the low-temperature–long-time (LTLT) treatments.

Since the heat treatment of pasteurization is not severe enough to render a product sterile, additional methods such as refrigeration, fermentation, or the addition of chemicals are often used to control microbial growth and to extend the shelf life of a product. For example, the pasteurization of milk does not kill thermoduric bacteria (those resistant to heat), such as Lactobacillus and Streptococcus, or thermophilic bacteria (those that grow at high temperatures), such as Bacillus and Clostridium. Therefore, pasteurized milk must be kept under refrigerated conditions.

Liquid foods such as milk, fruit juices, beers, wines, and liquid eggs are pasteurized using plate-type heat exchangers. Wine and fruit juices are normally deaerated prior to pasteurization in order to remove oxygen and minimize oxidative deterioration of the products. Plate-type heat exchangers consist of a large number of thin, vertical steel plates that are clamped together in a frame. The plates are separated by small gaskets that allow the liquid to flow between each successive plate. The liquid product and heating medium (e.g., hot water) are pumped through alternate channels, and the gaskets ensure that the liquid product and heating or cooling mediums are kept separate. Plate-type heat exchangers are effective in rapid heating and cooling applications. After the pasteurization process is completed, the product is packaged under aseptic conditions to prevent recontamination of the product.

13.13 ASEPTIC PROCESSING

The aseptic process involves placing a sterilized product into a sterilized package that is then sealed under sterile conditions. It began in 1914 with the development of sterile filters for use in the wine industry. However, because of unreliable machinery, it remained commercially unsuccessful until 1948 when William McKinley Martin helped develop the Martin system, which later became known as the Dole Aseptic Canning System. This system involved the sterilization of liquid foods by rapidly heating them in tubular heat exchangers, followed by holding and cooling steps. The cans and lids were sterilized with superheated steam, and the sterilized containers were filled with the sterile liquid food. The lids were then sealed in an atmosphere of superheated steam. By the 1980s hydrogen peroxide was being used throughout Europe and the United States for the sterilization of polyethylene surfaces.

13.13.1 Commercial Sterility

In aseptic processing the thermal process is based on achieving commercial sterility—i.e., no more than 1 nonsterile package for every 10,000 processed packages. The aseptic process uses the high-temperature–short-time (HTST) method in which foods are heated at a high temperature for a short period of time. The time and temperature conditions depend on several factors, such as size, shape, and type of food. The HTST method results in a higher retention of quality characteristics, such as vitamins, odour, flavour, and texture, while achieving the same level of sterility as the traditional canning process in which food is heated at a lower temperature for a longer period of time.

The heating and cooling of liquid foods can be performed using metal plate heat exchangers. These heat exchangers have large surface areas that result in improved heating and cooling rates. Other types of heat exchangers involve surrounding the food with steam or directly injecting steam into the food. Products sterilized with steam are then pumped into a vacuum chamber, where they are cooled rapidly.

Liquid foods that contain large solid particles are heated in scraped-surface heat exchangers. These heat exchangers use blades to continuously scrape the inside surface of the heating chamber. The scraping action protects highly viscous foods from being burned on the heating surface.

An alternate method for heating foods, called ohmic heating, passes a low-frequency electric current of 50 to 60 hertz directly through the food. A liquid food containing solids, such as diced fruit, is pumped through a pipe surrounded by electrodes. The product is heated as long as the electrical conductivity of the food is uniform throughout the entire volume. This uniform rate of heating prevents the overprocessing of any individual region of the food. Ohmic heating yields a food product of higher quality than those processed using conventional systems.

13.13.2 Packaging Aseptically Processed Products

The packaging containers used in aseptic processing are sterilized separately before they are used. The packaging machinery is sterilized using steam, sterile gases, or hydrogen peroxide. The sterilization process is generally monitored by culturing a test organism. For example, the remaining presence of the highly heat-resistant bacterium Bacillus subtilis globigii can be used as a marker to measure the completeness of sterilization. Packages must be sealed under sterile conditions, usually using high-temperature sealing plates. Foods that are aseptically processed do not require refrigeration for storage.

13.14 BLANCHING

Blanching is a thermal process used mostly for vegetable tissues prior to freezing, drying, or canning. Before canning, blanching serves several purposes, including cleaning of the product, reducing the microbial load, removing any entrapped gases, and wilting the tissues of leafy vegetables so that they can be easily put into the containers. Blanching also inactivates enzymes that cause deterioration of foods during frozen storage.

Blanching is carried out at temperatures close to 100 °C (212 °F) for two to five minutes in either a water bath or a steam chamber. Because steam blanchers use a minimal amount of water, extra care must be taken to ensure that the product is uniformly exposed to the steam. Steam blanching leafy vegetables is especially difficult because they tend to clump together. The effectiveness of the blanching treatment is usually determined by measuring the residual activity of an enzyme called peroxidase.

13.15 CONTROLLING WATER ACTIVITY

Foods containing high concentrations of water are generally more susceptible to deterioration by microbial contamination and enzymatic activity. The water content of foods can be controlled by removing water through dehydration or by adding solutes to the food. In both cases the concentration of solutes in the food increases and the concentration of water decreases.

13.15.1 Dehydration

Dehydration, or drying, of foods has long been practiced commercially in the production of spaghetti and other starch products. As a result of advances made during World War II, the technique has been applied to a growing list of food products, including fruits, vegetables, skim milk, potatoes, soup mixes, and meats.

Pathogenic (toxin-producing) bacteria occasionally withstand the unfavourable environment of dried foods, causing food poisoning when the product is rehydrated and eaten. Control of bacterial contaminants in dried foods requires high-quality raw materials having low contamination, adequate sanitation in the processing plant, pasteurization before drying, and storage conditions that protect from infection by dust, insects, and rodents or other animals.

Foodstuffs may be dried in air, superheated steam, vacuum, or inert gas or by direct application of heat. Air is the most generally used drying medium, because it is plentiful and convenient and permits gradual drying, allowing sufficient control to avoid overheating that might result in scorching and discoloration. Air may be used both to transport heat to the food being dried and to carry away liberated moisture vapour. The use of other gases requires special moisture recovery systems.

Loss of moisture content produced by drying results in increased concentration of nutrients in the remaining food mass. The proteins, fats, and carbohydrates in dried foods are present in larger amounts per unit weight than in their fresh counterparts, and the nutrient value of most reconstituted or rehydrated foods is comparable to that of fresh items. The biological value of dried protein is dependent, however, on the method of drying. Prolonged exposure to high temperatures can render the protein less useful in the diet. Low-temperature treatment, on the other hand, may increase the digestibility of protein. Some vitamins are sensitive to the dehydration process. For example, in dried meats significant amounts of vitamin C and the B vitamins—riboflavin, thiamine, and niacin—are lost during dehydration.

Dried eggs, meat, milk, and vegetables are ordinarily packaged in tin or aluminum containers. Fibreboard or other types of material may be employed but are less satisfactory than metal, which offers protection against insects and moisture loss or gain and which permits packaging with an inert gas.

In-package desiccants (drying agents) improve storage stability of dehydrated white potatoes, sweet potatoes, cabbage, carrots, beets, and onions and give substantial protection against browning. Retention of ascorbic acid (vitamin C) is markedly improved by packaging at temperatures up to 49 °C (120 °F); the packaging gas may be either nitrogen or air.

A related technique, freeze-drying, employs high vacuum conditions, permitting establishment of specific temperature and pressure conditions. The raw food is frozen, and the low pressure conditions cause the ice in the food to sublimate directly into vapour (i.e., it does not transit through the liquid state). Adequate control of processing conditions contributes to satisfactory rehydration, with substantial retention of nutrient, colour, flavour, and texture characteristics.

13.15.2 Concentration of Moist Foods

Foods with substantial acidity, when concentrated to 65 percent or more soluble solids, may be preserved by mild heat treatments. High acid content is not a requirement for preserving foods concentrated to over 70 percent solids.

Fruit jelly and preserve manufacture, an important fruit by-product industry, is based on the high-solids–high-acid principle, with its moderate heat-treatment requirements. Fruits that possess excellent qualities but are visually unattractive may be preserved and utilized in the form of concentrates, which have a pleasing taste and substantial nutritive value.

Jellies and other fruit preserves are prepared from fruit by adding sugar and concentrating by evaporation to a point where microbial spoilage cannot occur. The prepared product can be stored without hermetic sealing, although such protection is useful to control mold growth, moisture loss, and oxidation. In modern practice, vacuum sealing has replaced the use of a paraffin cover.

The jelly-forming characteristics of fruits and their extracts are due to pectin, a substance present in varying amounts in all fruits. The essential ingredients in a fruit gel are pectin, acid, sugar, and water. Flavouring and colouring agents may be added, and additional pectin and acid may be added to overcome any deficiencies in the fruit itself.

Candied and glacéed fruits are made by slow impregnation of the fruit with syrup until the concentration of sugar in the tissue is sufficiently high to prevent growth of spoilage microorganisms. The candying process is conducted by treating fruits with syrups of progressively increasing sugar concentrations, so that the fruit does not soften into jam or become tough and leathery. After sugar impregnation the fruit is washed and dried. The resulting candied fruit may be packaged and marketed in this condition or may be dipped into syrup, becoming coated with a thin glazing of sugar (glacéed) and again dried.

13.16 FERMENTATION AND PICKLING

Although microorganisms are usually thought of as causing spoilage, they are capable under certain conditions of producing desirable effects, including oxidative and alcoholic fermentation. The microorganisms that grow in a food product, and the changes they produce, are determined by acidity, available carbohydrates, oxygen, and temperature. An important food preservation method combines salting to control microorganisms selectively and fermentation to stabilize the treated tissues.

13.16.1 Pickled Fruits and Vegetables

Fresh fruits and vegetables soften after 24 hours in a watery solution and begin a slow, mixed fermentation-putrefaction. The addition of salt suppresses undesirable microbial activity, creating a favourable environment for the desired fermentation. Most green vegetables and fruit may be preserved by pickling.

When the pickling process is applied to a cucumber, its fermentable carbohydrate reserve is turned into acid, its colour changes from bright green to olive or yellow-green, and its tissue becomes translucent. The salt concentration is maintained at 8 to 10 percent during the first week and is increased 1 percent a week thereafter until the solution reaches 16 percent. Under properly controlled conditions the salted, fermented cucumber, called salt stock, may be held for several years.

Salt stock is not a consumer commodity. It must be freshened and prepared into consumer items. In cucumbers this is accomplished by leaching the salt from the cured cucumber with warm water (43–54 °C [110–130 °F]) for 10 to 14 hours. This process is repeated at least twice, and, in the final wash, alum may be added to firm the tissue and turmeric to improve the colour.

13.16.2 Pickled Meat

Meat may be preserved by dry curing or with a pickling solution. The ingredients used in curing and pickling are sodium nitrate, sodium nitrite, sodium chloride, sugar, and citric acid or vinegar.

Various methods are used: the meat may be mixed with dry ingredients; it may be soaked in pickling solution; pickling solution may be pumped or injected into the flesh; or a combination of these methods may be used.

Curing may be combined with smoking. Smoke acts as a dehydrating agent and coats the meat surfaces with various chemicals, including small amounts of formaldehyde.

13.16.3 Deterioration of Fermented and Pickled Products

Fermented foods and pickled products require protection against molds, which metabolize the acid developed and allow the advance of other microorganisms. Fermented and pickled food products placed in cool storage can be expected to remain stable for several months. Longer storage periods demand more complete protection, such as canning.

Nutrient retention in fermented and pickled products is about equal to retention for products preserved by other methods. Carbohydrates usually undergo conversion to acid or to alcohol, but these are also of nutritive value. In some instances, nutrient levels are increased because of the presence of yeasts.

13.17 CHEMICAL PRESERVATION

Chemical food preservatives are substances which, under certain conditions, either delay the growth of microorganisms without necessarily destroying them or prevent deterioration of quality during manufacture and distribution. The former group includes some natural food constituents which, when added to foods, retard or prevent the growth of microorganisms. Sugar is used partly for this purpose in making jams, jellies, and marmalades and in candying fruit. The use of vinegar and salt in pickling and of alcohol in brandying also falls in This Category. Some Chemicals Foreign To Foods Are added to prevent the growth of microorganisms. The latter group includes some natural food constituents such as ascorbic acid (vitamin C), which is added to frozen peaches to prevent browning, and a long list of chemical compounds foreign to foods and classified as antioxidants, bleaching agents, acidulants, neutralizers, stabilizers, firming agents, and humectants.

13.17.1 Organic Chemical Preservatives

Sodium benzoate and other benzoates are among the principal chemical preservatives. The use of benzoates in certain products in prescribed quantity (usually not exceeding 0.1 percent) is permitted in most countries, some of which require a declaration of its use on the label of the food container. Since free benzoic acid actually is the active agent, benzoates must be used in an acid medium in order to be effective. The ability of cranberries to resist rapid deterioration is attributed to their high benzoic acid content. Benzoic acid is more effective against yeasts than against molds and bacteria.

Other organic compounds used as preservatives include vanillic acid esters, monochloroacetic acid, propionates, sorbic acid, dehydroacetic acid, and glycols.

13.17.2 Inorganic chemical preservatives

Sulfur dioxide and sulfites are perhaps the most important inorganic chemical preservatives. Sulfites are more effective against molds than against yeasts and are widely used in the preservation of fruits and vegetables. Sulfur compounds are extensively used in wine making and, as in most other instances when this preservative is used, much care has to be exercised to keep the concentrations low in order to avoid undesirable effects on flavour.

Oxidizing agents such as nitrates and nitrites are commonly used in the curing of meats.

13.18 FOOD IRRADIATION

Food irradiation involves the use of either high-speed electron beams or high-energy radiation with wavelengths smaller than 200 nanometres, or 2000 angstroms (e.g., X-rays and gamma rays). These rays contain sufficient energy to break chemical bonds and ionize molecules that lie in their path. The two most common sources of high-energy radiation used in the food industry are cobalt-60 (60Co) and cesium-137 (137Cs). For the same level of energy, gamma rays have a greater penetrating power into foods than high-speed electrons.

The unit of absorbed dose of radiation by a material is denoted as the gray (Gy), one gray being equal to the absorption of one joule of energy by one kilogram of food. The energy possessed by an electron is called an electron volt (eV). One eV is the amount of kinetic energy gained by an electron as it accelerates through an electric potential difference of one volt. It is usually more convenient to use a larger unit such as megaelectron volt (MeV), which is equal to one million electron volts.

13.18.1 Biological Effects of Irradiation

Irradiation has both direct and indirect effects on biological materials. The direct effects are due to the collision of radiation with atoms, resulting in an ejection of electrons from the atoms. The indirect effects are due to the formation of free radicals (unstable molecules carrying an extra electron) during the radiolysis (radiation-induced splitting) of water molecules. The radiolysis of water molecules produces hydroxyl radicals, highly reactive species that interact with the organic molecules present in foods. The products of these interactions cause many of the characteristics associated with the spoilage of food, such as off-flavours and off-odours.

13.18.2 Positive Effects

The bactericidal (bacteria-killing) effect of ionizing radiation is due to damage of the biomolecules of bacterial cells. The free radicals produced during irradiation may destroy or change the structure of cellular membranes. In addition, radiation causes irreversible changes to the nucleic acid molecules (i.e., DNA and RNA) of bacterial cells, inhibiting their ability to grow. Pathogenic bacteria that are unable to produce resistant endospores in foods such as poultry, meats, and seafood can be eliminated by radiation doses of 3 to 10 kilograys. If the dose of radiation is too low, then the damaged DNA can be repaired by specialized enzymes. If oxygen is present during irradiation, the bacteria are more readily damaged. Doses in the range of 0.2 to 0.36 kilograys are required to stop the reproduction of Trichinella spiralis (the parasitic worm that causes trichinosis) in pork, although much higher doses are necessary to eliminate it from the meat. The dose of radiation used on food products is divided into three levels. Radappertization is a dose in the range of 20 to 30 kilograys, necessary to sterilize a food product. Radurization is a dose of 1 to 10 kilograys, that, like pasteurization, is useful for targeting specific pathogens. Radicidation involves doses of less than 1 kilogray for extending shelf life and inhibiting sprouting.

13.18.3 Negative Effects

In the absence of oxygen, radiolysis of lipids leads to cleavage of the interatomic bonds in the fat molecules, producing compounds such as carbon dioxide, alkanes, alkenes, and aldehydes. In addition, lipids are highly vulnerable to oxidation by free radicals, a process that yields peroxides, carbonyl compounds, alcohols, and lactones. The consequent rancidity, resulting from the irradiation of high-fat foods, is highly destructive to their sensory quality. To minimize such harmful effects, fatty foods must be vacuum-packaged and held at subfreezing temperatures during irradiation.

Proteins are not significantly degraded at the low doses of radiation employed in the food industry. For this reason irradiation does not inactivate enzymes involved in food spoilage, as most enzymes survive doses of up to 10 kilograys. On the other hand, the large carbohydrate molecules that provide structure to foods are depolymerized (broken down) by irradiation. This depolymerization reduces the gelling power of the long chains of structural carbohydrates. However, in most foods some protection against these deleterious effects is provided by other food constituents. Vitamins A, E, and B1 (thiamine) are also sensitive to irradiation. The nutritional losses of a food product are high if air is not excluded during irradiation.

13.18.4 Safety Concerns

Based on the beneficial effects of irradiation on certain foods, several countries have permitted its use for specific purposes, such as the inhibition of sprouting of potatoes, onions, and garlic; the extension of shelf life of strawberries, mangoes, pears, grapes, cherries, red currants, and cod and haddock fillets; and the insect disinfestation of pulses, peanuts, dried fruits, papayas, wheat, and ground-wheat products.

The processing room used for irradiation of foods is lined with lead or thick concrete walls to prevent radiation from escaping. The energy source, such as a radioactive element or a machine source of electrons, is located inside the room. (Radioactive elements such as 60Co are contained in stainless steel tubes. Because an isotope cannot be switched on or off, when not in use it is lowered into a large reservoir of water.) Prior to the irradiation treatment, personnel vacate the room. The food to be irradiated is then conveyed by remote means into the room and exposed to the radiation source for a predetermined time. The time of exposure and the distance between the radiation source and the food material determine the irradiation treatment. After treatment, the irradiated food is conveyed out of the room, and the radioactive element is again lowered into the water reservoir.

Large-scale studies conducted around the world have concluded that irradiation does not cause harmful reactions in foods. In 1980 a joint committee of the Food and Agriculture Organization (FAO), the International Atomic Energy Agency (IAEA), and the World Health Organization (WHO) declared that an overall average dose of radiation of 10 kilograys was safe for food products. The maximum energy emitted by 60Co and 137Cs is too low to induce radioactivity in food. The energy output of electron-beam generators is carefully regulated, and the recommended energy outputs are too low to cause radioactivity in foods.

13.19 PACKAGING

Because packaging helps to control the immediate environment of a food product, it is useful in creating conditions that extend the storage life of a food. Packaging materials commonly used for foods may be classified as flexible (paper, thin laminates, and plastic film), semi-rigid (aluminum foil, laminates, paperboard, and thermoformed plastic), and rigid (metal, glass, and thick plastic). Plastic materials are widely used in food packaging because they are relatively cheap, lightweight, and easy to form into desired shapes.

The selective permeability of polymer-based materials to gases, such as carbon dioxide and oxygen, as well as light and moisture, has led to the development of modified-atmosphere packaging. If the barrier properties are carefully selected, a packaging material can maintain a modified atmosphere inside the package and thus extend the shelf life of the food product.

Dehydrated foods must be protected from moisture during storage. Packaging materials such as polyvinyl chloride, polyvinylidene chloride, and polypropylene offer low moisture permeability. Similarly, packaging materials with low gas permeability are used for fatty foods in order to minimize oxidation reactions. Because fresh fruits and vegetables respire, they require packaging materials, such as polyethylene, that have high permeability to gases.

Smart packages offer properties that meet the special needs of certain foods. For example, packages made with oxygen-absorbing materials remove oxygen from the inside of the package, thus protecting oxygen-sensitive products from oxidation. Temperature-sensitive films exhibit an abrupt change in gas permeability when they are subjected to a temperature above or below a set constant. These films change from a crystalline structure to an amorphous structure at a set temperature, causing the gas permeability to change substantially.

13.20 STORAGE

Food storage is an important component of food preservation. Many reactions that may deteriorate the quality of a food product occur during storage. The nutrient content of foods may be adversely affected by improper storage. For example, a significant amount of vitamin C and thiamine may be lost from foods during storage. Other undesirable quality changes that may occur during storage include changes in colour, development of off-flavours, and loss of texture. A properly designed food storage system allows fresh or processed foods to be stored for extended duration while maintaining quality.

The most important storage parameter is temperature. Most foods benefit from storage at a constant, low temperature where the rates of most reactions decrease and quality losses are minimized. In addition, foods containing high concentrations of water must be stored in high-humidity environments in order to prevent the excessive loss of moisture. Careful control of atmospheric gases, such as oxygen, carbon dioxide, and ethylene, is important in extending the storage life of many products.

14

PLANT CRYOPRESERVATION: CURRENT ISSUES

14.1 INTRODUCTION

Plant cryopreservation involves the storage of plant tissues (usually seed or shoot tips) in liquid nitrogen (LN) at -196°C or in the vapour phase of LN at -135°C in such a way that the viability of stored tissues is retained following re-warming. Cryopreservation is usually applied to species with recalcitrant (i.e. dehydration sensitive) seeds that are not storable by any other means, or preservation of specific cultivars of vegetatively propagated crop plants like banana or potato, or for unique ornamental genotypes. Another reason to utilise cryostorage is to conserve endangered plant species, particularly where seeds may be extremely scarce or of doubtful quality and/or the species is threatened with imminent extinction.

The main advantage of cryopreservation is that once material has been successfully cooled to LN temperatures, it can be conserved in principle indefinitely, because at these ultra-low temperatures no metabolic processes occur. Replenishing a small volume of LN weekly in cryo-dewars is the only on-going maintenance operation usually required in cryostorage. There are further advantages to this approach: the low costs of storage, minimal space requirements and reduced labour maintenance compared to living collections and even when compared to maintenance of tissue cultures at room temperature. Once in storage, there is no risk of new contamination by fungus or bacteria, and cryogenically stored material has been reported to retain genetic stability. Depending on the species, small cryopreserved samples may take several weeks to re-establish shoot cultures, and several months to a year may be required to produce micropropagated plants capable of transfer to soil under greenhouse conditions and (following weaning) into the field.

Shoot tips (containing the apical meristem) are the most commonly used plant material for cryostorage. The apical meristem is composed of small unvacuolated cells served with a relatively small vascular system. The organised structure of apical meristems generally results in direct shoot formation after re-warming, thereby maintaining the genetic integrity of the resulting propagated material. While callus tissue (unorganized wound tissue) can also be cry stored, the risk of occurrence of genetic deviations may be higher when utilising the indirect organogenesis pathway. Besides shoot tips, callus cultures, cell cultures, somatic embryos, pollen or plant buds as well as recalcitrant and orthodox seeds can be used as explants in plant cryopreservation.

Plant cryopreservation began with research on the freezing of mulberry twigs in LN. Since then, a huge variety of plants and genotypes have been successfully cryostored for conservation of agriculture and horticultural genotypes, as well as for endangered and threatened plant species. This chapter reviewsand gives examples of different plant cryopreservation protocols that have been successfully applied. It will focus on free radical damage and membrane structure, both important topics in the cryopreservation of biological tissues. The topic of genetic and epigenetic stability inplant cryopreservation is also discussed. Recent reviews of plant cryopreservation have been written by Benson and Reed.

14.2 PLANT CRYOPRESERVATION METHODS

In the last three decades a number of different cryopreservation protocols, such as classical slow-cooling, vitrification, droplet vitrification, encapsulation/ dehydration and encapsulation/ vitrification protocols have been developed and utilised for germplasm storage. The choice of cryopreservation method to attain the highest survival rates is largely dependent on the plant species and tissue type that is being cryostored.

14.2.1 Slow Cooling or Controlled Rate Cooling

This technique involves the simple dehydration of plant material before cryogenic storage in LN. This is can be done by slow cooling of the plant tissue to a temperature of approximately -40°C. This forces the formation of extracellular ice ahead of intracellular ice, thus causing an outflow of water from the cells due to the resulting osmotic imbalance and, consequently, dehydration. Dehydration can also be brought about by incubation of tissue material on media containing a relatively high concentration of an osmo-regulant, commonly sucrose, although other compounds can also be used. Usually water concentrations must be decreased to between 10% and 20% of the fresh weight for optimal cryogenic survival. This has the aforementioned effects of reducing the extent of ice crystal formation due to the reduced water concentration and assisting in the achievement of the vitrified state of water as a result of the increased solute concentration. These techniques do not necessarily make use of cryoprotective agents (CPAs), however they can be used in conjunction with them to further improve dehydration, though these agents can be toxic to plant cells at high concentrations. Rapid re-warming rates are used after cryogenic storage to prevent ice crystal formation during thawing. This approach can result in extreme rates of dehydration, which can cause cell volume reductions that are potentially lethal. It has been suggested that slow-cooling is only suitable for non-organised tissues, as sufficient dehydration is more difficult to achieve in tissues with complex structures due to the different rates of water movement between and within plant cells with different characteristics.

14.2.2 Encapsulation-Dehydration

This method, developed by Fabre, involves encapsulating shoot tips, somatic embryos or callus cells within alginate beads. This is followed by incubation in media with high sugar concentrations in order to raise intracellular solute concentrations and promote desiccation. Finally, silica gel or airflow is used to dehydrate the beads until the moisture content drops to 20-30%, before they are immersed in LN. The encapsulating material is thought to promote a vitrified state in the tissue regardless of the cooling and re-warming rates, thus reducing damage from ice crystal formation. Mechanical stress is also reduced because the bead protects the explants from damage during handling. The benefits of this method include avoiding the use of high concentrations of (potentially toxic) CPAs (Reinhoud et al., 2000) and the presence of a nutritive bead, which may enhance post-regeneration survival or re-growth of the material following immersion in LN and re-warming.

14.2.3 Vitrification

Vitrification involves the treatment of tissues in a mixture of highly concentrated penetrating and non-penetrating CPAs applied at non-freezing temperatures, followed by rapid cooling in LN. The combinationof high intracellular solute concentrations (due to dehydration and some CPA penetration) and rapid cooling prevents the nucleation of water and the formation of ice crystals, both intracellularly and extracellularly, thus promoting the vitrification of water.

For plants sensitive to direct exposure to vitrification solutions, due to dehydration intolerance and osmotic stresses, a loading step of 10-20 minutes can be incorporated prior to incubation within the CPA solution. This is done by incubation of the samples within a less toxic/concentrated CPA solution (a media containing 0.4 M sucrose and 2 M glycerol proved highly effective, thereby improving dehydration tolerance. The CPAs used in vitrification usually contain high concentrations of glycerol, dimethyl sulfoxide (DMSO), ethylene glycol and various sugars. The most commonly used mixture of CPAs for vitrification is plant vitrification solution 2 (PVS2), which consists of 30% (w/v) glycerol, 15% (w/v) ethylene glycol and 15% (w/v) DMSO in basal culture medium containing 0.4 M sucrose.

Exposure time to cryoprotective solutions is a vital step in the cryostorage process. Volk and Walters demonstrated that the extent of penetration of PVS2 into mint and garlic shoot tips was directly proportional to exposure time. The water content of the shoot tips also significantly decreased with an increase in exposure time to PVS2. Greater penetration of CPAs can be useful as it helps to increase the internal solute concentration and may contribute to maintaining cell volume, thus preventing damage to the cells. However, overexposure to CPAs may cause damage to the cells owing to the toxic nature of the CPAs or excessive dehydration.

The vitrification protocol is a more widely applied cryopreservation method than slow cooling due to its ease of use, high reproducibility and the wide range of species with complex tissue structure (such as shoot tips and embryos) that have been successfully cryopreserved with this procedure.

14.2.3.1 Droplet-vitrification

The droplet-vitrification technique is a modification of the basic vitrification protocol that involves placing the sample within a droplet of 1-10 μl of cryoprotective solution on a pieceof aluminium foil before immersion in LN (as opposed to 1-2 mL of cryoprotective solution in the original protocol), as shown in Fig. 1.

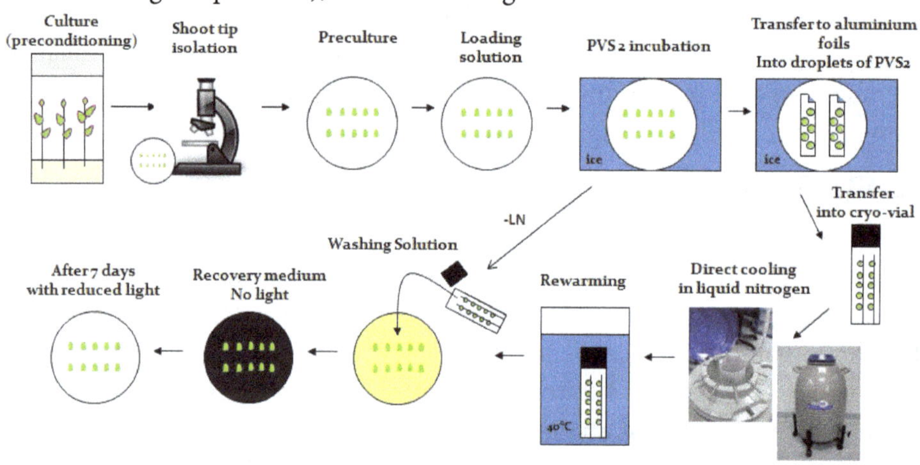

Fig. 1: Cryopreservation procedure for the droplet vitrification method.

This approach achieves higher cooling and re-warming rates, as the small volume of liquid allows a higher rate of heat transfer to and from the sample (Sakai & Engelmann, 2007). Cooling rates are increased to >130°C/sec (Panis & Lambardi, 2005), therefore facilitating the direct transition of intracellular water from a liquid state to a glassy state far more rapidly, thus minimising water crystallisation. The droplet-vitrification protocol has been successfully applied in the cryopreservation of garlic and chrysanthemum, yams, lily, potato and other plants.

14.2.3.2 Encapsulation-vitrification

Another modification of the vitrification approach termed encapsulation/ vitrification combines elements of the encapsulation/dehydration method with the vitrification method. As with the standard encapsulation method, shoot tips or calluses are first encapsulated in alginate beads and then the encapsulated material is incubated in a vitrification solution to promote sufficient dehydration and vitrification rather than dehydration under a constant airflow, which is very time consuming and relatively imprecise. Successful protocols have been established for potato, gentian, strawberry and pineapple.

Free radical and oxidative damage in plant cryopreservation

While achieving an optimum cryopreservation protocol that successfully avoids ice damage is important, there are various other factors that can affect post-cryogenic survival (Fig. 2). During the cryopreservation procedure plant tissues are susceptible to a variety of stresses, including oxidative stresses .

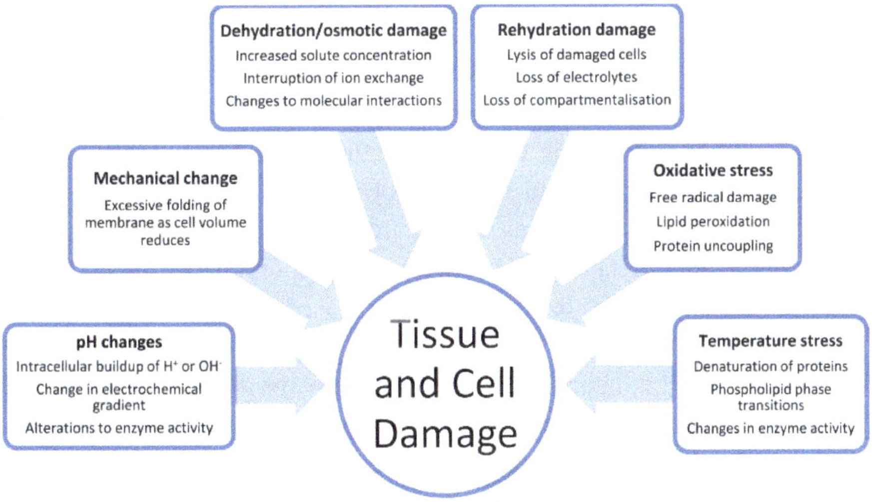

Fig. 2: Main causes of damage to plant tissues during cooling and cryopreservation (modified from).

Reactive oxygen species (ROS)

The formation of reactive oxygen species (ROS) during cryopreservation can occur during the many steps involved in this process. For example, ROS formation has been detected in photo- oxidative stress during tissue culture, during excision of shoot apices, osmotic injury and desiccation following application of CPAs, as well as during the rapid changes in temperature when the samples are first cryostored in LN and then re-warmed.

ROS are highly reactive molecules that can cause a wide range of damage in cells. There is a large variety of molecules that are classified as ROS, some of which include oxygen-free radical species and reactive oxygen non-radical derivatives (Table 1).

Table 1. Common reactive oxygen species (ROS).

Radicals	Non-Radicals
Superoxide (O2•-)	Hydrogen peroxide (H_2O_2)
Hydroxyl (OH•)	Peroxynitrite (ONOO-)
Hydroperoxyl (OOH•)	Peroxynitrite (ONOO-)
Peroxyl (ROO•)	Hypchlorous acid (HOCI)
Alkoxyl (RO•)	Hypobromous acid (HOBr)
Carbonate (CO3•-)	Ozone (O3)
Carbon dioxide (CO2•-)	Singlet oxygen (1 Δg) Singlet
Oxygen (1Σg+)	

Many of the more active ROS are free radicals, which are molecules that contain an unpaired electron, thus being able to react non-specifically with neighbouring molecules by removing electrons and causing a self-propagating chain reaction of radical formation. The removal of electrons can lead to a loss of function and structural alterations in macromolecules like proteins, lipids and DNA. ROS are frequently produced as by-products during cellular metabolism. The electron transport chain used in respiration and photosynthesis are the major producers of ROS, caused by the leakage of free electrons onto molecular oxygen (O_2), resulting in the formation of superoxide. The formation of ROS is controlled by a high concentration of antioxidants and proteins that can quench the ROS and fix the damage in these regions. Nevertheless, if there is a sudden increase in ROS formation then cellular repair processes can be overwhelmed and excessive damage can occur.

Temporarily reducing cryo-sample exposure to light immediately after cryopreservation has been shown to increase post-cryogenic survival due to the removal of photo-oxidative stress in the plants. Photo-oxidative stress in plants can result in high levels of singlet oxygen (1Σg+) and superoxide (O2•-) being produced, either from direct UV radiation on oxygen or the leakage of light energy onto oxygen from chlorophyll when the carotenoid pigments become saturated. Plants are highly susceptible to photo-oxidative stresses at low temperatures when exposed to strong light conditions. This can be demonstrated in alpine plants, many of which display adaptations (especially to leaves, i.e. the production of carotenoid pigments) that reduce photo-oxidation damage from enhanced UV-B radiation at high altitudes. When the ability of antioxidants to quench the formation of ROS and the recycling of antioxidants is reduced, greater damage occurs to the chloroplast through lipid peroxidation, inactivation of photosynthetic proteins and loss of pigments (i.e. bleaching). Damage to the chloroplast during chilling stress has been shown to severely impede growth rates; therefore, reducing the damage that occurs to plant cells due to low temperature oxidative stress is vitally important for improving survival and recovery in cryopreservation.

The most reactive ROS commonly found in plants include superoxide (O2•-), the hydroxyl radical (OH•), hydroperoxyl (OOH•) and singlet oxygen (1Σg+). Superoxide and singlet oxygen are often formed as by-products of the electron transport chain from both metabolism and photosynthesis, while hydroperoxyl and hydroxyl radicals are commonly formed in a process called Fenton's reaction, where hydrogen peroxide is converted into the hydroxyl or hydroperoxyl radical.

$$Fe^{2+} + H_2O_2 \rightarrow Fe^{3+} + OH\bullet + OH^-$$

$$Fe^{3+} + H_2O_2 \rightarrow Fe^{2+} + OOH\bullet + H^+ \qquad (1)$$

The formation of the hydroxyl radical is the major cause of lipid peroxidation in membranes, but can also cause a wide range of damage to all cellular components, including proteins and DNA. The addition of specific chelating agents (such as desferrioxamine) has been shown to reduce the levels of iron in cryopreserved tissues, with subsequent decreased levels of the hydroxyl radical observed.

Damage caused by ROS is difficult to quantify as these molecules are non-specific in their interactions, reacting freely with lipids, proteins and DNA. ROS are highly reactive (and therefore short-lived) and thus, direct measurement of the ROS present in cells is difficult due to time constraints, and does not reflect the damage that may be done prior to the ROS being quenched by antioxidants. Consequently, it is easier to measure the formation of by- products of oxidative damage or the antioxidant status of the cells. The ratio of oxidised to reduced antioxidants is a good indication of ROS formation and the ability of cells to regulate oxidative stress. Identifying end products of ROS oxidation is an indication of the damage caused and is a sign that cells have been unable to satisfactorily quench ROS activity.

Lipid peroxidation

The cell membrane represents one of the major areas where cryo-injury can occur. Any damage to the cell membrane can alter the delicate balance between intra and extracellular solutes, leading to cell death (Anchordoguy et al., 1987; Dowgert & Steponkus, 1984; Gordon-Kamm & Steponkus, 1984; Lynch & Steponkus, 1987). Lipid peroxidation of fatty acids (FA) in phospholipids can cause extensive damage to the cell membrane if the chain reaction is not controlled, leading to large areas where the semi-permeability of the membrane is altered and thus can no longer function normally. Lipid peroxidation is caused when specific ROS (hydroxyl radical, peroxyl radical and singlet oxygen) interact with a FA. Polyunsaturated fatty acids (PUFA) are the most susceptible to peroxidation. Glutathione peroxidase can detoxify lipid peroxides by reducing them back to their lipid alcohol form and, in the process, oxidising glutathione. The formation of the toxic end-products of lipid peroxidation, such as malondialdehyde (MDA) and 4- hydroxylnonenal (HNE), can also cause damage to cells. The toxicity of MDA is debatable; however, there is evidence that MDA can interact with proteins and DNA, causing loss of function in proteins and mutations in DNA. The formation of HNE has shown greater toxicity to cells as it can damage mitochondria, inhibit synthesis of DNA and proteins, and interfere with the action of repair proteins such as chaperones. Identifying the formation of MDA and HNE is commonly used in cryopreservation as an indicator of oxidative stress. High levels of MDA or HNE detected have correlated with decreased survival rates in rice cell suspensions, olive somatic embryos, flax and blackberry shoot tips.

Volatile headspace sampling (VHS) measures the formation of volatile compounds released from a sample in a non-destructive and non-invasive assay method. This provides an important tool for measuring oxidative stress. Free radical damage can cause the formation of volatile compounds such as methane, ethane, ethylene and pentane. Quantification of these compounds can be indicative of oxidative damage. The detection of ethylene in plants is of particular importance as ethylene is a vital hormone. Fang et al. (2008) observed that decreased levels of ethylene production correlated with decreased survival and growth. The production of the other volatile compounds is indicative of lipid peroxidation, where increased levels correlate to excessive oxidative damage. observed a large increase in volatile compounds produced after thawing that has since been observed in other species.

The use of DMSO as a free radical scavenger and probe for the hydroxyl radical can be utilised with VHS. The formation of methane when DMSO interacts with the hydroxyl radical can be measured if the sample is placed in an airtight container, with the levels of methane detected correlating with the formation of the hydroxyl radical. This technique has been used as a measurement of oxidative stress in multiple different plant species, such as rice, cocoa, Daucus carota and flax (Benson & Withers 1987; Benson et al., 1995; Obert et al., 2005; Fang et al., 2008). The production of methane is particularly strong during the initial phase of recovery, where it is predicted that antioxidant activity and production is reduced, resulting in increased ROS (Fang et al., 2008). The use of chelating agents to reduce the formation of hydroxyl radicals from the Fenton reaction has shown significant benefits in plant cryopreservation. Desferrioxamine is the most common chelating agent used. It binds to and reduces the amount of free iron. The addition of desferrioxamine to rice cells during cryopreservation showed a decreased recovery period after cryopreservation (Benson et al., 1995). Detection of methane formation was delayed in flax tissue when exposed to desferrioxamine, indicating a delayed production of hydroxyl radicals.

Antioxidants

An antioxidant can be defined as any molecule that "delays, prevents or removes oxidative damage to a target molecule" (Halliwell & Gutteridge, 2007). Antioxidants can be classified into two major groups: enzymes that catalytically remove ROS, or sacrificial antioxidant molecules that are preferentially oxidised to protect more important molecules by quenching ROS (Halliwell & Gutteridge, 2007). The addition of exogenous antioxidants during cryopreservation has been shown to result in increased survival in some cases (Uchendu et al., 2010; Wang & Deng, 2004). It is possible that the addition of exogenous antioxidants may also aid in reducing oxidative stresses immediately following re-warming, when cellular metabolism has not been restored to its original state.

Glutathione (GSH) is one of the main water soluble sacrificial antioxidants in plants. This low molecular weight thiol is converted to its oxidised form (GSSG) upon interaction with ROS. The ratio of reduced to oxidised GSH is a good indicator of oxidative stress, as an increased amount of GSSG indicates the inability of the plant to control oxidative damage, thereby triggering premature cell death. GSH is recycled by the enzyme GSH reductase using NADPH as the electron donor. Ascorbic acid is the most abundant water-soluble antioxidant in plant cells. This antioxidant is able to quench free radicals, forming the stable radical semidehydroascorbate, which can be further oxidised to dehydroascorbate by another free radical. Arabidopsis mutants, which were unable to express ascorbic acid activity, demonstrated the important role this antioxidant plays in reducing oxidative stress, as these plants were not viable following exposure to photo- oxidative stress. Tocopherol (Vitamin E) is a lipophilic antioxidant. This is the main antioxidant involved in membrane protection against lipid peroxidation, as it is preferentially oxidised instead of PUFAs. This antioxidant is thought to be reduced to its original, functional state by ascorbic acid. Carotenoid pigments are the main defence in chloroplasts, where large amounts of singlet oxygen can be produced if the activated chlorophylls transfer their energy onto oxygen. The carotenoid pigments can absorb the energy directly from the chlorophylls, thus suppressing the formation of singlet oxygen, and they can also quench the singlet oxygen directly.

The enzyme superoxide dismutase (SOD) catalytically removes the ROS superoxide, producing oxygen and hydrogen peroxide. The removal of superoxide is more important than the formation of hydrogen peroxide, as superoxide is a more reactive species, causing wider damage in the cells. There is potential for SOD to cause formation of ROS if levels of hydrogen peroxide are not controlled. SOD contains a metal cofactor that can cause Fenton reactions and the formation of the hydroxyl radical. Catalase is the main enzyme involved in removing hydrogen peroxide, resulting in the decomposition of hydrogen peroxide to water and oxygen. This enzyme is vital for the removal of hydrogen peroxide before it can damage the cell or be converted to the highly reactive hydroxyl radical through Fenton's reaction.

Plant cryopreservation and membrane structure

Membrane systems within cells are usually the site of freezing injury in plants. Membrane stability is therefore important for reducing such injury. There are four types of injury: (i) expansion-induced lysis, where the cells overexpand as a result of increased extracellular osmotic pressure during warming/thawing; (ii) loss of osmotic responsiveness, where there is no osmotic change during warming due to a slow cooling rate (cells remain dehydrated); (iii) altered osmotic behaviour, where cells membranes turn "leaky", resulting in the release of water and solutes into the surroundings; and (iv) intracellular ice formation, where rapid cooling causes membrane disruption due to the formation of ice crystals.

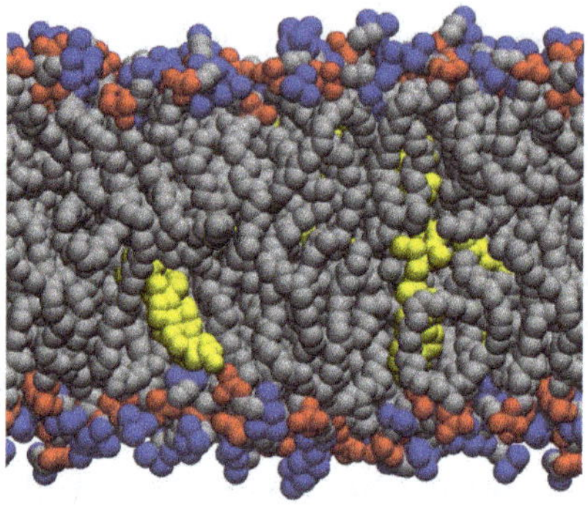

Fig. 3: Typical cell membrane structure consisting of a phospholipid bilayer with embedded sterols. Phospholipid chains are shown in grey, choline groups in blue and phosphate groups in red, while sterol molecules are shown in yellow.

The cell membrane is a bilayer consisting of different lipids and associated proteins (Fig. 3), where the lipids define the cell membrane structure and fluidity, and have a function in signal transduction. The three main classes of lipids found in cell membranes are glycerolipids (mostly phospholipids), sterols and sphingolipids.

Phospholipids

Phospholipids are amphiphilic molecules that form the core structure of the cell membrane lipid bilayer. They consist of a polar head (containing a phosphate group and simple organic molecule) and a (mostly) non-polar fatty acid tail. The most common phospholipid components of membranes include phosphatidylcholine (PC), phosphatidylethanolamine (PE), phosphatidylglycerol (PG), phosphatidylinositol (PI), phosphatidylserine (PS) and phosphatidic acid (PA). Phospholipids essentially maintain the structure, fluidity and permeability of the membrane, determined to some extent by the type of phospholipid present and the level of unsaturation in their fatty acid tails.

Sterols

Sterols are steroid alcohols that are integral components of living cells. They are formed as intermediate molecules during the metabolic production of hormones and are an integral component of cell membranes. Sterols are non-polar molecules with a polar hydroxyl (-OH) side chain that allows them to interact with the polarand non-polar groups of the phospholipid bilayer. They restrict the motion of the fatty acid chains in phospholipids, thus controlling the fluidity of the cell membrane. The major plant sterols are cholesterol, campesterol, sitosterol and stigmasterol. Sitosterol is usually found in higher concentration in apex tissue than stigmasterol. The composition and concentration of these sterols within the membrane modify its permeability and fluidity. Cholesterol has the greatest stabilising effect on membranes due to its small side chain.

The ability of sterols to alter stability and permeability of membranes can have a large effect on post-cryopreservation survival of plant tissue. The ratio of stigmasterol to sitosterol has been found to increase after sucrose preculture and was unfavourable to obtaining high percentages of shoot regeneration after cryopreservation of banana meristems. It is investigated the interactions of sitosterol and stigmasterol with phosphatidylcholine molecules in soybean and discovered, using neutron scattering experiments, that sitosterol has a greater effect than stigmasterol on the ordering of the fatty acyl chains of PC and increasing the hydrophobic thickness of PC bilayers. Cold acclimation of winter rye (Secale cereale) seedlings showed an increase in free sterol content, with β-sitosterol having the largest increase. It also found an increase of β-sitosterol with cold acclimation in winter rye; however, these results were not mimicked in spring oats (Avena sativa L. cv Ogle), where there was no significant change in β-sitosterol, but the stigmasterol proportion increased, whilst cholesterol decreased.

Soluble sugars

Intracellular soluble sugars and sugar alcohols, such as the ones used in preculture media, reduce damage sustained by cell membranes when the cells undergo desiccation and can improve membrane stability. Smaller sugar molecules help membranes to osmotically retain water and may enter the interlamellar space to maintain a degree of hydration and increase the separation between membranes, thus reducing compressive stresses and, consequently, reducing the chance of a fluid-gel phase transition (Wolfe & Bryant, 1999). Furthermore, the polar hydroxyl (-OH) groups on sugars and sugar alcohols have been thought to interact with the polar membrane phospholipid headgroups and stabilise the membranes by maintaining the separation of the phospholipid molecules. Turner et al. (2001) tested several sugars and sugar polyalcohols and determined that the small size of molecules such as glycerol and erythritol allowed them to 'pack' around membranes and have better bonding abilities. Additionally, the stereochemical arrangement of the -OH groups, particularly their orientation along one side of the molecules, imparted more stable hydrogen bonds with the membrane phospholipids and, consequently, resulted in more stable membranes. Nonetheless, recent biophysical studies have established that specific interactions of sugar molecules to phospholipid headgroups are not necessary to explain the stabilising effect of sugars on membrane gel to fluid transition temperatures and that sugars do not in fact insert between phospholipid headgroups but instead a solvation layer of water molecules separates them from the phospholipid headgroups.

Preconditioning and cold acclimation

Cold acclimation is the process in which plants being exposed to low non-freezing temperatures increase their freezing tolerance. Preconditioning of raspberry and blueberry plants at 22/-1°C alternating temperatures for four weeks was essential for optimal post-cryopreservation shoot-tip re- growth using encapsulation-dehydration and vitrification protocols. Similarly, better recovery rates were seen in mint and yams when they were cold acclimated for several weeks before cryostorage.

Cold acclimation is thought to activate genes that improve plant survival at low temperatures by improving stability in membranes. Cell membrane phospholipids and sterols have been observed to increase in proportion upon cold acclimation in winter rye, with a particular increase in di-unsaturated PC and sitosterol. Changes in phospholipid and sterol composition were found in Arabidopsis thaliana after cold acclimation at 2°C for one week, which increased its freezing tolerance from -2°C down to -10°C. Cell membrane phospholipid changes were also observed in oat leaves. These changes may be related to improved membrane stability in these plants.

Preconditioning of plants has also been shown to increase antioxidant levels in plants prior to cryopreservation Baek and Skinner (2003) analysed the expression of antioxidant genes in wheat species after cold acclimation and found increased expression of antioxidant enzymes, such as SOD and catalase. Increased levels of antioxidants may allow plants to better tolerate oxidative stress. Lynch and Steponkus (1987) observed an increase in the di- unsaturated levels of PC and PE in winter rye seedlings. Sucrose pre-treatment of banana meristems prior to cryopreservation increased survival after warming and was related in most cases to a decrease of the double bond index in phospholipids, free fatty acids, glycolipids and sphingolipids.

Genetic and epigenetic stability

The aim of successful cryopreservation is to maintain genetically stable plant material. While cryopreservation is now recognised as the method of choice for the long term preservation of plant material, the usefulness of cryostorage only applies if it does not lead to genetic changes in the plant species of interest. It is thus recommended to avoid the use of tissue in a non-organised dedifferentiated state, such as callus, and to use organised tissue like shoot tips instead to reduce the likelihood of non- desirable genetic mutations as well as due to their higher regrowth rates. Cryopreservation can cause injury at cellular level, but it is not clear if this injury can change the genetic composition of plants. The genetic stability of cryopreserved plants has nonetheless been confirmed for most of the analysed samples at morphological, histological and molecular level (Harding, 2004). Where differences between control and cryopreserved genotypes were found, it was suggested that the genetic changes might not be associated with the cryopreservation process itself, but rather that they are caused by the overall tissue culture, cryoprotection and regeneration process.

Comparisons of morphological development and analyses of the characteristics of control and cryopreserved plants have shown no differences in many species, such as Prunus, sugarcane, onion, kiwi, Eucalyptus, coffee, Dendrobium and Cosmos, as well as in the hybrid aspen (Populus tremula), an economically important woody plant and widely used model forest plant. Alterations in phenotype related to flower colouring have been observed in Chrysanthemum, which might be related to the chimeric structure of the plant. Morphological and phenotypic studies in potato, where shoot tips were used for cryopreservation, showed stable genetic integrity. Biometric studies examining morphological characters, agronomic traits or vegetal development descriptors in Dioscorea floribunda, sugarcane and banana revealed no significant differences between cryopreserved-derived and control plants. No abnormalities in chromosome number or cell structure were observed in cryopreserved Vanda pumila (Orchidaceae), with the regrown shoot primordia being able to induce new meristematic tissues like those of the non-cryopreserved controls. Long term storage of cryopreserved plants, such as strawberry, pea, Rubus and potato, did not result in any overall changes in regeneration capability and phenotype when regenerated explants were compared at the time point of storage as well as 10, 12 and 28 years later.

Histological studies and cytological analysis using flow cytometry have confirmed the genetic stability of plant species such as pea, oil palm, silver birch, Rubus, Solanum tuberosum and rice. Biochemical analyses have compared products of gene expression such as the formation and concentration of secondary metabolites. Examples of compounds compared in cryopreserved and control plants have been diosgenin in Dioscorea floribunda, chlorophyll and pyrethrin in Chrysanthemum, hypericin production in Hypericum perforatum L. betalanin pigments in Beta vulgaris and nicotin alkaloids in Nicotiana rustica, which were all unchanged in cryopreserved plants and thus confirmed the integrity of metabolic functions after cryostorage. Similar stability was observed upon comparison of proteins and enzymes.

A variety of different techniques and markers have been applied to compare genomic DNA patterns, such as restriction fragment length polymorphism (RFLP), randomly amplified polymorphic DNA (RAPDs) fragments, simple sequence repeat (SSR) analysis and amplified fragment length polymorphism (AFLP). Most studies have confirmed the presence of genetic stability and where changes in the genome have been found, such as in sugarcane and potato with RFLP markers, the changes could not be related to the process of cryopreservation itself.

In contrast to genetic variations manifested by DNA nucleotide sequence alterations, epigenetic changes do not change the original DNA sequence but can result in heritable changes of gene expression. Typical features of epigenetic characteristics are DNA methylation, histone modification and changes in chromatin structure. Epigenetic gene regulatory mechanisms have a function in plant development and might be influenced or changed by environmental conditions and osmotic stress during tissue culture and cryopreservation. Some recent studies have analysed epigenetic characteristics like DNA methylation in tissue culture and cryopreserved plants. Modifications in DNA methylation have been found in almond, papaya), chrysanthemum, Ribes, strawberry, citrus and potato. Changes in methylation might be caused by stressful in vitro conditions, osmotic dehydration and the application of cryoprotective agents.

15

IMPACT OF CRYOPRESERVATION ASSOCIATED DAMAGE ON DIFFERENT PARAMETERS OF HUMAN SPERMATOZOA

15.1 INTRODUCTION

The cryopreservation of gametes, as well as more complex multicellular organisms such as embryos, is an increasingly common practice in the field of assisted reproduction technology (ART) and reproductive medicine. Cryopreservation is described as the process of long-term cell preservation and storage at cryogenic temperatures (-196 °C), thus maintaining cell viability by means of hampering ageing and cell degradation processes. Recent advances in the field of cryobiology have not only led to the development of novel fertility treatments, such as fertility preservation and gamete/embryo donation, but also helped co-ordinate gamete retrieval and fertilisation measures. Since its early beginnings in the 1970s, human sperm cryopreservation has been demonstrated to be a crucial aid in ART. Currently, many male patients attend sperm cryopreservation programmes for fertility preservation purposes.

The cryopreservation and storage of human ejaculated spermatozoa is commonly facilitated in groups of male patients reporting difficulties in semen sample collection, such as in cases of erectile dysfunction and lack of or retrograde ejaculation. Additionally, to schedule oocyte retrieval and semen sample collection, sperm cryopreservation is counselled in those male patients who will not be able to collect the sample the day of ovarian puncture. Furthermore, sperm cryopreservation should also be considered in cases of azoospermia to prevent repeated sperm retrieval surgeries. Similarly, cryopreservation is also recommended for groups of male patients who have impaired semen parameters, or in cases in which there is an observed, progressive decrease in sperm quality, to prevent the risk of azoospermia.

Sperm cryopreservation has been widely used for assisted reproductive technology (ART). Indications for sperm cryopreservation include donor insemination, cryopreservation prior to surgical infertility treatment, and malignancies to avoid additional surgery in couples undergoing repeated in vitro fertilisation (IVF) or intracytoplasmic sperm injection (ICSI) cycles. However, dramatic changes during cryopreservation have detrimental effects on the sperm membrane, resulting in a large increase in the percentage of poorly motile sperm or sperm with abnormal morphology. The negative effects related to rapid temperature decrease, such as osmotic injury, cellular dehydration, intracellular ice crystal formation, and oxidative stress can also damage the sperm in ways that affect reproductive outcome. This comprehensive chapter focusses on describing the detrimental effects of the cryopreservation process on sperm and aims to clarify that not all impaired sperm parameters have the same impact on the clinical practice of ART. Regarding the parameters studied, some of the biomarkers used for sperm maturity, hyaluronic acid binding capacity, or damaged DNA have limited clinical significance compared to other semen parameters which provide more useful information for clinical practice and are often dismissed, such as total motility or total motile sperm count (TMSC). In the authors' experience, TMSC gives valuable quantitative information about the number of viable spermatozoa. Indeed, TMSC should be assessed specifically for groups of patients in which sample availability is limited.

Additional to these purposes, male fertility preservation is particularly relevant in oncologic patients, or in patients reporting chronic non-malignant diseases or autoimmune disorders. The gonadotoxic effect of treatments is the main reason why fertility preservation is counselled in such patients, since chemo/radiotherapy regimens as well as other medications for chronic diseases are a real threat to fertility potential. Indeed, in some cases sperm parameters are impaired even prior to any treatment. This could be in response to the direct impact of the tumour on the male genital tract, as is observed in testicular cancers. Additionally, indirect effects inherent to oncological processes could also be a plausible cause of sperm parameter impairment in these patients.

As the survival rates of patients living with malignant or non-malignant chronic disorders are increasing as a result of the new therapeutic options available to them,5 there is increasing interest from young male patients of reproductive age for fertility preservation. Since long-term quality of life following cancer treatment has become more prevalent in society, initiating semen cryopreservation before gonadotoxic treatments is seen as an effective means of maintaining male reproductive potential. A study investigated the growing trend to cryopreserve the sperm of pubertal boys and young adults being treated for leukaemia, lymphomas, or other types of germ cell tumours.Moreover, it is worth mentioning that the implementation of sperm cryopreservation protocols enables the establishment of donor sperm banks and, consequently, donor sperm programmes.

Unfortunately, despite the many strengths of sperm cryopreservation, important damages to the sperm cells are still reported. Furthermore, compared to other methods there is currently no standardised cryopreservation protocol for human spermatozoa that leads to higher viability yields after thawing. The chapter provide comprehensive information regarding the state-of-the-art human spermatozoa cryopreservation techniques, highlighting deleterious effects associated with freezing and vitrification protocols.

15.2 METHODS

15.2.1 Cryopreservation of Human Spermatozoa

The cryopreservation of sperm cells is not a harmless process; indeed, the significant damage of spermatozoa due to freezing and vitrification protocols is well-described. Cryoinjury, which is the damage, associated with cryopreservation, leads to an impairment of sperm parameters. In addition, as reported that the ageing process is not totally prevented by cryopreservation. They suggest that during cryopreservation the telomere-shortening phenomenon is observed, as well as an induction of cellular senescence.

15.2.2 Changes in Sperm Cell during Cryopreservation

The sperm cell membrane is a vital structure during the freezing and vitrification processes since cell survival is strongly dependent on its viability, especially during the temperature decrease and following the return to physiological temperature by thawing. The sperm cell membrane exhibits as a fluid mosaic-like structure, presenting a phospholipid bilayer with protein and carbohydrates dispersed throughout. The respective stiffness or fluidity of the sperm cell membrane is determined by the fatty acids present in the phospholipid bilayer.

The membranes of human spermatozoa are characterised by their large number of polyunsaturated fatty acids (PUFA), which are responsible for membrane fluidity. Unsaturated lipids, together with the presence of cholesterol, impair the phospholipid packaging, thereby decreasing the number of possible interactions between phospholipids. As a result, membrane fluidity is increased due to steric effects. Increased sperm cell membrane fluidity ensures a better response to the cryopreservation process,8,9 and therefore a large number of PUFA translates into a greater protection against cryoinjury during cryopreservation.10 As well as fluidity, permeability is another key feature of the sperm cell membrane. Permeability of the sperm cell membrane will continually decrease during the freezing stage of cryopreservation to the point where the membrane becomes an impermeable structure; hence, all fluids retained in cell cytoplasm will become vulnerable to crystalisation.

A crucial parameter during cryopreservation is the freezing rate. During the temperature decrease from physiologic temperature (37 °C) to 0 °C, a temperature shock occurs. This initial temperature decrease promotes the change of phase of membrane phospholipids from the fluid to gel phases. As the phase exchange is not simultaneous across the whole phospholipid bilayer, both phases can be observed before completion. The coexistence of both phases, fluid and gel, leads to an impaired phospholipid arrangement, increasing the cell's permeability to solutes. The use of cryoprotectant agents and a progressive freezing rate minimises this damage during the first freezing step.

The following temperature decrease below 0 °C promotes intracellular ice crystal formation. It is therefore paramount to induce a major degree of cell dehydration, so as to minimise the water content inside the sperm cell, because ice crystals cause physical damage to organelles and cell membranes. A specific, optimum freezing rate is associated to each cell type in which cryoinjury is minimised. Supra-optimal freezing rates promote ice crystal formation as a result of incomplete cell cytoplasm dehydration, whereas osmotic damage due to the high solute concentration in the extracellular medium is associated with sub-optimal freezing rates.

In parallel to the temperature decrease, a calcium influx towards the sperm cell cytoplasm is observed. Under this circumstance, phospholipids present in cryoprotectant agents contribute to membrane resistance to calcium shock, and therefore an appropriate balance in calcium levels is important to ensuring good post-thawing recovery. As well as the temperature decrease, high calcium levels in the extracellular media help promote cellular damage during the cryopreservation procedure.

15.2.3 Assessment of the Cryopreservation Process

Cryopreservation-associated damage to the spermatozoa not only alters the cell membrane structure, but also the cell's metabolism and mitochondrial bioenergetic processes.16,17 After thawing, protein degradation and phosphorylation have been reported.18 Interestingly, capacitation-like changes to spermatozoa are associated with protein phosphorylation. It is noteworthy that the lifespan of spermatozoa exhibiting capacitation-like features is shortened and, in combination with the decrease in spermatozoa with intact acrosomes after thawing, this fact would reinforce the hypothesis that the fertility potential of frozen-thawed spermatozoa is threatened.

All these cited changes have an impact on sperm parameters, such as decreasing motility or altered morphology. Furthermore, as well as mitochondrial damage, impaired motility, and morphology, sperm DNA integrity can be affected during cryopreservation. As Figure 1 shows, the cited parameters are the main targets of cryodamage.

15.2.4 Sperm Morphology

The effect of cryopreservation on sperm morphology has been the subject of a wide range of studies. Some studies demonstrate a decrease in normal morphological forms in the semen sample after thawing. When slow freezing is considered, it is suggested that the osmotic shock caused by cryoprotectant agents required in this protocol would be the main reason for observing altered morphological forms in the thawed samples, whereas when vitrification is the chosen cryopreservation protocol, the impairment of normal morphology would be associated to extracellular ice crystal formation.

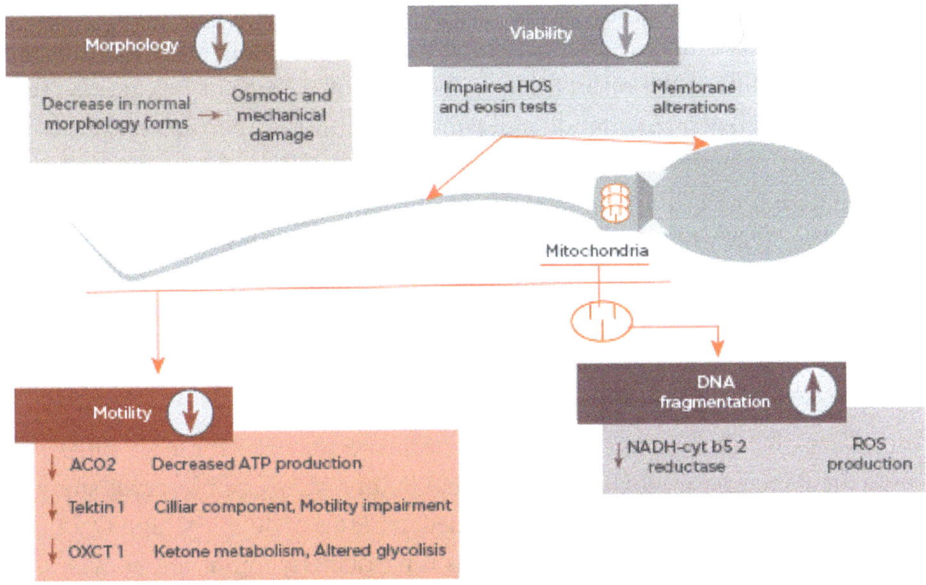

Fig. 1: Main sperm parameters affected by cryodamage and its causes. ATP: adenosine triphosphate; HOS: hypo-osmotic swelling; ROS: reactive oxygen species.

15.2.5 Sperm DNA Integrity

Cryopreservation-associated alterations to sperm DNA are mainly the result of two processes: abortive apoptosis and oxidative stress caused by reactive oxygen species (ROS). ROS production could be the main cause of damage to sperm DNA during cryopreservation. With regard to ROS production, endogenous ROS release is common in human spermatozoa because superoxide anion dismutation leads to hydrogen peroxide production. The cytotoxic effects of hydrogen peroxide are well-established, as seen by its high oxidant activity. In this sense, as has been suggested previously high levels of glutathione-peroxidases could be a biomarker for success after cryopreservation procedures since their activity consists of eliminating DNA damaging free radicals. In parallel, other components found in seminal plasma, such as antioxidant enzymes, prevent ROS release. Similarly, the protective role of albumin against endogenous ROS production has been reported. Albumin's protective role involves its capability for binding lipid peroxides and neutralising their cytotoxic effects. However, the clinical impact of cryopreservation-associated DNA fragmentation is still a debatable issue, as suggested in recent reviews and meta-analyses. Differences among samples regarding cryoresistance, protocols for cryopreservation, and the wide range of methods used to assess DNA integrity make the agreement of a consensus difficult.

The meta-analysis carried out puts forward the idea of an increased risk of miscarriage being associated with the use of high-fragmented spermatozoa. Unfortunately, as the authors also commented, there are design flaws in the studies included in the meta-analysis. Firstly, the threshold values for DNA fragmentation index varies among the studies, from 10–36%. Even using a high sperm fragmentation index did not observe any differences in clinical outcomes, regardless of the IVF technique used and the origin of the oocytes. In addition, the results clearly show a dependence of risk ratio's ranges on the test used for DNA fragmentation assessment, and thus important variations among the risk ratios are observed when different DNA fragmentation assays are used. Another common limitation regarding the assessment of sperm DNA fragmentation and clinical outcomes is the cohort of oocytes included. An elegant study defends the premise that oocyte quality is a major factor to consider due to the ability of good-quality oocytes to conditionally repair DNA damage. Therefore, this should be considered when it is stated that sperm DNA damage worsens reproductive outcomes through an increase in miscarriage rate in clinical studies including poor quality oocytes.

Addressing the issue of using frozen or fresh spermatozoa for ART purposes, it is reported a slight increase in single-stranded DNA lesions when cryopreserved spermatozoa were used, but no difference could be established, in terms of double strand DNA damage, between fresh and frozen sperm samples. This is particularly important because different sorts of DNA damage have different clinical outcomes. Consequently, double-stranded DNA breakages are associated with miscarriages, while single-stranded DNA lesions have no further clinical impact except a putative decrease in fertilisation rate and a delay on pregnancy achievement. Moreover, comparative studies of different cryopreservation protocols fail to find differences in sperm DNA integrity among the protocols studied. More remarkable is that most of these studies reported a high degree of DNA integrity, superior to 70%, after cryopreservation. In contrast, some of these studies observed lower sperm DNA integrity, in the range of 15%, in thawed samples. However, it is relevant to note that the sperm DNA integrity prior to cryopreservation is not available in these cases.

15.2.6 Sperm Maturity

To date, several maturity markers have been reported for sperm cells. In brief, high cytoplasmic levels of creatine kinase (CK) correlate with the spermatozoon's immaturity, suggesting an impairment in the last phase of spermatogenesis, when the cytoplasm extrusion occurs. Moreover, high expression of chaperonin HspA2, in combination with low levels of CK, are biomarkers of sperm maturity.

At least three hyaluronic acid binding proteins have been described in mature spermatozoa. These proteins are membrane receptors that are involved in relevant physiological processes, such as acrosome reaction, hyaluronidase release, and the binding to the oocyte's zona pellucida. Therefore, a sperm maturity profile has been developed based on the capability of spermatozoa to bind hyaluronic acid. According to evidences, hyaluronic acid binding sites in spermatozoa are not altered by cryopreservation; hence, hyaluronic acid-binding capability would not be a useful marker to assess cryopreservation outcomes.

15.3 SPERM SURVIVAL: VIABILITY AND MOTILITY

Sperm survival is the main parameter to consider when assessing cryopreservation outcomes. Traditionally, sperm survival is assessed by different assays with the same biological basis: sperm cell membrane integrity. Standardised viability tests are based on the assessment of sperm cell membrane's integrity as spermatozoa exhibiting altered cell membranes are associated with the loss of cell functionality. Eosin staining and the hypo-osmotic swelling test (HOS-test) allow to distinguish viable, but not motile, spermatozoa; hence the use of the cited tests is interesting in clinical cases in which sperm motility is difficult to assess. Several studies agree that, regarding the number of viable spermatozoa after thawing, it is because of physical and chemical stress that cryopreservation procedures worsen the results derived from those viability tests.

Despite strengths, these classic viability tests are not exempt from limitations. Eosin staining does not allow the use of the stained sperm sample in the following artificial reproduction techniques; hence the putative information derived from eosin staining results cannot be fully translated to the clinical practice when it comes to selecting the spermatozoon to be microinjected. Alternatively, after HOS-testing, the spermatozoa remain viable because the biological basis of HOS-test relies on the sperm cell's swelling capacity as a response to preserve osmotic balance under hypo-osmotic conditions. As hypo-osmotic conditions are established in vitro by exposing spermatozoa to a relatively harmless solution of sodium citrate and fructose, thus maintaining sperm viability, the sample could be used in the following ICSI, allowing the selection of viable spermatozoa exhibiting swelled tails.

However, some studies suggest that the spontaneously developed tail swellings occur in some cases under physiological conditions, resulting in an increased false positive rate because the tail swelling occurs in the absence of hypo-osmotic solution. It also reported that the spontaneously developed tail swelling phenomenon is particularly increased in thawed samples.

On the other hand, biologically, the movement of sperm cells is fuelled by means of energy from adenosine triphosphate (ATP). One of the sources of ATP can be the oxidative phosphorylation carried out by mitochondria located in the mid-piece of human spermatozoa. In spite of some concerns about the putative transference of ATP derived from mitochondrial activity through the flagellum, it could be hypothesised that ATP released from oxidative phosphorylation by mitochondria could be transported by means of the flux transfer chains, a model. The decrease in sperm motility in certain circumstances in which mitochondrial function is inhibited is additionally represented in robust evidences. Nevertheless, ATP used for the movement of spermatozoa is not exclusively produced by oxidative phosphorylation in the mitochondria. Indeed, glycolysis is another source of ATP that fuels sperm motility. In fact, a recent comprehensive review addressing the issue of the signalling pathways involved in sperm motility suggest that, possibly, both mentioned mechanisms co-operate to allow ATP production for sperm motility. The degree of contribution of each metabolic pathway could depend on the substrate's availability in the female reproductive tract or in the culture media.

Several authors agreed that a decreased motility rate can be observed, as well as decreases in other kinetic parameters, in thawed samples. The observed decrease in sperm motility could be due to the mitochondrial damage described after cryopreservation, as less functional mitochondria with reduced activity have been reported in thawed samples.

In this sense, sperm motility could provide a relatively accurate estimation of viability provided the premise that a motile spermatozoon is viable. The Computer Assisted Semen Analyzer (CASA) system can assess more accurately than manual observation the motility of a given sample, because this sort of software considers all types of motility that a spermatozoon can exhibit and is not only limited to progressive motility traditionally assessed by manual observations. Indeed, sperm total motility (TM) strongly correlates with HOS-test results in terms of sperm viability. Therefore, TM could be used as an estimation of the viability of a given sperm sample. TM assessed by CASA systems could be considered a rapid and objective method to estimate cryopreservation outcomes regarding sperm survival. However, there exist some exceptional clinical circumstances in which sperm viability tests, such as the HOS-test, cannot be replaced. In patients reporting an absolute structural alteration of the sperm flagellum, as in Kartagener syndrome, the HOS-test should be used rather than other assessments of sperm motility due to the physical incapability of the spermatozoa to move.

15.4 TOTAL MOTILE SPERM COUNT

Sperm count is also a paramount parameter that should be considered in certain clinical scenarios, particularly when the sample's availability is limited. The combination of TM percentage and sperm count in a single parameter provides useful information, not only about viability rate, but also quantitative valuable information regarding the total number of viable spermatozoa. A quantitative assessment of the number of viable spermatozoa is especially crucial in groups of sub-fertile patients reporting impaired sperm count or motility, as in these circumstances it is vital to maximise the number of motile spermatozoa achieved. Recent studies agree that using total motile sperm count (TMSC) as a sperm parameter contributes to achieving better outcomes in different clinical cases, such as in testicular cancer, severe male factor diagnosis, and in ICSI cycles. Differences in the clinical impact associated with each sperm parameter included, based on the main evidences included in this review.

Sperm Parameter	Main Outcome
Sperm Morphology	Decrease in normal morphology forms in thawed sperm samples. Normal morphology forms is not predictor fo better clinical outcomes of neither homologous artificial insemination nor IVF-ICSI treatments. Usefulness of sperm morphology assessment to improve chance of reproductive success in Art remains controversial. Intra- and interlaboratory differences in assessing sperm morphology hinder results interpretation.
Sperm Maturity	Poor correlation between sperm maturity (evaluated by hyaluron-binding assay) and sperm survival after thawing. No difference in maturity markers prior or affer sperm freezing. No correlation between sperm hyaluron binding assay and fertilisation rate after an IVF cycle
Sperm DNA integrity	Lack of consensus regarding which sperm DNA fragmentation test is more suitable for clinical practice. No agreement to establish a clinically relevant threshold value for sperm DNA fragmentation. Need of clinical trails addressing the issue of the effectiveness of assessing sperm DNA interity in ART.
Sperm DNA integrity	Evidence is not enough to recommend routinely assessment of DNA integrity. Increased risk ratio for miscarriage when highfragmented spermatozoa are used. Increase of 10% single strand DNA lesions in thawed sperm samples compared with fresh sperm samples. No impact on miscarriage rate because there is not any difference in double strand DNA lesions in thawed samples in comparison with fresh samples.
Sperm Viability: Classic Vitalty test	Impaired vitality test results after cryopreservation due to the sperm cell membrane damage. HOS-test false positive results sue to spontaneously developed tail swelling are increased after thawing.
Sperm viability: Motility	A decrease in spermatozoa kinetics occurs after thawing.
Total motile sperm count	TMSC is a predictor of recovery after thawing in patients with testicular cancer. There is a correlation between higher TMSC and outcomes. TMSC correlates better than WHO 2010 parameters with the severity of male factor subfertility. TMSC predicts better than WHO 2010 parameters the outcomes of ICSI cycles. TMSC is an accurate estimator of the number of viable spermatozoa after thawing.

16

CRYOPRESERVATION AND IN VITRO STORAGE

16.1 PRINCIPLES AND EXPERIMENTAL ACHIEVEMENTS

In vitro storage and cryopreservation of vegetative material is likely to have limited application to gene conservation activities for forest tree species, with minor use as a backup strategy for species displaying seed storage problems the most likely possibility. These technologies offer no solution to the urgent problem of genetic erosion of tropical forest gene pools. Cryopreservation of pollen is deserving of more attention as a method for medium and - maybe - in long-term gene conservation with industrial plantation species. Cryopreservation warrants much more attention as a means of maintaining juvenility and capturing gains offered by clonal forestry with industrial species, but only for plantation programmes where good breeding programmes are in place and clonal forestry is a realistic goal. In particular, this applies to some conifers and eucalypts. In the longer term, the use of in vitro germplasm transport procedures may become useful for species which are the subject of cooperative regional breeding programmes.

In Vitro Storage

As detailed elsewhere in this report, micropropagation procedures have been developed for over 1 000 plant species, many of which are micropropagated commercially. Rapid multiplication, involving rapid growth and frequent subculture, generally is an objective of commercial micropropagation. The basis of successful storage of cultures is an increase in the interval between subcultures by reduction of growth. Methods employed to this end have been well reviewed.

Alternatives are:

- Temperature reduction. This is the most common approach to storage of cultures. A limitation for tropical species is that many are intolerant of low temperatures. Culture temperature must be over 20°C for cassava, and over 18°C for oil palm.

- Reduction of light intensity.
- Alterations to the culture medium. Such alterations include reductions in the sugar and/or mineral composition, the addition of cryoprotective or osmotic agents such as mannitol, the addition of growth retardants such as abscisic acid, and the addition of activated charcoal. Additions to the medium have to be tested carefully - osmotic agents can have detrimental effects on the tissues of some species, and cultures of tropical species frequently respond poorly to the addition of abscisic acid.
- Modification of the gaseous environment, in particular the lowering of oxygen content (frequently achieved by covering the culture with a thin layer of oil).

Storage of cultures has involved explants of various types. Shoots with roots have demonstrated better survival than those without, and microtubers are suited to storage for potato. An innovation which holds promise for some species is the encapsulation of explants in alginate beads. Withers (1992) tabulates some 30 species for which successful slow growth storage, from several months to several years, has been reported. Success has been achieved with several woody species, including Pinus radiata, Alnus glutinosa, and species of Eucalyptus and Populus. Research aimed at developing in vitro storage protocols is largely empirical, involving the testing of various treatments and environmental conditions. It is likely though that, with sufficient effort, in vitro storage procedures could be developed for many of the tree species for which micropropagation protocols exist. Some tropical species may be difficult, however, to store in this way.

In vitro storage is now routinely used for germplasm storage of some crops. The in vitro cassava gene bank at CIAT, Colombia, comprises nearly 5 000 clones, in an area of 50 square metres, with transfer (subculture) intervals of 12–14 months. While in vitro storage thus offers some advantages over field genebanks, the management of large collections remains problematical, due to the requirement for periodic subculture. The possible introduction of genetic variants during culture may be a risk with some types of cultures.

16.2 CRYOPRESERVATION

Although some microorganisms and higher plant pollen can be freeze dried, suspension of growth in higher plant somatic tissues can be achieved only by transfer to ultra low temperatures such as that of liquid nitrogen. At these temperatures, metabolism is suspended, and free radical damage caused by ionizing background radiation is the only factor potentially causing deterioration. Results indicate that exposure to such temperatures need not be intrinsically damaging, although transitions must be carefully managed. Cryopreservation procedures thus involve the successive steps of choice of material, pretreatment, freezing, storage, thawing, and post-treatment handling, for which optima must be defined for each species.

Growth recovery in cryopreserved cell suspensions takes place if a certain percentage of cells survives, and reports of successful cryopreservation most commonly concern cell suspensions. Maintenance of structural integrity is more important with organized tissues such as meristem and somatic, pollinic or zygotic embryos, and reports of successful cryopreservation of these are less common. Meristematic cells are the most resilient, and young somatic embryos or immature zygotic embryos are preferred. In oil palm, somatic embryos of only a specific morphological type can be cryopreserved. Pretreatment comprise the application of a cryoprotectant, dehydration, or a combination of both. Pretreatment with cryoprotectants such as sucrose, sorbitol, mannitol, dimethylsulfoxide, or polyethylene glycol dehydrates tissues, but may also act by protecting membranes and enzymatic binding sites from injury. The appropriate type, concentration and duration of treatment varies with species . The optimal duration of desiccation treatment varies from two to four hours, during which water content is reduced to 10–16%.

A two stage cooling process is most commonly employed. Cultures are cooled gradually down to a specific temperature, and then immersed in liquid nitrogen. Species vary in their requirements with respect to both the rate at which this cooling is conducted, and the temperature to be reached, e.g. Citrus embryos have to be cooled at a very specific rate down to -42°C, while those of oil palm can withstand a wide range of freezing rates. With this approach, ice formation occurs first in the external medium, and water flows out of the cells. Cells reach a level of dehydration at which crystallization of the remaining water causes no damage. Programmable freezing apparatus is required for this work. For certain tissues of some species, a one step rapid freezing, accomplished by direct immersion in liquid nitrogen, is satisfactory. In this case, intracellular ice crystallizes as microcrystals which are not harmful to cell components. A third alternative is "vitrification", whereby water forms an amorphous glassy structure, without a crystallization phase. This is achieved by the use of very high concentrations of cryoprotectants and very rapid cooling. This process has been applied successfully to a range of cell suspension, protoplast and embryogenic cultures, and also to shoot cultures of asparagus, mint, orange, carnation, potato and white clover. Thawing after storage is generally rapid, to avoid recrystallization. After the freeze-thaw cycle, culture conditions different from those of the standard procedure may have to be defined in order to stimulate the regrowth of the material.

A recent innovation has been the incorporation of alginate bead encapsulation methods into cryopreservation procedures. This method offers protection from adverse effects of some of the treatments. Somatic embryos or shoot tips are suspended in culture medium supplemented with 3% Na-alginate. The mixture is then dispensed from a pipette. Beads of about 4 mm diameter containing 1 to 3 embryos or shoot tips are precultured for several days in media supplemented with sucrose. Coated organs are dried under sterile air flow at ambient temperature and humidity for up to six hours. After dehydration, artificial seeds are transferred into cryovials without liquid medium for cooling, using either a one or two step procedure. The encapsulation-dehydration procedure has been applied to cryopreservation of somatic embryos of carrot and shoot tips of pear, potato, grape and carnation. Cryopreservation has been applied successfully to over 70 plant species, including tropical crops such as coconut, rubber, cocoa, oil palm, peanut, rice, banana, cassava, and coffee. Forest tree material for which successful cryopreservation has been reported includes embryos of Quercus petraea, Fagus sylvatica, Aesculus hippocastanum , Araucaria excelsa, Castanea , Artocarpus, and Juglans ; embryogenic cell lines of Picea glauca, Acer pseudoplatanus , Picea abies and Pinus taeda , seeds of Abies alba, Sequoiadendron giganteum, Larix decidua, Pseudotsuga menziesii, Picea abies, Pinus sylvestris (but not Quercus petraea), pollen of Betula pendula, B.pubescens, Larix decidua, L.kaempferi, Pseudotsuga menziesii, Picea abies, Pinus sylvestris, Quercus petraea and Q. robur , and encapsulated shoot tips of Eucalyptus gunnii .

The expectation is that, once cryopreserved, material can be stored indefinitely in liquid nitrogen without further loss. Strawberry and peanut meristem were capable of regeneration after two years storage, and 1.5 years for oil palm, while potato and cassava meristem survived for four years. Cell lines of Picea glauca have shown no loss of viability after over two years of storage. The main emphasis of most cryopreservation research programmes lies in defining non-damaging procedures for cooling and rewarming the cultures, for which species have differing requirements. Rapid progress is being made with this work, and it seems likely that, given a sufficiently intensive research effort, prospects for developing a suitable protocol for tissue preservation for any chosen species within the short term would be reasonably good. One obstacle to application for many species lies in the need to regenerate plants from the tissues or organs after the period of cryopreservation.

16.3 APPLICATIONS IN FOREST TREE IMPROVEMENT

Three broad areas of application of cryopreservation and in vitro storage techniques to forest tree improvement warrant discussion:

16.3.1 Germplasm Preservation

In vitro storage is now routinely used for germplasm storage of cassava, at CIAT , Chavez et al. and for many Solanum genotypes at CIP, in Peru. ORSTOM, at Montpellier, is developing methods for the preservation of coffee by in vitro culture of immature zygotic embryos. Work aimed at the storage of somatic embryos of mango is underway. In the IBPGR in vitro program, high priority is being given to the development of good in vitro culture procedures for sweet potato, taro, cocoyam, banana and plantain, sugarcane, cocoa, citrus and forage grasses. With respect to cryopreservation, preservation of a substantial number of genotypes of oil palm by ORSTOM is the closest to routine application, while the development of routine procedures for banana and plantain (at CATIE), cassava (at CIAT) and coconut is underway.

A number of features characterize the above crops targeted for in vitro or cryopreservation approaches to gene conservation:

- They are limited in number, and are all of major, widespread importance.
- Most are characterized by a long history of domestication and use.
- Biological and agronomic features are relatively well known.
- Several are vegetatively propagated.
- Some have recalcitrant seeds.

Gene conservation in these crops is to some extent then directed towards:

- The collection of known genotypes from many different areas of use. These genotypes have been evaluated to some degree, and it is desirable gene combinations, rather than individual genes, which are the target of preservation.
- Preservation of threatened wild relatives, characteristics of which would also be partially known.

Under these circumstances, conservation priorities are reasonably easily set. For these crops, in vitro storage and cryopreservation offer alternatives to the traditional dependence on field genebanks. For agricultural crops where seed collection and storage are possible, however, seed storage is the recommended approach to gene conservation.

Undoubtedly, urgent gene conservation and preservation problems exist for forest trees species, mainly among the tropical hardwoods and non-industrial species. Compared to the crop species discussed above, the following features characterize these tree species:

- There is a large number of species of known or potential interest. Many may never be used commercially, and some will be important only locally.
- Biological features of many are poorly known.
- Species distributions, and those of relatives, are poorly known.
- Seeds of some species will be in the recalcitrant category, but many are orthodox.

Major impediments to the preservation of these gene pools are:

- Resources available are sufficient for only a very small fraction of the survey and collection work which would be required.
- Seed storage facilities in many tropical countries are unreliable, such that satisfactory storage of seed collected now cannot be guaranteed.

The potential use of in vitro storage and cryopreservation as a complementary method for the ex situ conservation of long-lived perennial species. This potential is undoubtedly real. It is of value though to carry this a little further, to consider specifically to which groups of tree species the technology might be applicable in the near future. It can be observed that replacement of existing seed storage facilities with more sophisticated technology is not likely to make a positive impact on the gene conservation problem with tropical tree species. Even for the recalcitrant species, activities should favour the establishment of ex situ plantings which will enable evaluation of the material. Cryopreservation and in vitro storage techniques thus offer no solution to erosion of tropical forest gene pools. More concerted efforts at documentation and in situ conservation are required.

Of established industrial species, cryopreservation and in vitro storage of vegetative material are not advantageous approaches to gene conservation in most of the conifers and eucalypts, for which seed can be stored satisfactorily. Poplars display some of the features characteristic of the fruit tree species described above - problematic seed storage, and breeding and commercial use based substantially on known clones. In vitro storage or cryopreservation may have a role as a back-up measure in large poplar breeding programmes. Similarly also for other species with seed production and storage problems, e.g. Juglans. It should be noted though that, for most forest tree species, introduction of in vitro or cryopreserved material into a breeding program will involve a delay of several years to flowering, and generally also for evaluation. Flowering genotypes in field clone banks can be much more quickly incorporated. An exception to the flowering obstacle is stored pollen, which can be rapidly incorporated into breeding programmes, and pollen storage should be more prominent in gene conservation programmes with industrial species. Cryopreservation may facilitate the broader use of pollen storage in these programmes. The use of cryopreservation and in vitro storage could involve selection for genotypes which tolerate the conditions imposed (e.g. freezing and thawing), and which are capable of regeneration following such treatment. The conserved material is thus not necessarily a random sample of the genetic material collected.

16.3.2 Maintenance of Juvenility

Cryopreservation is perhaps the most promising approach currently on offer for maintenance of the juvenile state and capture of genetic gains through clonal forestry with industrial species. The technology is applicable mainly in cases where good breeding programmes are in place and clonal forestry is a realistic goal. In particular, this applies to some programmes with conifers. Although some good breeding programmes are in place for the eucalypts, the need is not so urgent for these because rejuvenation can be achieved by other means. The approach ultimately may have some application to other tropical hardwoods, as good breeding programmes are established for these. Maturation is a much less important issue with many of the non- industrials, in particular those which flower precociously, can be selected early, and which can still be readily propagated at selection age.

16.3.3 Transport of Germplasm

The use of plants, organs or cells cultured in vitro circumvents to some extent the limitations which quarantine regulations place on the import of vegetative material (in particular) into many countries. International germplasm exchanges are becoming very important in the breeding of many crops, in particular to and from international genebanks at the International Agricultural Research Centres (IARCs) and other centres. In vitro exchanges are becoming common for some of these, e.g. for potato, coconut, banana and oil palm. In addition, in vitro collection, overcoming problems of inadequate or immature seeds and deterioration of vegetative material in transit, is now being used. Simple field procedures have been devised for cocoa, avocado, coconut, citrus, cassava, cotton, and forage grasses..

With the increasing emphasis on cooperative breeding of both industrial and non-industrial forest tree species in tropical countries, international movement of material will become increasingly important. In the short term, such movement is likely to be mainly in the form of seed for species and provenance testing. Movement of vegetative material, however, may become more important as breeding programmes advance, and in vitro approaches could find application in the longer term.

17

CRYOPRESERVATION FOR PRESERVATION OF POTATO GENETIC RESOURCES

17.1 INTRODUCTION

Cryopreservation is becoming an increasingly used method for the long-term storage of plant genetic resources (PGRs). Cryopreservation requires only a minimum of space and low level of maintenance. Methods for cryopreservation have been developed for a large number of plant species and further research is being conducted to enable adoption of this approach even more broadly. A timely book 'Plant Cryopreservation —A Practical Guide' was published to aid in the use of cryopreservation techniques globally, for the preservation of all forms of plant biodiversity. Also, an updated 'Genebank Standards for Plant Genetic Resources for Food and Agriculture' was issued from Commission on Genetic Resources for Food and Agriculture. In this book, one chapter discusses genebank requirements for in vitro culture and cryopreservation. Thus cryopreservation techniques using in vitro shoot tips are recognized as a long-term storage tool for PGR.

Cryopreservation is becoming a very important tool for the long-term storage of plant genetic resources and efficient cryopreservation protocols have been developed for a large number of plant species. Practical procedures, developed using in vitro tissue culture, can be a simple and reliable preservation option of potato genetic resources rather than maintaining by vegetative propagation in genebanks due their allogamous nature. Cryopreserved materials insure a long-term backup of field collections against loss of plant germplasm. Occurrence of genetic variation, in tissue culture cells during prolonged subcultures, can be avoided with suitable cryopreservation protocols that provide high regrowth, leading and facilitating a systematic and strategic cryo-banking of plant genetic resources. Cryopreservation protocols for potato reviewed here, can efficiently complement field and in vitro conservation, providing for preservation of genotypes difficult to preserve by other methods, wild types and other species decided as priority collections.

Recent research on cryopreservation has focused on practical procedures for genebank storage, thereby enabling cells and meristems to be cryopreserved by direct transfer into liquid nitrogen (LN). The development of simple and reliable methods for cryopreservation facilitates cryo-banking. Optimal cryopreservation conditions produced high levels of regrowth after LN storage. Potato is one of the most important food crops for food security. There are more than 4,500 varieties of Solanum tuberosum L. Preservation of potato genetic resources (GRs) in genebanks is mostly by vegetative propagation due their allogamous nature and many genebanks are maintaining potato GRs as field collections.

A review by discussed in detail recent advances in potato cryopreservation based on thermal, biochemical, genomic and ultrastructural analyses. There are some recent reviews of the development of potato cryopreservation protocols. This review introduces practical and successful cryopreservation protocols of in vitro grown potato shoot tips, which have been used and are being implemented for cryo-banking in institutions around the world. In addition, here new cryopreservation techniques using aluminium plates developed are described and the importance of cryopreservation techniques for long-term storage are discussed.

17.2 EX SITU PRESERVATION

Ex situ preservation of PGRs is the storage of seeds or plant materials under artificial conditions to maintain their long-term viability and availability for use. Globally, genebanks are employing seed storage, field collections, in vitro storage (tissue culture or cryopreservation) for ex situ preservation of PGR. Storage of PGR that produce orthodox seeds, which are tolerant to low moisture content and low temperatures, at appropriate temperature and humidity, is the most convenient ex situ preservation method. Many major seed crops are included this category. However, recalcitrant seeds, which are sensitive to low moisture content and low temperatures, do not survive if they are stored under the standard storage conditions used for orthodox seeds. This category of seeds includes several important tropical and sub-tropical tree species. There is one more category recognized as intermediates between orthodox and recalcitrant seeds and known as intermediate seeds, which can tolerate combinations of desiccation and low temperature.

Field genebanks maintain living plants. Field genebanks are used for the plants which produce non-orthodox seeds or no seeds and are vegetatively propagated. Vegetatively propagated plants comprise of two types, the perennial and annual/biennial types. The former can be maintained in the field for long period without replanting but for the latter replanting is necessary annually or biennially. The preservation of these PGRs requires an adequate area of land and continuous maintenance. Vegetatively maintained PGRs are vulnerable to loss from natural disasters and damage caused by pests and diseases.

In vitro genebanks are a means to overcome the disadvantage of the field genebanks and reflect progress in plant tissue culture techniques. Preserved in vitro germplasm can be propagated and regenerated into plantlets in a sterile and pathogen free environment. In vitro genebanks are used for species with an established tissue culture system. To maintain in vitro germplasm, it should be subcultured after specific periods of time under standard culture conditions to avoid deterioration and/or contamination of materials. Several slow growth (minimal growth) methods have been established for short (3 months) to middle (3 years) term storage using low temperature, minimal nutrition, growth retardant and so on, singularly or in combination. A drawback of tissue culture storage is the induction of genetic variation or mutation during prolonged subculturing. For this reason, minimal growth method is desirable for preservation of in vitro materials reduce the subculture interval. Selection of explant is also important in in vitro storage as somaclonal variation from cultured cells and callus may occur more easily compared to in vitro shoot cultures. Hence there is a preference for using shoots for in vitro storage by minimal growth method. Details regarding preservation methods can be found.

17.3 IN VITRO GENEBANKS OF POTATO

In vitro potato collection of 1,223 accessions are maintained at 25°C in the National Institute of Crop Science (NICS), Rural Development Administration (RDA), Rep. Korea. Also, in the National Center for Seeds and Seedlings (NCSS), Japan, an in vitro potato collection of 130 accessions is maintained at 25°C and normal subculture by node or shoot is performed every 3–4 months. The frequency of subculturing can be reduced by growing the micro plants on Murashige and Skoog medium supplemented with growth retardants or osmotic stress inducing polyols, incubating them under low temperature, low light intensity and varied photoperiod. The Leibniz Institute of Plant Genetics and Crop Plant Research (IPK), Germany maintains 2,855 potato accessions in vitro at 4°C as microtubers. The cycle of slow growth maintenance consists of a warm phase with long-day at 20°C for 2–3 months, a microtuber induction phase with short-day at 9°C for 2–4 months and a cold storage period, in which microtubers are stored at 4°C for 12–15 months. At the Central Potato Research Institute, Shimla, India, more than 1,500 parental lines and potato varieties are maintained in vitro on MS medium supplemented with 40 g/l sucrose and 20 g/l mannitol at 6–8°C and 16-h photoperiod. At the International Potato Center (CIP), Peru, the in vitro collection consists of 4,062 accessions that are maintained under slow growth conditions. The medium used contains MS salts, 40 g/l sorbitol, 20 g/l sucrose, 2 mg/l glycine, 0.5 mg/l nicotinic acid, 0.5 mg/l pyridoxine, 0.4 mg/l thiamine and 8 g/l agar. Cultures are maintained at 6–8°C under 22 µmol/m2s illumination and 16-h light. This allows in vitro plantlets to be stored for approximately 2 years without sub-culturing. Besides these Institutes, in vitro storage of potato GRs is conducted at many other Institutes, such as National Forestry, Crops and Live-stock Research Institute, Mexico and National Institute for Agricultural Research, Chile. Almost all Institutes mentioned above have started research on cryopreservation of potato GRs as an alternate of in vitro storage.

17.4 THE CONCEPT OF VITRIFICATION

Cryopreservation is based on the reduction and subsequent interruption of metabolic functions of biological materials by decreasing the temperature with LN (−196°C), while maintaining viability. At −196°C, almost all the cellular metabolic activities are quiescent and the cells can be preserved in such state for a long-term. It is essential to avoid lethal intracellular freezing that occurs during rapid cooling in LN and warming in order to maintain the viability of hydrated cells and tissues. Cells and tissues that are to be cryopreserved in LN, need to be sufficiently dehydrated before being immersed in LN.

There are two types of liquid-solid phase transitions in aqueous solutions. (a) Ice formation is the phase transition from liquid to ice crystals, and (b) vitrification is a phase transition from a liquid to amorphous glass that avoids crystallization. Water is very difficult to vitrify because the growth rate of crystals is very fast, even just below freezing point. However, highly concentrated cryoprotective solutions such as glycerol are very viscous and are easily supercooled below −70°C. This allows them to be vitrified on rapid cooling. Vitrification refers to the physical process by which a highly concentrated cryoprotective solution supercools to very low temperatures and finally solidifies into a metastable glass without crystallization. Vitrification had been proposed as a method for the cryopreservation of biological materials because of the potentially detrimental effects of extracellular and intracellular freezing might be avoided. Thus, vitrification is an effective freeze-avoidance mechanism for hydrated cells and tissues. As glass fills space in a tissue, it may contribute to the prevention of additional tissue collapse, solute concentration, and pH alteration during dehydration. Operationally, glass is expected to exhibit a lower water vapor pressure than the corresponding crystalline solid, thereby preventing further dehydration. Because glass is exceedingly viscous and stops all chemical reactions that require molecular diffusion, its formation leads to dormancy and stability over time. In any cryopreservation method, whole specimens or partial parts of specimens, which are in sufficient concentration of cytosol, can vitrify by rapid cooling into LN. In the plant vitrification method, plant vitrification solution (PVS) is used which is an extremely concentrated solution (7–8 M) of cryoprotectants. The most applied PVS is PVS2 solution which contains 30% (w/v) glycerol, 15% (w/v) ethylene glycol, 15% (w/v) dimethyl sulfoxide (DMSO) and 0.4 M sucrose in basal MS medium. This solution is supercooled below −70°C and vitrified at about −115°C without any detectable freezing events.

17.5 CRYOPRESERVATION

Over the last 25 years, several countries have been conserving PGR by cryopreservation. The cryopreservation technique is an effective approach for storage of plant cells, tissues, seeds and embryos. The development of new techniques for cryopreservation such as droplet vitrification, DMSO droplet and V cryo-plate methods, is facilitating the systematic and strategic cryobanking of PGRs. Cryo-storage of potato germplasm is at the cutting-edge of cryopreservation research. Many experiences obtained from potato cryo-banking in IPK, CIP, NAC RDA, CAES HRO and NIAS provide insights into the storage, principles and pitfalls governing the operations of a cryo-bank. These new techniques have facilitated the cryo-banking of other plant species that are established tissue culture techniques like Musa germplasm. summarized practical issues that need to be resolved before initiation of a cryo-bank. These issues include the plant materials to be preserved, storage records, storage forms, quantity to store, protocol testing, storage controls and recovery, as well as facilities and equipment. In the case of potato where huge diversity exists the genotype needs to be considered in the routine protocol of cryopreservation. It is important to have many choices of protocol for cryopreservation, because there are many types of plant propagules and plant species to be cryopreserved. The first thing that should be done is to determine how to make materials for cryopreservation, which are not only healthy and vigorous shoots, but also at a uniform stage and size. Despite research conducted to date some genotypes remain 'recalcitrant materials'. Secondly, it is necessary to determine whether the specimens are sensitive to some chemicals such as PVS or excess dehydration. Thirdly, the unloading step and regeneration medium should be reexamined. Current new protocols apply rapid cooling and warming by direct immersion in LN and unloading solution. Several papers report that optimal exposure time to PVS2 shows a wide spectrum of efficiency. This means high regrowth might be obtained by rapid cooling and warming, even though the water content of specimen has not reached optimal. Cryopreservation should be considered as a backup to field collections to insure against loss of plant germplasm. Priority of collections to be cryopreserved should be given to the 'at risk' plants that have an increased chance of being lost from a collection. Some minor and endangered crops have genotypes that are difficult to establish in in vitro culture. It is crucial for successful cryopreservation to develop efficient micro propagation systems. Developing tissue culture systems which allow rapid multiplication and which stimulate regrowth after retrieving samples from LN will be the next challenge for cryopreservation research.

Preservation of in vitro shoot tips and somatic embryos at cryogenic temperatures is considered to be a suitable alternative that can ensure the long-term security of vegetatively maintained germplasm. Once stored in LN, germplasm can be kept for apparently almost unlimited periods, and as a result cryopreservation is the most appropriate for long-term storage of base collections. Cryopreservation is often combined with tissue culture preservation for in vitro storage. Because the cryopreservation procedure is usually proceeded by tissue culture, except when preserving seeds, pollen and dormant buds.

In the vitrification method, cells and tissues must be sufficiently dehydrated with plant vitrification solution without causing injury to be capable of vitrifying upon rapid cooling into LN. High survival of in vitro grown materials is determined not only by the cryogenic protocol itself, but also by the physiological conditions of the materials to be cryopreserved. This means that some steps for acquisition dehydration tolerance or low temperature tolerance are crucial in the cryopreservation procedure. Following procedures such as preconditioning, preculture and osmoprotection (loading treatment) are vital for successful cryopreservation. The materials must be in an optimal physiological and morphological state to ensure high recovery and vigorous regrowth after LN exposure. In preconditioning, homogeneous specimens in terms of size, cellular composition, physiological state and growth stage, increase the chances of a positive and uniform response to treatment with loading solution (LS) and PVS2. Cold-hardening and preculture of shoot tips with sucrose-enriched media is effective for improving the post-thaw survival of some temperate and tropical species, due to increased membrane stability. During preculture on sucrose-enriched medium, concentrations of sugar, starch and proline are greatly increased in the shoot tips and may enhance the stability of membranes under conditions of severe dehydration. In addition, a cryoprotective or osmoprotective treatment with LS solution appears promising as a means of enhancing the dehydration tolerance of shoot tips of several species. The protective effect of this solution in cellular peri-protoplasmic spaces may be due to mitigation of the large osmotic stress from exposure to PVS2, as well as to some mechanisms that minimize the injurious membrane changes from severe dehydration.

The optimal dehydration time with PVS2 or air dehydration is also important. An accurate control of the dehydration procedures and prevention of chemical toxicity injuries or excess osmotic stresses during dehydration treatment are indispensable for successful cryopreservation. Optimal dehydration time is species-specific, and may vary with the size, stage and morphological state of the shoot tips. Oxidative processes involved in cryopreservation protocols may be responsible for the reduced viability of explants after LN exposure. Adding antioxidants in the cryoprotectant or the recovery media that counteract these reactions may improve recovery. Polyvinylpyrrolidone (PVP) and plant regulators supplementation in the recovery medium also increase regrowth significantly compared with no supplementation. PVP may be involved in adsorbing the phenolic compounds produced by dead cells.

17.6 PRACTICAL CRYO-STORAGE

During the last 25 years, several cryopreservation techniques have been established based on the conventional slow freezing method. These techniques such as the vitrification method, encapsulation/dehydration method and encapsulation/ vitrification method, involve the steps of extraction of freezable water from the tissue cells before cooling. As a result, vitrification of internal solutes takes place during cooling. Modified techniques have been developed which further reduce the chance for lethal ice-crystal formation through the application of ultra-fast cooling and rewarming rates. These techniques are called the droplet vitrification method, V cryo-plate method and D cryo-plate method. Detailed descriptions of these cryopreservation protocols can be found in.

The current status of the main cryo-stored germplasm, apart from potato, is shown. Seed preservation at super low temperature (by vapor or liquid phase of LN) has been successfully achieved for a wide range of crop species by the standard seed bank protocol. There are several large cryopreserved collections of orthodox seeds. In National Center for Genetic Resources Preservation (NCGRP), USA approximately 10% of the seed accessions preserved (over 37,000 accessions) have been cryopreserved. Whereas, more than 1,200 seed accessions of 50 species have been cryopreserved at the National Bureau of Plant Genetic Resources (NBPGR), India and 400 Panax ginseng seed accessions have been cryopreserved in National Agrobiodiversity Center, Rural Development Administration (NAC, RDA), Rep. Korea. At the Institute de Recherche pour le Développement (IRD), France, a cryopreserved collection of coffee germplasm have been also stored safely in LN even though it is a non-orthodox seed.

Table 1 Current status of main cryo-storage in the world except potato germplasm

Institute	Materials	Plants	Cryo-storage accessions (No.)	Cryopreservation methods
NCGRP, USA	Orthodox seeds	10% seeds of accession	over 37,000	Desiccation
NBPGR, India	Orthodox seeds	50 species	1,200	
NAB, RDA, Rep. Korea	Orthodox seeds	Panax ginseng	400	
IRD, France	Non orthodox seeds	Coffea spp.	500	
NCGRP, USA	Dormant buds	Malus spp.	2,200	Cryopreservation using dormant buds (Slow freezing)
NIAS, Japan	Dormant buds	Morus spp.	1,236	
NBPGR, India	Dormant buds	Morus spp.	329	
AFOCEL, France	Dormant buds	Ulmus spp.	440	
NCGRP/NCGR, USA	In vitro shoot tips	Pyrus spp.	100	Slow freezing
NCGRP/NCG, USA	In vitro shoot tips	Rubus spp. et al.	57	Slow freezing
CIAT, Colombia	In vitro shoot tips	Manihot esculenta	480	Droplet vitrification
INIBAP, Belgium	In vitro shoot tips	Musa spp.	700	Droplet vitrification

Institute	Materials	Plants	Cryo-storage accessions (No.)	Cryopreservation methods
NICS, RDA, Rep. Korea	Shoot from cloves	Allium sativum L.	300	Droplet vitrification
IPK, Germany	In vitro shoot tips	Allium sativum L.	101	Vitrification
IPK, Germany	In vitro shoot tips	Mentha L.	86	Droplet vitrification
SARC, Japan	In vitro shoot tips	Wasabia japonica M.	40	Vitrification
NIAS, Japan	In vitro shoot tips	Juncus effusus	50	D cryo-plate

These information obtained in 2nd International Symposium on Plant Cryopreservation, Fort Collins, Colorado, USA, except seeds. NCGRP (National Center for Genetic Resources Preservation); NCGR (National Clonal Germplasm Repository); NBPGR (National Bureau of Plant Genetic Resources); NAC, RDA (National Agrobiodiversity Center, Rural Development Administration); IRD (Institute de Recherche pour le Développement); NCGR (National Clonal Germplasm Repository); NIAS (National Institute of Agrobiological Sciences); AFOCEL (Association Forêt Cellulose); CIAT (International Center for Tropical Agriculture); INIBAP (International Network for the Improvement of Banana and Plantain); NICS RDA (National Institute of Crop Science, Rural Development Administration); IPK (Leibniz Institute of Plant Genetics and Crop Plant Research); SARC (Shimane Agriculture Research Center).

Some temperate woody plants can be cryopreserved by using dormant buds and this cryopreservation method is called 'Cryopreservation of dormant buds'. This method is now applied to Malus spp., Morus spp. and Ulmus spp. at four different Institutes having a large scale cryo-storage infrastructure (Table 1). Cryopreservation protocol for mulberry dormant buds developed by National Institute of Agrobiological Sciences (NIAS), Japan is as follows. Mulberry branches with axillary buds are collected during the winter season when the buds are quiescent. After harvest, dormant buds with vascular tissue are removed and packed in polypropylene cryotube (8 ml) which are then cooled down to 0°C, −5°C, −10°C, −15°C and −20°C at successive one day intervals as a pre-freezing dehydration process. The final day when buds are at −20°C, the cryotube is removed from the cooling unit and quickly transferred into the vapor phase of LN tank (ca. −160°C) for long-term storage. The regrowth of 'Kenmochi' mulberry buds stored for 11.5 years in vapor phase of LN tank was 98% by tissue culture after rewarming.

The large scale cryo-storage of in vitro shoot tips has been accomplished at several Institutes by optimizing cryopreservation protocols (Table 1). The International Network for the Improvement of Banana and Plantain (INIBAP) has been maintaining the Musa spp. cryo-bank collection of over 700 accessions by the droplet vitrification method.

The other crops which have been cryopreserved in cryo-banks, are Pyrus spp., Rubus spp., Manihot esculenta , Allium sativum Menta L., Wasabia japonica , Juncus effuses (Table 1).

17.7 PRACTICAL CRYO-BANKING OF IN VITRO GROWN POTATO SHOOT TIPS

The current status of potato cryo-banks of in vitro grown shoot tips globally is shown (Table 2). IPK, Germany, and CIP, Peru, are two of the largest potato genebanks. Both institutes have been applying cryo-storage to potato and achieved large cryo-bank collections with over 1,456 and 869 accessions, respectively. The cryo-storage of the potato shoot tips have been also established at NCGRP, USA, NAC RDA, Rep. Korea, Central Agricultural Experiment Station, Hokkaido Research Organization (CAES HRO), Japan, and NIAS, Japan (Table 2). The cryopreservation methods used in these cryo-banks are DMSO droplet (IPK), droplet vitrification (CIP, NCGRP and NAC RDA), encapsulation vitrification (CAES HRO) and V cryo-plate (NIAS). With these methods, except encapsulation vitrification method, shoot tips are directly immersed in LN and in the rewarming solution on aluminum foil strips or cryo-plates. Cooling and warming rates are about 4,000–5,000°C/min and about 3,000–4,000°C/min, respectively, resulting in little or no crystallization and high regrowth after rewarming. In contrast, conventional methods such as vitrification and encapsulation vitrification use capped cryotubes for immersion into LN and retrieval from LN. When using capped cryotubes, cooling and warming rates are about 100–200°C/min and about 80–120°C/min, respectively, which are far less than new methods and this has a great impact on regrowth of cryopreserved materials.

Table 2 Current cryo-storage status of *in vitro* grown shoot tips of potato

Institute, country	Total accessions	Number of accessions				Cryopreservation methods
		Field preservation	Seed storage	In vitro storage	Cryo-storage	
IPK/GLKS, Germany	6,124 (2,846)	89	2,846 (2,846)	2,855	1,436	DMSO droplet vitrification
CIP, Peru	6,768 (2,414)	3,931	6,125 (2,289)	4,062 (49)	869	Droplet vitrification & Vitrification
Northern Region 6, USA	5,808					
NCGRP, USA					247	Droplet vitrification
NICS, RDA, Rep. Korea	1,223	670		1,223		
NAC, RDA, Rep. Korea					130	Droplet vitrification
NIAS, Japan	1,964	1,964		20	20	V cryo-plate
NCSS, Japan				130		
KAES HRO, Japan	500	500				Encapsulation vitrification
CAES HRO, Japan					100	

( ) means number of wild potato accessions.

IPK (Leibniz Institute of Plant Genetics and Crop Plant Research); GLKS (The Groß Lüsewitz Potato Collection); CIP (International Potato Center); NR6 (The US Potato Center); NCGRP (National Center for Genetic Resources Preservation); NICS RDA (National Institute of Crop Science, Rural Development Administration); NAC, RDA (National Agrobiodiversity Center, Rural Development Administration); NIAS (National Institute of Agrobiological Sciences); NCSS ((National Center of Seeds and Seedlings); KAES HRO (Kitami Agricultural Experiment Station, Hokkaido Research Organization); CAES HRO (Central Agricultural Experiment Station, Hokkaido Research Organization).

17.7.1 DMSO Droplet in IPK

Developed the ultra-rapid cooling method for in vitro shoot tips of potato called the DMSO droplet method, by using aluminium foil. Shoot tips are treated with a 10% DMSO in liquid MS medium with 30 g/l sucrose, 0.5 mg/l zeatin riboside, 0.2 mg/l GA3 and 0.5 mg/l IAA at room temperature (RT) for 2 h and frozen ultra-rapidly by direct immersion into LN in a 2.5 µl droplet of the same solution on a small piece of aluminium foil. The procedure is shown (Table 3). This protocol is very simple because only 10% DMSO is used as cryoprotectant solution. The explants (2–3 mm) are incubated in MSTo medium overnight at 22°C and treated with cryoprotectant solution (10% DMSO in MSTo medium) for 1–3 h at RT followed by transfer into droplets of 2.5 µl cryoprotectant solution one by one on aluminium foil. Afterwards the aluminium foil is immersed directly into cryotube filled with LN. The explants are rewarmed quickly by putting aluminium foils in liquid MS medium with 30 g/l sucrose at RT for regeneration. The shoot tips are plated on solid MSTo medium. Currently 1,436 potato accessions are stored at IPK using this protocol with a mean regeneration percentage of 46%. The DMSO droplet method is currently being improved. One improvement is the application of alternating preculture temperature for cold accumulation of shoots under 8-h photoperiod at 21/8°C day/night temperature for 7 days. The other improvement is the adoption of solid medium for regeneration. It is emphasized that a critical aspect in potato cryopreservation is the diverse response between different genotypes in terms of their regeneration capacities after cryopreservation. Therefore, genotypes not responsive to cryopreservation still need to be maintained using in vitro storage or field storage. In addition, certain genotypes may not even have tissue culture ability and need to be preserved as tubers. Finally, none of the conservation strategies (cryopreservation, in vitro storage, field storage) can be completely safe, because materials may always be lost regardless of the conservation technique.

Table 3 Five practical cryopreservation methods for *in vitro* grown shoot tips of potato

Procedure	**DMSO droplet method in IPK, Germany**	**Droplet vitrification method in CIP, Peru**	**Droplet vitrification method in NAC, RDA, Rep. Korea**	**Encapsulation vitrification method in CAES, HRO, Japan**	**V cryo-plate method in NIAS, Japan**
Culture	Culture on solid Murashige and Skoog medium (MS) with 20 g/l sucrose under 16 h photoperiod at 22°C for 3–4 weeks.	Culture on solid MS with 2 mg/l glycine, nicotinic acid, 0.5 mg/l pyridoxine, 0.1 mg/l thiamine, 25 g/l sucrose and 2.8 g/l phytagel at 22°C, 45 µmol/m2/s illumination and 16 h light.	Culture nodal segments on solid MS with 30 g/l sucrose, 2.2 g/l phytagel (Sigma Co.) at 24°C, under 16 h light, 100–140 µmol/m2/s illumination for 3–7 weeks.	Culture nodal segments on solid MS with 0.5 g/l casamino acid, 30 g/l sucrose, 2.5 g/l gellan gum at 23°C and 16 h photoperiod with light intensity of 50 µmol/m2/s every 2 weeks.	Culture on solid MS with 30 g/l sucrose, 0.3 g/l CaCl2 and 8 g/l agar at 20°C and 16 h photoperiod with light intensity of 104 µmol/m2/s every 3 months.
Preconditioning	Cold preculture under 8 h photoperiod at 21/8°C day/night temperature for 7 days.	Shoot multiply on the MS using apical cutting (~0.7 cm) for 3 weeks. If need, cold preculture on MS medium with 24 g/l sucrose at 6°C for 7 days.		Culture nodal segments on the MS for 7 days under standard condition or for 1 day under standard condition, then transfer at 4°C and 12 h photoperiod with light intensity of 20 µmol/m2/s and for 3 weeks.	Culture nodal segments with a lateral bud (about 5 mm) on the solid MS for about 2 weeks in the standard condition.
Shoot tips	Explants of 2–3 mm in length and 0.5–1 mm width.	Explants of 1.8–2.5 mm length apical shoot tips.	Axillary shoot tips (1.0–2.0 mm in size) are isolated by dissection from the upper to middle part of the mother-plants.	Excision shoot tips with 2 to 3 leaflets (about 1 mm length) from axillary buds.	Excision shoot tips with 2 leaflets (about 1.5 mm length) from shoots.
Preculture Mounting shoot tips on cryo-plate/ Encapsulation	Incubate in liquid MS with 30 g/l sucrose, 0.5 mg/l zeatin riboside, 0.2 mg/l GA3 and 0.5 mg/l IAA (MSTo) overnight at 22°C.	Place specimens on potato meristem medium, which is MS with 0.04 mg/l kinetin, 0.1 mg/l gibberellic acid, 24 g/l sucrose and 2.8 g/l phytagel (MMP) and incubated at RT for about 1 h.	Isolated shoot tips were precultured in liquid MS with 0.3 M sucrose for 8 h under standard conditions. Shoot tips were further precultured in liquid MS with 0.7 M sucrose for 18 h under the same conditions.	Preculture on MS with 0.3 M sucrose, 1 mg/l GA3, 0.01 mg/l BA and 1 µg/l NAA for 16 hr at 23°C. Encapsulation of shoot tip (one shoot tip/ bead); mixture of shoot tips and calcium-free MS with 2% Na-alginate and 0.4 M sucrose are dropped into 0.1 M CaCl2 soution with 30 g/l sucrose for 30 min, at 25°C.	Preculture on MS with 0.3 M sucrose overnight (16 hr) at 25°C. Pour 2% Na-alginates solution in the wells of cryo-plate. Place the precultured shoot tips one by one in the wells. Then, drop over them 0.1 M CaCl2 soution and wait for 15 min at 25°C for complete polymerization.

Procedure	DMSO droplet method in IPK, Germany	Droplet vitrification method in CIP, Peru	Droplet vitrification method in NAC, RDA, Rep. Korea	Encapsulation vitrification method in CAES, HRO, Japan	V cryo–plate method in NIAS, Japan
Osmoprotection	Treatment with cryoprotectant solution (10%DMSO in MSTo at RT for 1–3 h.	Treatment with osmoprotective solution (LS) with 2 M glycerol and 0.4 M sucrose at RT for 15 min.	Dehydration in 10 ml PVS2 solution for 20 min with continuous shaking (60 rpm).	Treatment beads with MS solution with 0.6 M sucrose, 2 M glycerol and the same plant hormones in preculture medium on rotary shaker (60 rpm) at 25°C for 90 min.	Treatment with LS solution contains 2 M glycerol + 0.8 M sucrose by transfer the cryo-plates in a 25 ml reservoir filled with 20 ml LS at 25°C for 30 min.
Dehydration		PVS2 treatment on ice. Shoot tips were exposed for 50 min to 2 ml ice-cooled PVS2 in glass vials.		Dehydration with PVS2 at 0°C for 3 hr. The beads in PVS2 are shaken (20 rpm) in a water bath.	Dehydration with PVS2 by placing cryo-plates in a 25 ml filled with 20 ml PVS2 at 25°C for 30 min.
Transfer the explants	Transfer the explants into droplets of 2.5 µl cryoprotectant solution one by one on aluminium foils (5 × 25 × 0.03 mm).	The 3 min before the end of each PVS2 treatment, the specimen were transfered to a PVS2 drop (10–15 µl) on an aluminum foil strip (0.5 × 2 cm).	A few min before plunging in LN, seven drops (2.5 µl each) of PVS2 solution are placed on an aluminum foil strip (7 × 20 mm) using a dispenser. One shoot tip is put in each of the seven PVS2 drops.		
Storage	Put the aluminium foils directly into cryotube filled with LN. Close the cryotube refilled with LN and stock into the Dewar with LN.	The strip holding the shoots is then rapidly plunged into a LN filled cryo-tube in a Petri dish on ice. Close the cryotube refilled with LN and stock into LN tanks.	The foil strip is immediately plunged in LN. After a few min, two foil strips is transferred in one 2 ml cryovial, which have been previously filled with LN and stored in a LN tank.	Cryotube containing beads and PVS2, is directly plunged in LN.	Transfer the cryo-plate in 2 ml cryotube, which is held on cryo-cane, and plunge it directly in LN.

Procedure	DMSO droplet method in IPK, Germany	Droplet vitrification method in CIP, Peru	Droplet vitrification method in NAC, RDA, Rep. Korea	Encapsulation vitrification method in CAES, HRO, Japan	V cryo–plate method in NIAS, Japan
Rewarming	Putting aluminium foils quickly in liquid MS with 30 g/l sucrose at room temperature (RT).	Dropping the strips quickly in liquid MS with 1.2 M sucrose at RT and incubate for 20 min.	Foil strips are taken out of the cryovials and immediately plunged in 6 ml 0.8 M sucrose solution at 40°C for 30 s and 6 ml of the solution are added. Shoot tips are further incubated in the unloading solution at RT for 30 min.	Cryotube is put into water bath at 38°C for 3 min. After removing the PVS2 solution, the beads are washed with 1 ml of 1.2 M sucrose solution for 10 min.	Cryo-plate is retrieved from the cryotube in LN and immersed in 2 ml cryotube containing 2 ml MS basal medium with 1 M sucrose, in which it is incubated for 15 min at RT.
Regeneration	Shoot tips are plated on solid MSTo (10 g/l agar) under a 16 h photoperiod at 22°C.	Post-cryo culture in the dark on MMP with progressively decreased sucrose levels (daily transfers from 0.3, to 0.2, to 0.1 M and maintained on 0.07 M). One week after warming, shoot tips were transferred from the filter paper to fresh MMP (0.07 M sucrose) and incubated at 22°C under standard condition.	Shoot tips are post-cultured on semi solid MS with 0.05 mg/l IAA, 0.3 mg/l zeatin, 0.05 mg/l GA3, 30 g/l sucrose and 1.8 mg/l phytagel at 24°C under low light intensity for 7 days and then transferred to standard culture conditions.	Beads with shoot tips are plated on the solid MS with 30 g/l sucrose and the same plant hormones in preculture medium for 1 day, then transferred on the solid MS with 30 g/l sucrose and 0.5 µg/l GA3 under standard condition.	Shoot tips are plated on the solid MS and cultured under standard condition.

17.7.2 Droplet Vitrification in CIP

There are several variations of the vitrification method. The discrimination of these methods is that the standard vitrification protocol takes place in a cryotube while the droplet vitrification protocol is done on aluminium foil strips, and encapsulation vitrification adds an encapsulation step to the standard protocol. The droplet vitrification method was developed in banana cryopreservation at first resulting increase in a regrowth rate of 40–50% compared to standard vitrification. In the droplet vitrification method, shoot tips are plated on aluminium foil during cooling and dehydrated with highly concentrated vitrification solution such as PVS2. Rewarming after cryopreservation resulted in regrowth of 46–51%, 8–47%. These results confirm that regrowth ability of cryopreserved potato shoot tips is genotype-dependent as indicated by Genotype dependency was also observed as the results showed that higher linoleic acid content in frost resistant genotypes of potato, are positively correlated with higher regeneration rates after cryopreservation.

The procedure of the latest droplet vitrification in CIP are shown (Table 3). The shoot tips (1.3–2.5 mm) are dissected from the shoots preconditioned and incubated at RT for about 1 h on potato meristem medium containing 24 g/l sucrose, 0.04 mg/l kinetin and 0.1 mg/l gibberellic acid. Osmoprotection is performed by LS (MS medium with 2 M glycerol and 0.4 M sucrose) for 15 min at RT. After that shoot tips are dehydrated by PVS2 for 50 min at 0°C and 3 min before the end of each PVS2 treatment, the shoot tips are transferred to a PVS2 drop (10–15 μl) on an aluminum foil strip (0.5 × 2 cm). Then, the strip holding the shoots are rapidly immersed into a LN filled cryo-tube in a Petri dish on ice. The strips are then rewarmed quickly by dropping them in liquid MS medium with 1.2 M sucrose at RT and incubated for 20 min for regeneration. Post-cryo cultures are kept in the dark on MMP with a progressive decrease in sucrose levels (daily transfers from 0.3, to 0.2, to 0.1 M and maintained on medium with 0.07 M sucrose). One week after plating, shoot tips are transferred from the filter paper to fresh MMP (0.07 M sucrose) and incubated at 22°C under standard conditions. This protocol was successfully applied to 4 genotypes showing different reactions to abiotic stress, which had high regrowth levels ranging from 23% to 76%. The response to cryopreservation is strongly genotype and species- specific, limiting the use of the current protocols to large diverse collections such as the CIP's potato collection. To overcome this issue, it is crucial not only to optimize the protocol for different genotypes, but also to make a uniform, healthy and robust shoot tips able to tolerate cryopreservation procedures and the regeneration system. Also, it is important that the method is simple and a suitable protocol for large scale application to potato cryopreservation. In the schematic procedure, the treatment combining 15 min LS at RT, 50 min PVS2 at 0°C, apical shoot tips from 3 week old mother plantlets, explant size (1.8–2.5 mm), cooling and warming on aluminium foil strips, produced positive result in the basic protocol for improving potato PVS2 vitrification.

17.7.3 Droplet Vitrification in NAC, RDA

The optimization of the protocol for diverse genotypes is a prerequisite for implementing large scale cryopreservation of potato collections in genebanks. The irregular survival levels observed within different potato species could be related to many factors such as subculture conditions, size of the shoot tips and their location on the plantlet axis, sucrose concentration of the preculture medium, preculture time, dehydration, cooling and warming, and unloading step. The subculture conditions, light intensity, aeration and planting density significantly affected survival of cryopreserved shoot tips. Also, the subculture duration and the position of the shoot tips on the axis of the in vitro plantlets have a significant effect on survival of cryopreserved shoot tips.

The procedure of droplet vitrification in NAC, RDA is shown (Table 3). The shoot tips (1.0–2.0 mm) are dissected from the subcultured shoots. The optimal duration and size of shoot tips were different by genotypes; 7 weeks and 1.5–2.0 mm for 'Dejima' and 5 weeks, 1.0–1.5 mm for 'STN13'. The shoot tips are precultured with 0.3 M sucrose for 8 h followed by 0.7 M sucrose for 18 h. The precultured shoot tips are dehydrated with PVS2 for 20 min without osmoprotection treatment. A few min before the end of each PVS2 treatment, the shoot tips are transferred to a PVS2 drop (2.5 μl each) on an aluminum foil strip (7 × 20 mm). Then, the strip holding the shoots are rapidly immersed into LN. For rapid warming, the strips are directly dipped in 6 ml 0.8 M sucrose solution at 40°C for 30 s and 6 ml of the solution are added. Shoot tips are further incubated in the unloading solution at RT for 30 min. Shoot tips are post-cultured on semi solid MS with 0.05 mg/l IAA, 0.3 mg/l zeatin, 0.05 mg/l GA3, 30 g/l sucrose and 1.8 mg/l phytagel at 24°C under low light intensity for 7 days and then transferred to standard culture conditions. This optimized protocol was successfully applied to 12 potato accessions, including wild species, resulting 64–94% regrowth. The same protocol was applied to the Korean in vitro potato collection at NAC RDA for cryo-storage of 130 accessions.

17.7.4 Encapsulation Vitrification in CAES HRO Japan

Originally, the encapsulation-dehydration technique was developed for easy handling of a large number of meristems at the same time using the vitrification method. applied this method (Table 3) to various potato cultivars. In this method, uniform shoot tips from axillary shoots with 2 to 3 leaflets (approximately 1 mm in size) are used which are induced from nodal segments at 23°C for 7 days or at 4°C for 3 weeks. Shoot tips are precultured on MS medium supplemented with 0.3 M sucrose, 1 mg/l GA3, 0.01 mg/l BA and 1 μg/l NAA overnight at 23°C. After that shoot tips are trapped within alginate gel beads (one shoot tip/one bead, diameter: 4–5 mm) with 0.4 M sucrose. Encapsulated shoot tips are treated with a mixture of 2 M glycerol and 0.6 M sucrose on a rotary shaker (60 rpm) at 25°C for 90 min to induce dehydration tolerance. They are then dehydrated with PVS2 at 0°C for 3 h in a water bath with shaking (20 rpm). The 10–15 beads are suspended in 1 ml PVS2 solution per cryotube and plunged directly into LN. For regeneration, cryotube are put into a water bath at 38°C for 3 min. After removing the PVS2 solution, rewarmed beads are washed with 1 ml of 1.2 M sucrose solution for 10 min. Then, beads with shoot tips are plated on the solid MS medium having 3% sucrose and the same plant hormones as in preculture medium for 1 day, then transferred on the solid MS medium with 30 g/l sucrose and 0.5 µg/l GA3 under standard condition. This protocol was successfully applied to 14 potato accessions resulting in 47–71% regrowth. The protocol was applied to Hokkaido Research Organization (HRO) potato collection and in CAES HRO, 100 accessions have been cryopreserved in LN tank by using this method.

17.7.5 V Cryo–Plate in NIAS

In conventional vitrification procedures, including droplet vitrification, small size shoot tips are suspended in various solutions employed, which have to be removed and added by repeated pipetting. This often results in damage and loss of shoot tips during the course of the cryopreservation protocol. Moreover, vitrification procedures require a precise control of duration of treatment with vitrification solutions due to the narrow range of optimal treatment durations. In the droplet vitrification procedure, dehydrated shoot tips also need to be transferred on aluminium strips just before immersion in LN. The vitrification method using aluminium plates, named V cryoplate method was developed in order to establish a simple, reproducible and reliable protocol using aluminium cryoplates. This new method is now adopted for several plant species such as strawberry, Dalmatian chrysanthemum, mint, mulberry, carnation, mat rush and sugarcane. It is a user-friendly procedure and permits high cooling and warming rates of treated materials. As a result, superior regrowth is obtained after cryopreservation for the plant species which have been tested by this method.

Currently, the V cryo-plate method was successfully applied to in vitro grown potato. The procedure is shown in Table 3. The shoots from nodal segments are cultured on solid MS medium containing 30 g/l sucrose and 0.3 g/l CaCl2 at 20°C for 2 weeks. The shoot tips (about 1.5 mm) are excised from the in vitro grown shoots and precultured on MS medium containing 0.3 M sucrose at 25°C for overnight. The precultured shoot tips are placed on aluminium cryo-plates (7 mm × 37 mm × 0.5 mm) with ten wells (diameter 1.5 mm, depth 0.75 mm) and embedded with calcium alginate gel. Osmoprotection is performed by immersing the cryo-plates for 30 min at 25°C in 25 ml pipetting reservoirs filled with LS (MS medium with 2 M glycerol and 0.8 M sucrose). For dehydration, the cryoplates are transferred and immersed in another reservoirs filled with PVS2 for 30 min at 25°C. Then, the cryo-plate is transferred in an uncapped 2 ml cryotube and directly plunged into LN. For rewarming, the cryo-plate is retrieved from the cryotube in LN and immersed in a 2 ml cryotube containing 2 ml MS basal medium with 1 M sucrose, in which it is incubated for 15 min at RT. Rewarmed shoot tips are placed on solid MS medium and cultured under standard conditions. This protocol was successfully applied to 16 cultivars and 4 wild potato accessions, resulting in high regrowth levels ranging from 93% to 100%.

In genebank NIAS Japan, 1,964 accessions of potato GRs are currently maintained in the field. Now, cryopreservation is a preservation option of potato GRs, as a long-term back-up system for the field collections. Currently, an air-dehydration method is being applied using aluminium cryo-plates named D cryo-plate method, which combines the encapsulation-dehydration method and V cryo-plate method. In this method, shoot tips/ buds attached to the cryoplates are dehydrated under the laminar air flow cabinet's air current after treating with LS. Air dehydration can minimize damage to specimens by avoiding the use of PVS2, using materials with comparatively higher moisture content (MC) and performing minimal excision of young leaves and/or sheaths. In in vitro mat rush buds, the D cryo-plate method overcame problems associated with sensitivity to PVS2, insufficient or excessive dehydration, damage to and loss of material during excision and manipulations. This method is going to be applied for in vitro grown shoots of potatoes.

17.7.6 Longevity and Genetic Stability

Germplasm will survive for a long time in cryogenic storage, it is not known for exactly how long. Estimates of the actual shelf life of cryogenically stored material are critical for efficient gene-banking, but are difficult to obtain because of instrument limitations or the extended times required for measurements. The seeds of Brassica oleracea cryo-stored in LN for 20 years maintained their viability up to 97% after storage, but the seeds stored at −18°C for 25 years had lower viability (11%). Cryogenic storage clearly prolonged shelf life of lettuce seeds with half-lives projected as ~450 and ~2600 years for fresh lettuce seeds stored in the vapor and liquid phases of liquid nitrogen, respectively.

Maintaining viability and genetic stability during storage is also important for cryopreserved in vitro shoot tips. After the development of vitrification methods, a few research publications appeared to suggest the exact viability and genetic stability of materials after long-term cryo-storage.It is demonstrated that it was possible to cryopreserve in vitro strawberry and pea meristems in LN for 28 years. In the case of in vitro grown strawberry meristems, there was no decrease in the percentage of viable meristems stored for 8 weeks or 28 years. This result significantly extends the reported duration of successful cryopreservation of plant meristems and provides corroborating evidence to the theory that plant meristems may be cryopreserved indefinitely. Also, for in vitro grown wasabi shoots, there was no significant differences of regrowth and morphological characteristics among 10 year cryo-storage, 2 h cryo-storage, treated control and control by the vitrification method. In biochemical analyses of sinigrin, which is a chemical precursor of the mustard oil, there was no significant difference of concentration level among them. All restriction fragment length polymorphism (RAPD) fragment patterns of 10 year cryo-storage tested were identical to those of 2 h cryo-storage. From these results, concluded that wasabi plants derived from shoot tips cryopreserved for 10 years by vitrification method are genetically stable. Suggested that callus formation might increase the frequency of genetic variants. However, optimized cryogenic techniques with suitable conditions can provide high survival after rewarming. High survival is attributed to a lower degree of injury incurred by explants during cooling and rewarming. This indicates that a high level of survival after cryopreservation is necessary to reducing genetic changes.

In potato, many studies on genetic integrity after cryopreservation have been reported. Morphological and phenotypic, cytological and molecular comparisons were conducted revealing that plant material was genetically stable as a result of cryopreservation. Potato shoot tips cryopreserved by the DMSO droplet method and stored in LN for several years was found to have no adverse effect on the regeneration rates. The genetic stability was also confirmed using morphological parameters, flow cytometric measurements and RFLP analyses, concluding that the cryopreservation technique may not induce somaclonal variation.

To date, many molecular and morphological study approaches for the genetic stability of cryopreserved plants had been reported. No observable significant differences have been observed in the regenerated material. Using a biochemical approach, the diosgenin content in Dioscorea deltoidea and the ginsenoside content in Panax ginseng were analyzed and found to be the same as those of controls. However, the possibility of some genetic changes may occur in cryopreserved plants. It is thus necessary to monitor the genetic stability of regenerated plants (true-to-type status) as a result of cryopreservation.

18

OOCYTE CRYOPRESERVATION

18.1 HISTORY

Oocyte cryopreservation is a method to preserve fertility for young cancer patients. Its indications have been extended to include the quarantine and storage of donor oocytes in egg donation programs, women with medical conditions that may culminate in premature ovarian insufficiency, and women who wish to safeguard fertility decline associated with ageing. In this review, we discuss the history of oocyte cryopreservation and its various clinical applications, with a focus on the safety of the procedure for cancer patients, especially those with hormone-dependent cancers such as breast cancers. We also discuss ethical considerations for women who are cryopreserving their oocytes to protect against age-related fertility loss, the optimal age to undergo oocyte cryopreservation, and the optimal number of oocytes to freeze.

Early attempts to cryopreserve human oocytes were hindered by a high incidence of aneuploidy and digynic polyploidy in the cryopreserved mammalian oocytes because of damage of the sensitive meiotic spindle at the metaphase II stage when mature oocytes were frozen. In 1986, a breakthrough was made using oocytes cryopreserved with the slow-freezing method1. However, oocyte survival on thawing was low, and safety of the children born from cryopreserved oocytes was a concern. In 2004, the ban on zygote and embryo cryopreservation in Italy provided an incentive to optimise oocyte cryopreservation. The first baby born from vitrified oocytes was reported in 1992.

The major obstacle restricting the clinical application of oocytes vitrification was the lack of an appropriate carrier.

The introduction of Cryotop enabled an extremely rapidcooling rate with minimal fluid volume and achieved an oocyte survival rate of >90% and the establishment of live births. Vitrification is preferred to slow freezing, with higher rates of oocyte survival, fertilisation, embryo cleavage, and clinical pregnancy6. Comparisons between fresh and vitrified oocytes showed comparable oocyte survival and clinical pregnancy rates. In view of the growing evidence on efficacy, the European Society for Human Reproduction and Embryology and the American Society for Reproductive Medicine affirmed that oocyte cryopreservation should no longer be considered as experimental in 2012 and 2013, respectively.

18.2 OOCYTE CRYOPRESERVATION

Oocyte cryopreservation is an option for women with various medical conditions to preserve fertility. It is also widely applied in oocyte donation and has been extended to women who wish to preserve their fertility against agerelated fertility loss. As the number and quality of oocytes decrease with advanced reproductive age, women who wish to cryopreserve oocytes should preferably consider this procedure before 37 years of age. The procedures of ovarian stimulation and egg retrieval are generally safe for women. The safety of children born from cryopreserved oocytes is reassuring, but the long-term outcome is lacking.

Controlled ovarian stimulation is the first step. Among the various protocols adopted from in vitro fertilisation (IVF) treatment cycles, the antagonist protocol is preferred because of its flexibility and shorter duration. Treatment is initiated in the early follicular phase of a spontaneous or combined oral contraceptive pill-induced menstrual cycle with daily injections of follicle-stimulating hormone. Daily gonadotropin releasing hormone (GnRH) antagonist injections are added on day 6 (fixed protocol) or when the leading follicles are $\geq$14 mm in diameter (flexible protocol), in order to prevent premature luteinising hormone surge. An ovulation trigger injection is administered when the leading follicle exceeds 18 mm in diameter. An GnRH agonist is almost always used to minimise the risk of ovarian hyperstimulation syndrome (OHSS). The oocytes are then retrieved by aspirating these follicles under transvaginal ultrasound guidance.

18.3 CLINICAL APPLICATIONS

In the early days, oocyte cryopreservation was largely reserved for women undergoing gonadotoxic chemotherapy or radiotherapy. As the technology evolves, it has become the standard protocol for cryopreserving donor oocytes to establish egg banks. Indications for oocyte cryopreservation have further been extended to include medical conditions other than cancers and to thosewho wish to delay childbearing for various reasons – often known as social reasons, elective oocyte cryopreservation.

18.3.1 Fertility Preservation for Cancer Patients

Cancer treatments often have detrimental effects on female fertility when involving irradiation to the pelvic organs, surgical removal of the ovaries, or systemic gonadotoxic agents. The extent of damage to the ovarian function depends on the type and dose of chemotherapeutic agent used, the patient's age, and the ovarian reserve at baseline. Alkylating agents such as cyclophosphamide are the most gonadotoxic, causing depletion of the primordial follicle pool, and thus compromising the ovarian reserve.

Advances in cancer treatments have substantially improved patient survival. Professional organisations such as the American Society for Reproductive Medicine14 and the American Society of Clinical Oncologyrecommend oncologists to discuss with their patients the impacts of chemotherapy on fertility during cancer treatment planning and refer patients to reproductive specialists to discuss the possibility of fertility preservation. Despite this, only 3036 (44%) of 6976 patients in the United States were counselled regarding the risk of infertility associated with chemotherapy.

In cancer patients undergoing ovarian stimulation for oocyte cryopreservation, there is a concern that supraphysiological levels of oestrogen during ovarian stimulation may stimulate growth of hormone-dependent cancers such as breast cancers. The addition of an aromatase inhibitor in combination with gonadotropins has been proposed.

Letrozole is an aromatase inhibitor that effectively reduces the peak oestradiol level and does not affect the oocyte yield. It is usually administered orally starting on the second or third day of a spontaneous cycle until the day of ovulation trigger, and then restarted after oocyte retrieval until menstruation returns19. Final egg maturation is achieved with a GnRH agonist instead of the conventional human chorionic gonadotropin. GnRH agonist triggering results in significantly decreased oestradiol level on the day of retrieval and a faster drop of oestradiol levels in subsequent days. GnRH agonist also reduces the risk of OHSS.

Breast cancer patients who underwent combined letrozole-gonadotropin ovarian stimulation showed no significant difference with controls in terms of short-term recurrence rate and relapse-free survival, as well as longer term follow-up of 5 (range, 1-13) years22 and 6.3 years (range, 3 months to 23.6 years). For cancer patients, there is a time constraint before the commencement of chemotherapy or radiotherapy. Novel ovarian stimulation protocols can shorten the time interval to oocyte retrieval. The random-start protocol initiates ovarian stimulation at the time of patient presentation rather than waiting for spontaneous menstruation. It is equally effective as conventional start protocol in terms of the total number of mature oocytes retrieved, oocyte maturity rate and fertilisation rate, irrespective of whether the stimulation is started in the late follicular phase or the luteal phase.

The number of oocytes retrieved is important in determining the probability of the patient having a successful live birth in the future. Cancer patients do not have much time to undergo repeated ovarian stimulation and oocyte retrieval cycles. The double stimulation or DuoStim protocol combines conventional follicular phase stimulation together with luteal phase stimulation, so that two oocyte retrieval procedures can be performed within the same ovarian cycle, maximising the total number of oocytes that can be retrieved for an individual patient.

The oocytes collected from the luteal phase stimulation have comparable rates of fertilisation, blastulation, euploid embryo, and pregnancy after embryo transfer, compared with oocytes collected from the follicular phase stimulation in IVF patients. Because of the low utilisation rate of cryopreserved oocytes in cancer patients, data regarding pregnancy and live birth rates from these two novel ovarian stimulation protocols are scarce and inconclusive. Nonetheless, the European Society of Human Reproduction and Embryology recognised these as optio s when there is urgency in cryopreserving oocytes.

18.3.2 Other Medical Indications

Other medical and iatrogenic conditions causing premature ovarian insufficiency include autoimmune, genetic and epigenetic, environmental, metabolic, and gynaecological conditions. The impact of endometriotic cysts (both the occurrence and removal) on the ovarian reserve is often overlooked. Women with endometriomas have a faster depletion of ovarian follicles and early (premature) ovarian insufficiency. It is pertinent that doctors looking after these women discuss or refer these women to an appropriate specialist who can offer the option of fertility preservation including oocyte or embryo cryopreservation.

18.4 OOCYTE DONATION PROGRAMMES

Oocyte donation is an alternative to adoption for women with premature ovarian failure who desire to bear infertility owing to the reduction in the number and the quality of oocytes or simply diminished ovarian reserve. Women who are carriers of a known genetic disease who wish to avoid passing the abnormality to the next generation can also benefit from oocyte donation. In the early days, fresh donor oocytes were used, and the menstrual cycles of the donor and the recipient had to be synchronised to allow transfer of the resultant (fresh) embryo in the same cycle. In addition, donor oocytes cannot be quarantined for infectious diseases such as HIV. Cryopreserved oocytes can be quarantined for the incubation period and kept in an oocyte bank. Recipients can have the embryos replaced at their 'convenience'.

Oocytes from a pool of donors can be allocated to more than one recipient, potentially improving the efficiency and reducing the cost and waiting time of oocyte donation programmes. Oocyte donation programmes enable study of the efficacy of the oocyte cryopreservation process, as frozen and fresh oocytes can be compared with regard to their capacity to be fertilised, cleaved, implant, and ultimately the live birth rate. The oocyte survival rate was reported to be 92% on thawing, with a comparable ongoing pregnancy rate between vitrified and fresh oocytes (43.7% vs 41.7%) in a single-centre prospective randomised study. However, the Society for Assisted Reproductive Technology reported a lower live birth rate in recipients of cryopreserved donor oocytes (43.2% vs 49.6%). A follow-up study reporting two additional years (2103-2015) of the national outcome data using the same database37 confirmed a lower live birth rate per recipient cycle started (39.7% vs 51.1%) and per embryo transfer (45.3% vs 56.4%), despite a similar number of embryos transferred. Reasons for the lower live birth rates in the US remain speculative. It may be attributable to the allotment of oocytes from one donor to several recipients rather than giving the entire cohort to a single recipient in order to reduce cycle costs for each recipient. However, it is impossible to evenly divide a cohort based on quality. This explains cancellation in rccipient cycles where all allocated oocytes fail to survive the thawing process. Although this may lower the live birth rate per recipient cycle started, it should not affect the live birth rate per embryo transfer.

Another possible explanation may be related to the freezing and thawing process in each centre. It can negatively affect the developmental potential of an oocyte and that of the subsequent embryo. Vitrification technique is difficult to master, as is the thawing process7-9,36-38. Embryologists in different centres may not have the same level of skill and experience. Finally, donor selection by commercial donor oocyte banks may not be as rigid as selection for fresh donor cycles by fertility centres.

Cryopreserved donor oocytes offer advantages over fresh ones. These include simplified access to a larger pool of oocytes particularly for ethnic minorities, ability to transport frozen oocytes over long distances thereby reducing the need for reproductive tourism and lowering the cost per treatment cycle. Although cryopreservation of donor oocytes has become a routine practice for some IVF centres, the American Society for Reproductive Medicine does not recommend routine donor oocyte banking until clinical data on safety and equivalent efficacy of oocyte cryopreservation become available.

18.5 ELECTIVE FERTILITY PRESERVATION

Women's fertility declines with age and the decline accelerates after the age of 35 years. The decline is due to reduction in oocyte quantity and quality , reflected by an increased miscarriage rate43 and a higher risk of carrying a fetus with chromosomal abnormalities. However, a study using shared oocytes between the donors and recipients showed that oocyte recipients had comparable pregnancy and delivery rates to their donors, indicating that uterine or endometrial factors do not seemcto be reduced in women of advanced reproductive age, and that the age-related decline in fertility can be largely overcome by using younger oocytes. When women of advanced reproductive age fail to conceive, they have the option to undergo IVF with donor oocytes. If they have previously cryopreserved their own oocytes, they can use their own oocytes for IVF and have their own genetic offspring. Compared with using donor oocytes, elective oocyte cryopreservation may potentially reduce the cost of multiple cycles of IVF. Thus, EPP may offer a solution to prevent unavoidable age-related infertility.

18.6 ETHICAL ISSUES

Society is divided on whether oocyte cryopreservation should be made available to women who wish to postpone child-bearing. This issue can be examined from the perspectives of autonomy, beneficence, nonmaleficence, and justice. Oocyte cryopreservation enhances women's reproductive autonomy by enabling them to decide whether, when, and with whom they wish to start a family.

It allows them to divert their energies towards alternative life goals such as education and career plans and not to rush to start a family because of the biological clock pressure. Reproductive autonomy is further enhanced by granting women, particularly single women, the control of their destiny. To generate embryos, the couple has to be legally married and those contributing the gametes will have a stake. Conflicts may ensue when the partner changes his mind and decides against having children, or when they separate, divorce, or posthumously. The Hong Kong Human Reproductive Technology Ordinance prohibits the transfer of embryos to persons who are not the parties to a marriage. It prohibits the posthumous use of embryos and any stored embryos have to be disposed of when the partner passes away. Furthermore, oocytes are generally not afforded the same status as embryos. The latter may conjure emotional or religious connotations upon disposal. Cryopreserving oocytes thus provides a more flexible option for single women and for those who prefer not to generate and then cryopreserve embryos. Women have a narrower reproductive window than men; their optimal fecundity spans less than two decades and is drastically reduced 5 to 10 years before the menopause.

Historically, women had to choose between childbearing or education and career development. This biological inequity can be partially offset by oocyte cryopreservation. Women can pursue other life goals or career plans without losing their natural reproductive potential and/or before they are able to find a suitable partner. Oocyte cryopreservation can thus foster gender equality. In regard to beneficence, oocyte cryopreservation, strictly speaking, is not fertility preservation; rather it preserves gametes for future attempts at reproduction. There is no guarantee that one or more live births will result from the cryopreserved oocytes. In fact, it may do harm by giving women a false sense of security so that they may delay childbearing until it is too late. The process is not without risks (non-maleficence). Controlled ovarian stimulation can lead to severe OHSS and its attendant complications. Oocyte retrieval is painful and invasive; it can be complicated by substantial internal haemorrhage and pelvic infection. These can result in infertility and even mortality. Nonetheless, if society accepts oocyte donors to undergo a medical intervention for no personal benefit, there is no reason why the same risks become unacceptable when a woman chooses to cryopreserve her own oocytes. Current adoption of the antagonist (stimulation) protocol, use of a GnRH agonist ovulation trigger and withholding embryo transfer can practically reduce the risk of severe OHSS to near zero.

Delaying childbearing until women are in their fifth decade or beyond may also do more harm than good, because older women have more obstetric and neonatal complications49. Their offspring may face negative psychosocial consequences of being born to a mother of advanced age, and may lose a parent relatively early in his/ her life. Children as caregivers are more likely to suffer from depression and behavioural problems, and they have less time for school activities and to make friends. They live in constant fear of losing one or both parents50. The long-term impact of early (before the age of 18 years) parental death has shown a negative impact in adulthood with regards to trust, relationships, self-esteem, loneliness, and isolation.

In terms of justice, oocyte cryopreservation is expensive and often not covered by health insurance and thus not every woman has access to this option, although Apple and Facebook offer EPP to their women employees as health benefits. There are social, racial and ethnic disparities in women's access to this option52. It is also important that women are not pressured into delaying childbearing just because their company is providing insurance coverage for oocyte cryopreservation, and that they will not be considered as non-committal to their career if they choose to have children early rather than cryopreserving their oocytes and defer motherhood to a later age46, thereby undermining the whole concept of reproductive autonomy.

18.7 OPTIMAL AGE FOR EPP

The conception rate, natural or via reproductive technology treatment, diminishes rapidly with advancing maternal age. This is largely due to the age-related decline in the quantity and quality of oocytes in a woman's ovaries. In a study of associations between maternal age and the prevalence of embryonic aneuploidy in over 15 000 consecutive trophectoderm biopsies, the lowest risk was seen in women in their mid to late twenties, and the risk of having no euploid embryo was lowest in women aged 26 to 37 years.

In a retrospective analysis of IVF patients, the chance of having a live birth for each fresh oocyte reduced gradually from 8.67% for women aged <30 years to 7.33% for those aged 35 to 37 years, and then rapidly to 1.06% for those aged ≥43 years. In women who used their autologous cryopreserved oocytes, the chance of having a live birth showed a similar downward trend. In another retrospective study of women who underwent EPP, the live birth rate was significantly higher at the age cutoff of 35 years (50% [95% CI=32.7-67.3] vs 22.9% [95% CI=14.9-30.9]).

Younger women may be able to maximise the number of 'good quality' oocytes for storage, but they may be less likely to use these oocytes in the future. The procedure and expense of oocyte cryopreservation may become unnecessary if they never have to use these oocytes. Cryopreserving oocytes at a later age may yield fewer and poorer-quality oocytes per cycle, and women may need multiple cycles to bank an adequate number of oocytes to have a reasonable chance of a live birth and this increases the cost. Using a mathematical model, in women who plan to delay childbearing until the age of 40 years, oocyte cryopreservation before the age of 38 years reduces the cost to achieve a live birth56. In a decision-tree model, the highest probability of live birth is seen when oocyte cryopreservation is performed at the age of <34 years (>74%), and that oocyte cryopreservation versus no action has the largest benefit at the age of 37 years and is most cost-effective. However, there is little benefit to cryopreserve oocytes for younger women aged 25 to years, because they may not need to use these oocytes in the end. Nonetheless, young women at risk of premature ovarian insufficiency should be counselled of the option of oocyte cryopreservation at an earlier age. In Hong Kong, the Council on Human Reproductive Technology specifies that "the maximum storage period for gametes or embryos stored for patients' own use in a reproductive technology procedure should not exceed 10 years". This means that for women younger than 32 years, their cryopreserved oocytes would have to be disposed of before the age of 42 years, thereby defeating the intention of EPP. Therefore, the optimal age for EPP – at least in Hong Kong – is between 33 and 37 years of age.

18.8 OPTIMAL NUMBER OF OOCYTES TO FREEZE

Every vitrified-warmed oocyte has a 5% to 7.4% chance of a live birth with an overall efficiency of 6.4%. The number of oocytes required varies with the woman's age at the time of cryopreservation. In a study of IVF outcome using vitrified oocytes, in women aged ≤35 years, the cumulative live birth rate increased sharply from five (15.4%) to eight (40.8%) oocytes, with an 8.4% gain for each additional oocyte banked, and the rate of increase plateaued at 10 to 15 oocytes (85.2%)55. This contrasted with a milder increase for women aged >36 years, their cumulative live birth rate was 5.1% (5 oocytes) and 19.9% (8 oocytes), reaching a plateau of 35.6% with 11 oocytes.

For women aged <38 years, 15 to 20 oocytes should be frozen to produce a 70% to 80% chance of having at least one live birth; and 25 to 30 oocytes should be frozen for women aged 38 to 40 years to produce a 65% to 75% chance of having at least one live birth54. Based on a mathematical model, women aged 34, 37, or 42 years, each with 20 mature oocytes frozen, are expected to have a 90%, 75%, and 37% chance of having at least one live birth, respectively; and 10, 20, and 61 oocytes should be frozen to produce a 75% likelihood of having at least one live birth58. All these studies are of single-centre, retrospective, and have not been validated or reproduced. In a study from Reprogenetics regarding the euploidy rates in donor egg cycles among 42 fertility clinics in the United States, the average euploidy rate per centre ranged from 39.5% to 82.5%, whereas the mean expected rate of euploidy was 68.4%. The implication of these findings is that centre-specific assisted reproductive technology practices and outcomes can vary considerably, including oocyte cryopreservation.

18.9 SAFETY FOR WOMEN

The risks associated with oocyte cryopreservation involve controlled ovarian stimulation and oocyte retrieval. The risks of oocyte retrieval include pelvic infection, internal bleeding, inadvertent damage to other intra-abdominal organs, and ovarian torsion. OHSS is the most serious complication. OHSS can be classified as mild, moderate, severe, and critical60. Mild and moderate OHSS is characterised by abdominal pain, enlarged ovaries, and weight gain, with an incidence of 3% to 6%; it can be managed conservatively, as it is self-limiting and generally resolves upon resumption of menstruation. Severe and critical OHSS can occur in 1% to 3% of IVF cycle. These women have fluid retention in the form of ascites and sometimes pleural effusion, massive ovarian enlargement, haemoconcentration and oliguria, and venous thrombo-embolism. It is potentially life-threatening and worsens by an ensuing pregnancy following fresh embryo transfer. The rise in endogenous human chorionic gonadotropin can exacerbate its symptoms and duration.

Elective cryopreservation of embryos can prevent OHSS61,62. Women who are cryopreserving oocytes will not have embryo transfer, and hence late-onset OHSS can be avoided. A systematic review and meta-analysis demonstrated a significantly lower incidence of OHSS in the GnRH antagonist protocol compared with the GnRH agonist protocol (OR=0.59, 95% CI=0.42-0.82)63. The risk can be further reduced with the concomitant use of a GnRH agonist for final oocyte maturation instead of the traditional human chorionic gonadotropin. A Cochrane review showed that the incidence of moderate to severe OHSS was significantly lower in the GnRH agonist trigger group compared with the human chorionic gonadotropin group (OR=0.10, 95% CI=0.01-0.82)64. Therefore, the antagonist stimulation protocol coupled with GnRH agonist trigger is recommended, as it minimises the risk of OHSS.

18.10 FOR CHILDREN

In the early days of oocyte cryopreservation, there were concerns about the risk of meiotic spindle damage leading to an increased aneuploidy and digynic triploidy in the subsequent embryos derived from cryopreserved oocytes. The aneuploidy rate by fluorescence in situ hybridisation showed that the percentage of embryos with aneuploidy in the cryopreservation group was comparable to that observed in the controls. A review of over 900 live births derived from cryopreserved oocytes also showed no increased risk of congenital anomalies, compared with the general population. In a retrospective study of 2252 live babies born from cryopreserved oocyte in the Italian National Register, only 0.9% had congenital malformations reported.

Despite these, the safety of long-term cryopreservation of oocytes is lacking. Cryopreserving oocytes for up to 4 years did not seem to affect the IVF success outcomes. The euploid rate of blastocysts was similar to those derived from fresh oocytes, after a median of 3.5 years (maximum, 6 years). There is no study reporting the long-term follow-up of children born with oocyte cryopreservation, especially when the oocytes are cryopreserved for a prolonged period.

19

POTATO SHOOT TIP CRYOPRESERVATION

19.1 INTRODUCTION

Potato belongs to the botanical family Solanaceae and within it to the genus Solanum, which consists of more than 2,000 species. Apart from the cultivated potato Solanum tuberosum L. ssp. tuberosum, seven other cultivated and 228 wild potato species have been identified. Taxonomically, they all belong to the section Petota containing the tuber-bearing potatoes. Spooner and Salas have recently revised the taxonomy of potato species by combining the seven cultivated species into one single potato species (S. tuberosum) with eight cultivar groups. The cultivated potato originated from wild Solanum species in the Andes of South America and the coastal strip of central to southern Chile. After the first potatoes were brought to Europe in the late sixteenth century, this crop plant was rapidly distributed and is now utilised all over the world, mostly in temperate regions. Today, there are more than 4,500 varieties of S. tuberosum ssp. tuberosum. Potato is known to have the richest genetic diversity of any cultivated plant because of its high number of varieties and related species. To prevent the loss of potato genetic resources, long-term conservation of plant material is accomplished in genebanks, genetic resource centres and private collections worldwide.

Potato is highly heterozygous, and the sexually produced seeds are, therefore, not true to type. Thus, maintenance of cultivated potato accessions is not possible by true potato seeds. Genebank stocks are therefore derived from vegetative material. To preserve the genetic resources of this crop plant for the long term, cryopreservation is the best current option. Cryopreservation is the storage of biological material at −196 °C in liquid nitrogen (LN) in such a way that viability is maintained and plant regeneration is obtained after rewarming. Potato cryopreservation was started in 1977 with two-step cooling procedures and ultra-rapid freezing techniques. The first protocols were time consuming because of the need to use programmable freezers and resulted in low survival. Later on protocols were refined and new methods developed, such as the dimethyl sulfoxide (DMSO) droplet method, vitrification, encapsulation/dehydration, encapsulation/ vitrification and droplet vitrification methods. These protocols improved the results in potato cryopreservation in terms of survival and regeneration frequencies and they are now routinely applied in several institutes to build up cryopreserved potato collections. This is concentrated primarily on currently used potato cryopreservation protocols. In this chapter, we illustrate in detail (a) historically important, (b) currently used and (c) the most recent advances in potato shoot tip cryopreservation with various potato species and varieties. We compare and contrast evolving trends in potato cryogenic research based on analysis of original and contemporary techniques, such as proteomics and differential scanning calorimetry (DSC) and explain how advances in understanding of cryogenic science have shaped improve- ments in techniques. We conclude with a discussion as to how the most recent advances (e.g., proteomics and DSC) may shape the future of potato cryogenic research.

19.2 CRYOPRESERVATION TECHNIQUES

19.2.1 Two-Step Cooling

The first cryopreservation studies on potato were carried out by, who used a two-step cooling procedure. In this method, cell water content was reduced by freeze-induced dehydration. Firstly, tuber sprouts and axillary buds were cryoprotected with various glycerol and/or sucrose solutions. Then they were cooled slowly in the vapour phase of LN and later directly in LN. obtained between 7% and 18% survival of tuber sprouts and axillary buds with this protocol. Regeneration percentages of up to 21% and 33% were obtained for 'Norland' and 'Red Pontiac', applying two-step cooling. Explants were cooled slowly with a speed of0.2–0.3 °C min−1 until −35 °C followed by immersion in LN. applied quite a similar protocol using 0.3 °C min−1 until −30 °C and then direct cooling in LN. Shoot regeneration after rewarming ranged between 4% and 85% depending on species. Overall, the method described is, however, rather complicated and time consuming and needs special equipment.

19.2.2 Ultra-Rapid Cooling

In comparison to the previously mentioned two-step cooling method, ultra-rapid cooling is easier in handling. In ultra-rapid cooling technique, shoot tips were isolated, precultured and cryoprotected with MS medium containing sucrose and 10% *v*/*v* DMSO. Subsequently,each single shoot tip was directly plunged into LN on a tip of a hypodermic needle. Rewarming was carried out quickly through transferring the needles directly into a vial of MS medium with benzylaminopurine (BAP) at 35 °C. After that shoot tips were removed from solution and were transferred to the regeneration medium. Presented one of the first successful cryopreservation protocolsfor potato shoot tips with as high survival as 20% and shoot regeneration capacity of10%. A significant step in this method is the quick cooling, so that water is transformed directly to amorphous ice, and the reverse occurs with rapid rewarming, so that devitrification does not occur on return to room temperature regime.

Table. 1 Examples of cryopreservation techniques applied for potato shoot tips

Method	Species	Number of cultivars of accessions	Mean regeneration (%)
Two-step cooling	Solanum chaocense	1	4-5
	Solanum chaocense	1	67
	Solanum chaocense	1	0-93
	Solanum huancabambense	1	8
	Solanum iopetalum	1	0
	Solanum jamesii	1	73
	Solanum phureja	3	15
	Solanum pinnatisectum	1	13
	Solanum polydenium	1	6
	Solanum sparsipilum	1	71-89
	Solanum stenotomum	1	77
	Solanum stoloniferum	1	15
	Solanum tubersum ssp. andigena	1	70
	S. tuberosum	2	9-8
	S. tubersum	2	0-1
	S. tubersum	5	21-85
	S. tubersum	2	0-33
	S. tubersum	14	22
	Solanum goniocalyx	1	10.6
	Solanum microdontum	1	48.0
	Solanum pinnatisectum	1	5.5
	S. tubersum	1	19-27
Ultra-rapid cooling	S. tubersum	2	1.3-18.0
	S. tubersum	2	1.3-42.7
	S. tubersum	4	4.9-20.0
Vitrification	S. tubersum	35	23
	Solanum chaucha	1	30
	S. goniocalyx	9	29
	S. goniocalyx x S.	10	47
	stenotomum	38	41
	S. phureja	57	33

Method	Species	Number of cultivars of accessions	Mean regeneration (%)
DMSO droplet method (droplet feezing)	S. stenotomum	4	45
	S. stenotomum X		
	S. goniocalyx	61	29
	S. tuberosum ssp.	4	21.3-58.0
	S. tuberosum	5	5-56
	S. tubersoum	2	2.3-71.0
	S. Suberosum	1	25.3-55.8
	Solanum acaule	1	62.4-85.6
	Solanum demissum	10	2.5-22.0
	S. tuberosum	>1,000	58
	S. tubersosum	2	12.2-75.0
	S. tubersosum	260	48
	S. tubersosum	4	0-29.6
	S. tubersosum	200	41
	S. tubersosum	219	40
	S. tubersosum	2	68.2-87.1
Droplet vitrification	Solanum chacoense	2	n.d.
	Solanum commersonii	2	64.0-91.7
	S. goniocalyx	2	72.9-87.7
	S. stentomu	1	15.0-87.7
	S. stentomu	1	n.d.
	S. tuberosum ssp. andigena		
	S. tubersosum	1	50-80
	S. tubersosum	3	8.8-55.5
	S. tubersosum	6	78.7-94.4
	S. tubersosum	4	8-47
	S. tubersosum	1	n.d.
	S. tubersosum	1	33.4-88.7
	Solanum x juzepczukii	1	n.d.
Encapsulation/ dehydration	Solanum x ajanhuiri	1	n.d.
	Solanum acaule	1	25-37
	Solanum \brachycarpum	1	0-36
	Solanum guerreroense	1	9-29
Emcapsulation/ dehydration	Solanum iopetalum	1	0-13
	Solanum phureja	2	42-78
	S. phureja	1	0.0-9.3

Method	Species	Number of cultivars of accessions	Mean regeneration (%)
	S. phureja	1	0-73
	S. tuberosum	2	12-71
	S. tuberosum	1	0-32
	S. tubersum	1	2.9-58.8
	S. tuberosum	2	0-61
Encapsulation/	S. tubersoum		
Vitrification		14	0-70

Another ultra-rapid cooling method (developed for the potato cryopreservation collection in the Czech Republic) involved culturing in vitro donor plants on semi-solid hormone-free MS medium containing a reduced amount of nitrogen. Four-day-old nodal cuttings were incubated in 2 M sucrose for 5 days; then shoot tips were excised and put on sterile filter paper moistened with 0.7 M sucrose for 20 h. Shoot tips were then transferred to aluminium foils and desiccated over silica gel for 1.5 h with a subsequent direct cooling in LN. Shoot tips were thawed in sterile water at room temperature and transferred to regeneration medium with growth regulators (0.5 mg l^{-1} indole acetic acid (IAA), 0.5 mg l^{-1} kinetin and 0.2 mg l^{-1} gibberellic acid (GA_3)). The average regeneration percentage was 23% in the 35 accessions tested. The main difference between this method compared to other rapid cooling protocols is the avoidance of DMSO and the use of aluminium foils for supporting the shoot tips.

19.2.3 Vitrification

The process of vitrification is described as solidification of liquids without crystallisation. When the solutions within the cells are vitrified, they are in a "glassy state" and amorphous. They lack organised structures but possess the mechanical and physical properties of a solid. Vitrification is one of the main and most widely applied plant cryopreservation methods. Most vitrification protocols apply the plant vitrification solution 2 (PVS2). It is a mixture of cryoprotectants containing 30% w/w glycerol, 15% w/w ethylene glycol, 15% w/w DMSO in MS medium supplemented with 0.4 M sucrose. Published the first vitrification protocol for S. tuberosum cultivars 'Kufri Badsha', 'Kufri Chandramukhi', 'Kufri Lalima', 'Kufri Lauvkar' and 'Kufri Sindhuri'. In this vitrification method, apical shoot tips of 0.5–0.7 mm in size were first isolated from 30-day-old plantlets and precultured for 2 days on filter paper discs over half strength liquid MS medium supplemented with 8.7 μM GA3 and different combinations of sucrose (0.3, 0.5 and 0.7 M) and mannitol (0, 0.2 and 0.4 M) under a 16 h photoperiod with approx. 40 μmol m−2 s−1 light intensity, at 24 °C. Shoot tips were then loaded in 20% PVS2 (30 min, 24 °C), afterwards in 60% PVS2 (15 min, ice bath) followed by 100% ice cold PVS2 (5 min) or directly in 100% PVS2 (5 min). Five shoot tips each were transferred into one cryovial filled with 0.7 ml of PVS2 and finally plunged into LN. After quick rewarming at 35 °C in a water bath for 1 min, cryovials were flushed with dilution medium (MS with 1.2 M sucrose) and incubated for 30 min. Then shoot tips were cultivated on semi-solid regeneration medium [MS with 0.2 M sucrose, 5.8 μM GA3, 1.0 μM BAP and 6 g/l Nobel agar (HiMedia, India)] and cultured under diffuse light (16 h photoperiod, approx. 6 μmol m−2 s−1) at 24 °C for 1 week. Subse- quently, shoot tips were transferred to standard medium (MS medium with 0.09 M sucrose, 2.9 μM GA3 and 6 g l-1 agar) under light conditions of approximately 40 μmol m−2 s−1 with 16 h photoperiod, at 24 °C. For successful regeneration, preculture with sucrose in combination with mannitol was necessary. Concentrations of mannitol above 0.2 M were detrimental. The highest response was found applying a preculture with 0.3 M sucrose and 0.2 M mannitol followed by sequential vitrification with highest shoot regeneration rate of 56%.

In the International Potato Center (CIP), cryopreservation research was initiated in 1995 using the vitrification method developed by Steponkus and co-operators. In this method, the vitrification solution contained ethylene glycol: sorbitol: BSA (bovine serum albumin) in the ratio of 50:15:6% *w*/*w* and samples were cooled in polypropylenestraws. A number of genotypes (197) were cryopreserved with an average regeneration rate of 46%. Using this protocol, approximately 400 potato landraces were tested with highly variable responses. Of the different genotypes, 30% did not show any survival, and in 30% of the genotypes, the regeneration capacity was less than 15%. In 2004, this protocol was further optimised, and CIP (in cooperation with the Laboratory of Tropical Crop Improvement, Katholieke Universiteit Leuven, Belgium) adopted the droplet vitrification method (see below) initially designed for cryopreserving banana shoot tips also for potato. Applied the vitrification method to four *S. tuberosum* cultivars ('Ackersegen','Blaue Schweden', 'Carnea' and 'Désirée'). In their study, one-week-old nodal cultures were used to isolate apical explants from the sprouted nodal buds (1 mm in length). Material was incubated overnight in MS medium with 30 g/l sucrose. Next day shoot tips were placed in loading solution (MS medium with 0.4 M sucrose and 2 M glycerol) for 20 min followed by incubation in PVS2 for 30 min. Cryovials filled with 1.5 ml of PVS2 together with ten shoot tips each were directly cooled in LN. Rewarming was performed in a water bath at 38 °C for 2 min. Then explants were washed three times with MS solution containing 1.2 M sucrose and transferredto regeneration medium, solidified with 0.8% agar). Average shoot regeneration capacity of the described original protocol was 58.0%, which was higher compared to experiments applying cold preculture at 21/8 °C for 7 days (45.9% regeneration capacity), liquid regeneration medium (21.3% regeneration capacity) or filter-sterilised DMSO (57.5% regeneration capacity). Using vitrifica- tion, significantly higher shoot tip regeneration frequency was obtained (58.0%) in comparison with DMSO droplet method. Incontrast to this, treatment at 10 °C for 3 weeks and application of the ice-blocking agent Supercool X1000 within the PVS2 resulted in regeneration of 71% and 55% in *S. tuberosum* cultivars 'Atlantic' and 'Superior', respectively.

19.2.4 DMSO Droplet Method (Droplet Freezing)

DMSO droplet method is an optimisation of the ultra-rapid cooling method. The expression 'droplet' refers to droplets of cryoprotectant on an aluminium foil, onto which the shoot tips are placed for coolingand rewarming. Also droplets of agarose together with a single shoot tip each and liquid medium have been used for regeneration. The innovative idea of usingaluminium foils originated from Who cryopreserved cassava shoot tips on foils using a two-step cooling method. adopted the idea of a fast cooling method also for potato shoot tips. The foils are a good carrier in transferring a large number of shoot tips at once quickly into and outof LN compared with using hypodermic needles, where only one shoot tip at a time can be transferred. Furthermore, aluminium is a very good heat conductor, which is significant for quick cooling and also for rewarming of explants. used foils of 20×7×0.03 mm in size, so that two foils fit well into one cryovial. Regarding the state of water during cooling, no information about the stateof the water (ice or glass) in the droplets was given in their study. However, as observed by using cassava, it is highly probable that vitrification may occur.

In the original droplet freezing protocol, shoot tips were isolated from 10 cm long in vitro plantlets to explants of 2–3 mm in length and 0.5–1 mm width, and then incubated in MS medium with 30 g/l-1 of sucrose, 0.5 mg/l-1 zeatin riboside, 0.2 mg/l-1 GA3 and 0.5 mg/l-1 IAA overnight. On the second day, shoot tips were placed in cryoprotectant solution (MSTo medium with 10% DMSO) at room temperature for 2 h. They were then transferred into droplets of 2.5 µl cryoprotectant solution and placed onto small heat-sterilised aluminium foils and directly cooled in LN. Shoot tips were rewarmed quickly by putting aluminium foils in liquid MS medium at room temperature. For regeneration, shoot tips were placed in drops of warmed agarose in Petri dishes. After solidification of drops, liquid MSTo solution was added. The average survival of the cryopreserved 219 accessions was 80%, and plant regeneration capacity was 40% analysed the influence of phytohormones (0.5 mg/l-1 zeatin riboside, 0.2 mg/l-1 GA3 and 0.5 mg/l-1 IAA) in the agarose drops on plant regeneration for 10 potato accessions. The original protocol according without any phytohormones in the agarose drops showed less plant regeneration frequency in comparison with the application of agarose drops having phytohormones. Plant regeneration was between 2.5% and 22% with a mean of 13.2% (Table 2). By applying the DMSO droplet method, compared the original DMSO droplet protocol with slightly modified versions for S. tuberosum cultivars 'Ackersegen', 'Blaue Schweden', 'Carnea' and 'Désirée'. In their study, no survival and regeneration was seen at all, if autoclaved DMSO was used instead of filter-sterilised DMSO. Application of solid medium during regeneration enhanced regeneration rate to an average of 15% in comparison with the original protocol using liquid medium (average regeneration rate of 13.8%). The best results with an average of 29.6% plant regeneration were given as a result of applying cold preculture of the potato donor plants (21/8 °C day/night temperature, 8 h photoperiod, 7 days) prior to shoot tip isolation. Similar results were subsequently observed. Improvements in regeneration results from 34.6% to 45.2% were found, when an alternating temperature preculture (22/8 °C day/night temperature, 8 h photoperiod, 7 days) was applied for the donor in vitro plants prior to shoot tip isolation carried on by a comparison between the original method of applying agarose drops and liquid medium for regeneration and by using solid regeneration medium (MSTo with 10 g/l-1 agar). It was found that the regeneration frequency was increased on solid medium in explants of S. tuberosum 'Désirée' and 'King Edward' and the regenerated shoots showed better plantlet structure with less callus formation. The combination of alternating temperature preculture and solid regeneration medium caused the highest average regeneration rate of 58.7%. As a result of those positive findings in the IPK Germany, the DMSO droplet method, which is the routine cryopreservation method, was improved by the application of alternating temperature preculture and solid medium for regeneration. In the IPK, up to date 1,119 accessions have been stored (Table 1,) with a mean regeneration percentage of 46%. This collection includes the accessions formerly stored at German Collection of Microorganisms and Cell Cultures (DSMZ, Germany) and the Institute of Crop Science of the Federal Agricultural Research Centre (FAL, Braunschweig, Germany) which have subsequently been integrated into the IPK collection.

The DMSO droplet method is a simple freezing protocol, which is now applicableto a wide variety of potato accessions. The method is quick and easy and the procedure is also relatively cheap. Problematic is still the low regeneration frequency of certain potato genotypes. Therefore, additional studiesare needed to improve the results for accessions showing low regeneration capacity.

19.2.5 Droplet Vitrification

Droplet vitrification, which combines elements of the DMSO droplet method and vitrification, is a relatively new method in potato cryopreservation. In the droplet vitrification method, shoot tips are plated on aluminium foils during cooling, like in the DMSO droplet method, but other types of cryoprotectants, such as PVS2 are used. The use of aluminium foils makes the transfer of shoot tips into and out of LN easier and faster. This can be important during application of PVS2, because slightly longer incubation times can be toxic for shoot tips. Using the droplet vitrification, first successful results were presented in three potato cultivars ('Désirée', 'Ostara' and 'Santé'). The plants were multiplied on hormone-free MS medium and maintained at 24 °C with photosynthetically active radiation of 39.06 μmol m−2 s−1 and 16 h photoperiod. Shoot tips with different sizes (1–6 mm in length) were isolated with hypodermic needles from four- to eight-week-old plants and incubated on filter paper with liquid MS medium containing 0.4 mg/l-1 of GA3, 0.5 mg/l-1 of zeatin and 0.2 mg/l-1 of IAA at 24 °C for 24 h. Afterwards various types of precultures with different sugar or sugar alcohols (sucrose, glucose, mannitol and sorbitol) were performed at 24 °C for 24 h. Then shoot tips were dehydrated in droplets of 4 μl with PVS2 on aluminium foil (0.6×1.5 cm) at room temperature for 10–30 min, before their transfer into pre-cooled cryovials and then into LN. Rapid warming was carried out by shaking aluminium foils in liquid medium at room temperature. For regeneration shoot tips were put onto semi-solid medium with 3.5 g/l 1 agar at 24 °C for 20 days. Afterwards elongated shoots were transferred to solidified medium with 7 g/l-1 of agar. The highest shoot regeneration frequencies were 55% in 'Désirée', 51% in 'Ostara' and 46% in 'Santé'. Used plantlets that were precultured at 6 °C on MS medium with 0.07 M sucrose for 3 weeks. Afterwards shoot tips were isolated (2–3 mm in size) and incubated in loading solution (MS with 2 M glycerol and 0.4 M sucrose) for 20 min. Shoot tips were transferred to PVS2 at 0 °C for 50 min, plated on aluminium foil strips, put into cryovials and plunged into LN. Rewarming was done rapidly in MS medium enriched with 1.2 M sucrose for 20 min. Explants were then cultivated on MS medium with 0.04 mg/l-1 kinetin and 0.1 mg/l-1 GA3 and 0.3– 0.1 M sucrose concentrations, reducing the sucrose level by 0.1 M per day. Regeneration capacity of shoot tips was genotype-dependent and ranged between 8% in S. tuberosum 'Wila Yari' to 47% in 'Désirée' also analysed certain biochemical compounds, such as phospholipids, glycolipids, aromatic amines and polyamines to find links to cryopreservation ability. They found that frost resistant genotypes had significantly higher regeneration rates after cryopreservation and that the linoleic acid content was positively correlated with tolerance to cryopreservation. It was also concluded that addition of putrescine to preculture medium can further enhance recovery rates for poorly regenerating genotypes. In recent years, the number of species successfully cryopreserved using the combined droplet vitrification protocol with rapid cooling and rewarming is continuously increasing, whereas two-step cooling methods are now used less often and then in most cases only for callus and suspension cultures. The reason for the increasing use of droplet vitrification lies in its fast and easy application without expensive cooling device which is necessary for two-step cooling.

19.2.6 Encapsulation/Dehydration

In 1990, a new cryopreservation technique for plants called encapsulation/ dehydration was developed by Fabre and Dereuddre using potato species Solanum phureja. In this method, donor plants were subcultured on medium A (MS minerals with Morel and Wetmore vitamins, Na-Fe- EDTA, 30 g/l-1 sucrose, 8 g/l-1 agar, pH 5.85). Apical sections of eight- to ten-week- old plantlets were transferred to medium A for 7 days. Shoot tips with three to four leaf primordia (approximately 0.5 mm in length) were excised and incubated on medium B (Morel macronutrients), Heller micronutrients, vitamins, Na-Fe-EDTA, 8 g/l-1 agar, 0.1 M sucrose and growth regulators, pH 5.8). Donor plants and shoot tips were cultured at 20 °C with 70% relative humidity and light intensity of 50 μmol m−2 s−1. Afterwards shoot tips were suspended in liquid calcium-free medium B supple- mented with 3% (w/v) sodium alginate. This mixture was dropped with the explants into liquid medium B containing 100 mM calcium chloride to form alginate beads. Beads of approximately 3 mm in diameter containing one or two shoot tips inside each bead were precultured in liquid medium B supplemented with sucrose. The beads were then dehydrated under sterile air flow at room temperature for 4 h. Encapsulated shoot tips were cooled directly by LN or they were progressively cooled from 20 to −40 °C before immersion into LN. Beads were slowly warmed in air at room temperature and transferred to medium B for further regeneration. The best results (9.3% regeneration) were found using a combination of preculture with 0.5 M sucrose and two-step freezing (Table 2).

Optimization of the above described protocol was made using three different diploid accessions of S. phureja and two tetraploid S. tuberosum cultivars ('Charly' and 'Hertha'). Donor materials were cultured at temperature of 21± 1 °C with a photoperiod of 16 h per day and light intensity of 25– 30 μmol m−2 s−1. Shoot tips were excised from nodal cuttings, grown for 3, 7, 14 and 21 days, and then incubated on culture medium. After encapsulation, preculture in liquid medium was applied during the next 2 days with stepwise increasing sucrose (0.3, 0.5, 0.75 and 1 M) to obtain a final concentration of 1 M. Beads were kept in 1 M sucrose medium for 1–7 days. In addition, direct preculture of encapsulated shoot tips in sucrose-enriched medium (0.1, 0.3, 0.5, 0.75, 1, 1.25 and 1.5 M) was tested for 2 days. After preculture, shoot tips were dried and dehydrated for 4.5 h in airtight boxes over silica gel to reach a water content of 0.20–0.22 g g-1 dry weight. Encapsulated shoot tips were transferred to cryovials and directly immersed in LN. Rewarming was done slowly in air at room temperature. For regeneration, shoot tips were plated on medium with growth regulators for 1 week and then later plated on medium supplemented only with GA3 to avoid callus formation. The highest response was found with direct preculture of encapsulated shoot tips in 0.75 M sucrose medium for 2 days prior to cooling in LN. Regeneration was between 50% and 64% in two S. tuberosum accessions tested, whereas diploid S. phureja was more tolerant to dehydration and had higher or similar shoot regeneration rates (64%, 69%, and 78%; Bouafia et al. Applying encapsulation/dehydration to six different Solanum species, found considerable variation in early and late recovery responses and between experimental repetitions. However, plant regeneration was attained for all cryopreserved species with the highest average regeneration percentages in Solanum acaule (32%) and the lowest in S. iopetalum (7%, Table 2). Differences in regeneration between diploid (2n=2x =24) (S. phureja) and tetraploid (2n =4x=48) species (S. tuberosum, S. acaule) were not found in contrast to the observations of only hexaploid (2n =6x = 72) species showed reduced percentages in regeneration capacity (between 7% and 19%) in comparison to diploid and tetraploid accessions (29–31%). Variations between experiments were also found in S. tuberosum 'Brodick' and 'Golden Wonder', when the encapsulation/ dehydration method was applied. Shoot regeneration was between 7% and 41% ('Brodick') and 0–61% ('Golden Wonder') in six repetitions. Different preculture treatments were performed to increase plant regeneration after cryopreservation using encapsulation/dehydration. These precultures were: (1) sucrose preculture of isolated shoot tips [1 day on MS-P medium (MS without casein and myo-inositol, and with reduced N content 25% NH4NO3 and 50% of original KNO3) with 0.7 M sucrose], (2) sucrose- glycerol preculture of isolated shoot tips (1 day on MS-P medium with 0.3 M sucrose and 0.5 M glycerol), (3) drought hardening of donor plants (4 ml of 2 M sucrose solution were added to donor plants for 5 days followed by shoot tip isolation) and (4) drought hardening plus sucrose preculture (as in case 3 and additional shoot tip preculture for 1 day in MS-P medium with 0.7 M sucrose). The highest survival (79%) and shoot regeneration responses (59%) were found in S. tuberosum 'Désirée' as a result of drought hardening plus sucrose preculture. Those results can be explained by the function of sucrose as cryoprotectant to stabilise membranes, to decrease the phase transition temperature of dehydrated phospholipid bilayer and to prevent protein denaturation during dehydration. Overall, a clear progress in optimising the encapsulation/dehydration method can be seen from the first experiments with two-step cooling and limited growth recovery to more recent experiments with high survival after rapid cooling.

19.2.7 Encapsulation/Vitrification

Encapsulation/vitrification methods combine certain elements of vitrification and encapsulation/dehydration. Analysed 14 potato cultivars applying this method. Plantlets were grown on solidified basal medium (MS medium with 0.5 g/l casamino acid, 30 g/l sucrose, 2.5 g/l gellan gum) at 23 °C and 16 h photoperiod with light intensity of 96 mol m−2 s−1. For cryopreservation, axillary meristems from nodal segments with 5 leaf primordia (approximately 1 mm in size) were isolated and precultured on basal medium with 0.3 M sucrose, 1 mg/l-1 GA3, 0.01 mg/l-1 BAP and 0.001 mg/l-1 NAA. Meristems were suspended in calcium-free MS medium supplemented with 2% (w/v) Na-alginate and 0.4 M sucrose. The mixture containing the explants was transferred into 0.4 M sucrose solution plus 0.1 M CaCl2 for 30 min, at 25 °C, to form beads. Beads were osmoprotected with MS medium with 0.4–1.6 M sucrose and 2 M glycerol and plant hormones as in basal medium on a rotary shaker (60 rpm) at 25 °C for 90 min. Encapsulated meristems were dehydrated with PVS2 at 0 °C for different lengths of time. Beads with PVS2 were shaken (45 rpm) in a water bath. In total, 10–15 beads were suspended in a volume of 1 ml PVS2 solution per cryovial and plunged directly in LN. For rewarming, cryovials were put into water bath at 38 °C for 3 min. After removing the PVS2 solution, rewarmed beads were washed with 1 ml of 1.2 M sucrose solution for 10 min. For regeneration, beads with shoot tips were plated on basal medium for 1 day. Then they were transferred to basal medium with 0.0005 mg/l-1 GA3. Regeneration capacity was determined for all 14 analysed potato accessions and it was the highest (70%) as a result of osmoprotection with 0.6 M sucrose and 2 M glycerol for 90 min, followed by 3 h incubation with PVS2 (Table 2). No improvement was found in cold treated nodal segments (4 °C, 3 weeks) in comparison with the control explants (without cold treatment). Rate of shoot formation and speed of growth were much higher and faster using encapsulation/vitrification in comparison with encapsulation/dehydration. In conclusion, the encapsulation/vitrification method is easy to handle (less time for bead dehydration by PVS2 is required instead of placement in laminar airflow or silica gel) and a large number of explants can be cryopreserved in the same time. Therefore, it seems to be relatively suitable for cryopreservation even on a large scale.

19.3 PARAMETERS AFFECTING CRYOPRESERVATION

Many elements which encompass cryopreservation, such as condition of donor cultures, shoot tips, preculture etc. play a significant role in the success of potato shoot tip cryopreservation. Some of these elements and their influence on regeneration capacity and plant vigour with their flow-on effects on the success of cryopreservation are reviewed below.

19.3.1 Donor Cultures

Cold preculture of donor plants prior to cryopreservation is known to improve results for woody and herbal species, which are naturally able to cold acclimate at low temperatures. Also in other species like Zoysia matrella, Zoysia japonica and yam (Dioscorea spp.), which are usually not able to acclimate to cold temperatures, cryopreservation results could be improved after exposure to low temperatures. In potato, low temperature preculture was applied before cryopreservation, when encapsulation/vitrification and droplet vitrification methods were used, but with weaker results when compared with the original protocols . Cold preculture at 4 °C for 3 weeks at 4 °C for 1 week or alternating temperature of 21/8 °C (day/night) for 1 week have been applied. However, with the DMSO droplet method, used an alternating temperature preculture of 21/8 °C (day/night) for 1 week and were able to improve cryopreservation results in four potato accessions. In a similar experiment applying also the DMSO droplet method and alternating temperature preculture at 22/8 °C (day/night) for 1 week, found significant improvement in regeneration ability of S. tuberosum 'Désirée', 'King Edward' and the wild species Solanum demissum using a solid regeneration medium. Based on the results of the conclusion can be drawn that the improvement in regeneration after alternating temperature preculture is not simply based on cold acclimation. The increase of soluble sugars found in the tissue is surely a step towards an acclimation response. However, since no complete cold acclimation has been found in the analysed potato samples, this better response should rather be explained as the result of a general increase of vigour of the in vitro plantlets due to the alternating temperature regime. Improvement in regeneration rate was also found applying cold preculture at 10 °C for 3 weeks or at 6 °C for 3 weeks. Based on those results, it could be concluded that cold preculture conditions need to be warmer than 6 °C for potato in vitro cultures to receive better cryopreservation response. Alternating temperature during preculture can also enhance recovery. Sometimes the effect of cold acclimation can even be replaced by incubation in sucrose solutions. Since the cultivated potato is not able to cold acclimate at all, drought hardening like incubation in sucrose-enriched media can replace the cold preculture. In 'Désirée' using the encapsulation/dehydration method, noticed that incubation of donor plants in medium with 2 M sucrose for 5 days followed by shoot tip preculture in medium with 0.7 M sucrose for 1 day together increased the plant regeneration levels after cryopreservation. Observed an increase in shoot regeneration from 50% to 80% after preculture of donor plants in medium with 0.055 M up to as high as 0.22 M sorbitol for 21 days followed by droplet vitrification.

The subculture duration can also have a significant impact on cryopreservation results. Different subculture durations were tested by Yoon et al. (2006). They found that optimum subculture duration was genotype-dependent and the best recovery results were obtained with subcultures of 7 weeks in S. tuberosum 'Dejima' and 5 weeks in S. stenotomum 'STN13'. Wide variations of subculture methods have been applied for different cryopreservation methods and genotypes. For example explants were taken 4 weeks after subculture or alternatively completely new cultures were established and explants were taken 1 week after each transfer onto fresh medium. Culture vessels and the place of cultivation can also have an influence on the cryopreservation response. Plants grown in certain incubators showed significantly higher survival and regeneration percentages in comparison to potato plants grown in culture rooms together with other in vitro cultures like garlic and shallot. Many other environmental subculture conditions, such as light intensity, aeration and also planting density significantly influence both, regeneration of non-cryopreserved and cryopreserved shoot tips. Yoon et al. (2006) attained the highest shoot regeneration percentages with a combination of high light intensity, ventilation of culture vessels and low planting density. Influence of culture vessel type for the donor plants was analysed by Keller et al. (2006). On average, plants grown in jars with cotton plugs in the lid (two similar types) gave best results in comparison with 1,000 ml WECK jars, 300 ml Erlenmeyer flasks or SIGMA tubes. Plant quality is also an important factor. Superficially clean looking plants can bear infections by endogenous bacteria which may break out after cryopreservation and thereby reduce regeneration levels.

19.3.2 Shoot Tips

The physiological state of shoot tips can influence the regeneration results. Higher regeneration levels in three *S. tuberosum* cultivars when apical shoot tips instead of axillary shoot tips were used. The regeneration rate was decreasing with bud position from apical to lower axillary buds. Similar results werefound. Percentages of regeneration from apical vs. nodal shoot tips were compared in four *S. tuberosum* accessions resulting in four- toeightfold better results, when apical shoot tips were used. Also found that the apical shoot tips of *S. tuberosum* 'Dejima' or *S. stenotomum* 'STN13' subcultured every 3 or 4 weeks showed better regeneration response than those below the apex. In contrast, apical shoot tips reacted weaker than upper or middle axillary shoot tips when donor plants were subcultured after longer periods in a frame, every 5 or 7 weeks. Incontrast found little or no differences in cryopreservation results between axillary shoot tips of the first to third node counted from the apex. Next to the physiological state, also the explant size seems to have a significant effect. Observed the highest regeneration level after cryopreservation with shoot tips of 3–4 mm in length, whereas smaller (1–2 mm) or larger (5–6 mm) shoot tips showed weaker results. Also reported that larger shoot tips (2–3 mm long, 0.5–1 mm thick) produced the bestregeneration rates. Such explants include the meristem together with a few enveloping leaf bases. Optimum length of the shoot tip in *S. tuberosum* 'Dejima'was 1.5–2 mm and in *S. stenotomum* 'STN13' 1.–1.5 mm, showing that the most favourable sizes of shoot tips can be genotype- or species- dependent. In addition, the shoot tips size within each genotype is dependent on the physiological age and environmental culture conditions.

19.3.3 Preculture of Shoot Tips

Most vitrification and encapsulation/dehydration protocols require a preculture step to increase shoot tip tolerance to dehydration and cooling stress conditions caused by submerging in LN. Sucrose is the most widely used compound to induce such tolerance. In addition, glucose, mannitol and sorbitol have been applied. Preculture of shoot tips can be made in one step or alternatively in several steps by increasing sugar concentration. Sucrose is most successfully applied in concentrations ranging from 0.3 to 0.75 M. Comparing different sucrose concentrations ranging from 0.3 to 1.5 M for 24 h after shoot tips isolation, discovered the optimum regeneration response in 'Désirée', 'Ostara' and 'Santé' through applying 0.5 M sucrose in preculture. By comparing sucrose, glucose, mannitol and sorbitol at their concentrations of 0.5 M they revealed the best results with sucrose. Sarkar and Naik (1998) observed improved survival rates of shoot tips after preculture on medium supplemented with combination of 0.3 M sucrose and 0.2 M mannitol. In contrast, stepwise increase of sucrose concentration in the preculture medium, i.e. 0.3 M sucrose for 24 h followed by 0.7 M sucrose for 7 h caused the highest response in two tested accessions. Preculture of donor plants with 2 M sucrose for 5 days followed by shoot tip incubation in 0.7 M sucrose for 1 day significantly increased internal soluble sugars and proline contents and thereby also finally the cryopreservation results.

19.3.4 Loading and Cryoprotection

Loading or osmoprotection is needed in protocols which rely on vitrification. In the loading process, usually sucrose in combination with glycerol is used to increase osmolarity within shoot tips and to reduce the free water content. Glycerol being a cell membrane penetrating cryoprotectant reduces the concentration of salt by reducing the amount of water frozen on a simple colligative basis. In contrast sucrose can only penetrate the cell wall. Sucrose acts as cryoprotectant by withdrawing water from the cell over plasmolysis and by protecting the plasmalemma against freeze-induced dehydration. Using sucrose alone, very low success in cryopreservation was obtained. In contrast, using sucrose in combination with glycerol, a higher plant regeneration capacity was observed and the optimum results were foundby applying 0.6 M sucrose with 2 M glycerol in the preculture medium. The same solution was successfully used for loading of 'Atlantic' and 'Superior' in a vitrification protocol. Supercool X1000 ice- blocking agent added to the PVS2 solution further improved the cryopreservation response. In vitrification-based protocols, PVS2 is the most commonly applied cryoprotectant. Shoot tips are either directly put into 100% PVS2 solution, or the PVS2 concentration is stepwise increased. Incubation time in PVS2 depends on working temperature (room temperature or 0 °C), shoot tip size and genotype. A separatemethod of cryoprotection is performed in the DMSO droplet method. After isolationthe shoot tips are incubated in MSTo medium overnight, and the next day they are cryoprotected in MSTo medium with 10% DMSO.

19.3.5 Cooling

First applied potato cryopreservation protocols utilised the two-step cooling system. Slowly cooled the explants by tying cryovials together with the explants to a cord and by gradually sinking them into LN, firstly by exposing the samples to the vapour of LN and then directly transferring them into LN. Towill used cooling speed of 0.3 °C min−1 until −40 °C before plunging the explants directly into LN. Similar speed of cooling was used. Explants in cryovials were then submerged into a methanol bath at −5 °C for 15 min. Freezing of cryoprotectant solution in the cryovials was induced by touching the walls with pre-cooled forceps. After 15 min, the methanol bath was cooled down at speed of 0.3 °C min−1 until −30 °C. Explants were afterwards plunged directly in LN. Used a programmable freezing unit, were shoot tips were cooled first at 1 °C min−1 to 0 °C then with 0.5 °C min−1 until −10 °C before transfer to LN. In comparison, two-step cooling had lower shoot regeneration than ultra-rapid cooling. Additional disadvantages of two-step cooling are difficult handling, longer processing time of the method and need of special equipment. Recently developed methods, such as DMSO droplet method, vitrification, encapsulation/dehydration, encapsulation/vitrification and droplet vitrification apply rapid cooling by plunging precultured and cryoprotected shoot tips directly in LN. The fastest cooling rate (24,000 °C min−1; is implemented by using partially solidified nitrogen (LN slush, approximately −208 °C, which was also applied in potato. The cooling rate of 360 °C min−1 is reached using PVS2 solution within cryovials and can be further increased by using aluminium foil strips (7,800 °C min−1). Explants are exposed directly to LN on aluminium foils, whereas explants in cryovials are less fast cooled because of the air or cryoprotectant barrier to LN.

19.3.6 Rewarming

In potat shoot tip cryopreservation, either slow, medium or rapid rewarming has been applied. Slow rewarming was carried out by putting the explants from LN to room temperature. Medium rewarming speed was achieved by plunging and shaking explants from LN into room temperatured medium with rewarming speeds of approximately 3,000 °C min−1. In most cases, rapid rewarming was carried out by transferring explants from LN into a 35–40 °C water bath for 1–3 min. Comparing different warming rates,

Found higher regeneration response in shoot tips warmed by dipping the foils with explants directly in preheated (40 °C) 0.8 M sucrose medium for 30 s than by warming the explants in 40 °C water bath for 5 s and then adding 0.8 M sucrose (40 °C). In the vitrification- based protocols, unloading with high sucrose concentrations is applied to dilute and remove PVS2 from explants, which could be toxic to cells and thereby reduce shoot regeneration capacity. In the successfully applied potato cryopreservation protocols unloading solutions with 0.8–1.2 M sucrose for 10–30 min have been introduced.

19.4 FURTHER CULTURE AND VIABILITY ASSESSMENT

The loss of regenerating plantlets does often occur in cryopreservation because of not surviving or damaged shoot tips. After recovery, MS culture media with 3% sucrose and plant hormones have generally been used. Absence of phytohormones resulted in reduced numbers, sometime even no regenerated plants (Marion Grübe, IPK, Gatersleben, personal communication). In the DMSO droplet method, a combination of 0.5 mg/l-1 zeatin riboside, 0.2 mg/l-1 GA3, and 0.5 mg/l-1 IAA was used. In certain protocols higher phytohormone concentrations were applied in the first weeks of the regeneration phase. Subsequently during the culture, lower levels or no phytohormones at all were supplied. employed medium supplemented with 0.1 mg/l-1 BAP, 0.01 mg/l-1 NAA and 5 mg/l-1 GA3 for a period of 1 week. In the same protocol, the shoot tips were later transferred to a medium with only 0.001 mg/l-1 GA3 to avoid callus formation. Similar procedure, first by regenerating shoot tips on medium with 2 mg/l-1 GA3 and 0.23 mg/l-1 BAP for 1 week and then on medium with 1 mg/l-1 GA3. Yoon et al. (2006) used medium with 0.05 mg/l-1 IAA, 0.3 mg/l-1 zeatin and 0.05 mg/l-1 GA3 for 1 week followed by a culture on medium without phytohormones. Increase in plant regeneration was obtained by addition of polyoxy ethylene (POE)-polyoxypropylene (trade name PluronicF-68) at concentrations of 0.0005% to the regeneration media. The consistency of the regeneration medium can also have an influence on cryopreservation results. Observed the highest regeneration rates in four cultivated potato varieties using a vitrification protocol and solid medium, whereas the highest recovery using the DMSO droplet method was attained with liquid regeneration medium. In contrast, also used the DMSO droplet method in two cultivated potato varieties to compare liquid and solid regeneration media obtaining better response on the solid medium in terms of plant regeneration and plantlet structure. Light regime also plays an important role in shoot tip regeneration. As proven 'Désirée' showed only high survival rate, if the light intensity was low (15 μmol m−2 s−1) during the first week of recovery. Afterwards it was necessary to cultivate plants in higher light conditions (45 μmol m−2 s−1) to ensure regular morphological development.

19.4.1 Ultrastructure

Ultrastructural studies on potato applying transmission electron microscopy have revealed structural damage on both surviving and killed explants resulting from different cryopreservation methods. First ultrastructural studies on potato shoot tips were conducted by Grout and Henshaw (1980) using *Solanum goniocalyx* and ultra-rapid freezing. After rewarming of shoot tips it was not possible visually to discriminate between killed and surviving explants during thc first 24 h. After rewarming and culture for 24 h, shoot tips showed areas of massive damage often associated with rupture of the epidermis that was also visible under light microscope. Ultrastructural analyses revealed considerable cellular damage in certain areas with total breakdown of protoplasts and sometimes cell wallrupture. The influence of the cell wall in freezing tolerance of cultured potato cells was analysed. In this study, thecell wall of cultured potato cells accounted for freezing injury because of mechanicalstress between cell wall and plasma membrane during a freeze-thaw cycle. Survivingexplants were characterised by single cells or small groups of cells with electron- dense cytoplasm which were located in primordia and dome regions of the explants. Survival involved either regrowth of the shoot apex or callus proliferation, although it was sometimes difficult to be certain whether the original apical meristem resumed growth or whether it was replaced by another meristem rapidly regenerated from the callus. Some studies indicate that the original meristem does not survive cryopreservation and thatonly small groups of cells are enough for regeneration of complete shoots. Insufficient numbersof undamaged cells could lead to callus formation instead of shoot development.

In a later study, four accessions of *S. stenotomum* were studied, which had been cryopreserved by the vitrification method. Aftercryopreservation, abnormal cytoplasm, cell plasmolysis at different stages and a large number of small vesicles were found. Further to a lesser extent cell wall rupture, protoplast outflow and anomalous nucleus shape were also discovered. Could observe a negative correlation between the percentage of survival and cellular damage meaning the genotype with the highest survival capacity had the least ultrastructural damage.

Analysed the influence of sugar preculture on ultrastructure ofshoot tips of *Asparagus officinalis*. They found surviving cells after sucrose preculture with many developed plastids that contained starch grains andextensive rough endoplasmic reticulum (rER) in the cytosol. Obtained similar results. They observed drastic changes in cell ultrastructure with the appearance of rER after sugar incubation. These ultrastructural changes resemble those occurring in cold-acclimated cells, thus, being a characteristic feature of freezing tolerance. Sucrose preculture could, therefore, substitute cold acclimation and increase freezing tolerance in tissue subjected to cryopreservation.

Ultrastructural studies on *S. tuberosum* 'Désirée' using the DMSO droplet method, showed changes, when shoot tips had been incubated in the cryoprotectant solution for 2 h (10% DMSO in MSTo medium,. After this time period, cells contained numerous small vesicles. Mitochondria and chloroplasts were increased in size and vacuoles assumed an irregular shape. Directly after cooling in LN and rewarming in MS medium, cells had been unchanged in comparison to the conditions prior to freezing. However, first visible changes were found 2 days after rewarming, when parts of the meristematic dome and epidermis showed signs of extensive damage. Rupture of plasmalemma, plasmolysis and destruction of cell organelles as well as strong heterochromatisationof nuclei were observed. Survival and regeneration of cells were mainly found in leaf primordial regions, where cells were physiologically very active and contained many mitochondria and intact or regenerating chloroplasts.

19.5 GENETIC AND EPIGENETIC FACTORS

19.5.1 Genomic DNA Stability

In theory, all metabolic activities at temperatures of LN are reduced to zero, so that after rewarming from cryopreservation, true-to-type plants are expected. Since cryopreservation protocols do not only involve cooling in andewarming from LN but also in vitro culture and regeneration processes, phenotypicand genomic changes can occur due to somaclonal variation. Therefore, the verification of true-to-type plants after cryopreservation is necessary. In potato, many studies on genetic integrity after cryopreservation have been performed. Morphological and phenotypic,cytological and molecular comparisons were conducted revealing that plant material was genetically stable as a result of cryopreservation.

Shoot tips of diverse potato species with different ploidy levels (diploid, tetraploid and hexaploid) exhibited normal developmental patterns (flowering, berry set and tuber formation) after regeneration from cryopreservation which had been carried outthrough the encapsulation/dehydration method. The ploidy status of the regenerated plants was maintained and chromosomal abnormalities were not observed based on cytological studies. Harding and Benson also verified the stability at the molecular level through the analyses of the ribosomal RNA genes (rDNA) as well as the nuclear- chloroplast DNA in the chloroplast genome in potato plants regenerated from cryopreserved shoot apices.

Using microsatellite fragments, regenerated plants from cryopreserved shoot tips of 'Brodick' and 'Golden Wonder' were compared to the control materials. Microsatellite profiles were identical in the regenerated material to those of the parental plants and their progeny. This demonstrates the stable somatic inheritance ofgenomic regions containing the measured simple sequences.

Applying the DMSO droplet method, genetic stability was also confirmed in a random sample of the collection using morphological parameters, flow cytometric measurements and restriction fragment length polymorphism (RFLP) analyses. The comparison of morphological parameters displayed only one abnormal and a few weak plants in a set of approximately 1,000 regenerated plants from 98 varieties tested. The vigour was not increasingly reduced compared to the regular variation occuring in the control plants, which had not been cryopreserved. RFLP analysis and flow cytometry did not show genetic abnormalities within 161 samples of regrown plants. Storage in LN for up to 10 years was found to have no adverse effect on the regeneration rates. Concluding from these results, it seems likely that the cryopreservation technique does not induce somaclonal variation. However, certain wild species, such as *Solanum microdentum* and *S. pinnatisectum* and dihaploid (2*n*=2*x* =24) *S. tuberosum* lines 'PDH 40', 'PDH 135', 'PDH 634' and 'PDH 638' produced more than one shoot after rewarming from cryopreservation. The authors pointed out that a certain portion of these shoots may have been originating from callus and could, therefore, have shown evidence of somaclonal variation. Plants recovered from cryopreserva- tion were tested concerning their ploidy levels, but only one tetraploid plant was found, which could be due to spontaneous chromosome doubling. Since the numberof regenerated shoots revealing abnormality was low, it was concluded that cryopreservation did not induce ploidy changes. The capacityto generate tubers was not affected by the cryopreservation process in the cryopreserved plant materials, but the development of flowers was impaired in comparison to tuber-derived control plants. Performed phenotypical comparisons between material cryopre- served for 1 year and the original clones maintained in vitro. A number of 18 accessions were analysed belonging to the following potato species and interspecific hybrids: *S. stenotomum, S. tuberosum* ssp. *andigena, S. goniocalyx, Solanum chaucha, S. tuberosum* ssp. *andigena* x *S. tuberosum, S. phureja,* and *Solanum curtilobum.* Even 23 different morphological descriptors were compared. Differences were found in six of the 18 accessions. Of these, there were twomisidentifications. The remaining four accessions displayed small morphological differences, which were attributable to pathological and environmental factors. Phenotypic integrity between cryopreserved and control stock plants of *S. tuberosum* 'Agria' and 'Marfona' was found, when the encapsulation/dehydration method was applied. In the same study, genetic stability was found using AFLP analyses and by using seven primer pairs. In addition, ploidy comparison did not reveal any differences between control and cryopreserved plants. Applying the encapsu- lation/vitrification method, Compared control plants with cryopreserved materials using RAPD analyses with 17 primers. Genetic integrity was also found for each analysed plant that had been cryopreserved.

19.5.2 Epigenetic Factors

Genetic changes are characterised through various alterations in the DNA nucleotide sequences (single copy sequences, low copy and repetitive sequences). In contrast, epigenetic modifications do not change the original DNA sequence, but may, nevertheless, result in heritable changes of gene expression potential. Typical features associated with epigenetic processes such as regulation of gene expression, activity of transposable elements, defence against foreign DNA and inheritance of specific gene expression patterns are DNA methylation, particular histone modifications and alterations in chromatin structure. DNA methylation is required for regular developmentof plants and may have an impact on vigour and morphogenesis. Some studies have discussed putative effects of changes in DNA methylation on genomic DNA stability and rearrangements in micropropagated and cryopreserved stocks. Modification of epigenetic features due to the stress during in vitro culture could possibly result in altered phenotypes in the recovered plants. First comparative studies of epigenetic marks in plants before and after cryopreservation have already been done. Modifications in DNA methylation have been determined in almond, papaya, chrysanthemum, *Ribes*, strawberry and citrus. In potato, no studies on epigenetic modifications in connection with cryopreservation have been published so far, but first analyses at IPK Gatersleben pointed to methylation changes, although it is not clear yet, whether these changes aredue to cryopreservation or long-term in vitro storage (A Kaczmarczyk, unpublished results). Analysis of epigenetic changes induced by ex situ storage systems like in vitro culture has been carried out in potato so far in only one study. DNA methylation changes have been found in mature greenhouse plants of *S. tuberosum* 'Désirée' and 'Golden Wonder' after they had grown on MS with 6% (*w*/*v*)mannitol for 6 months. The growth on mannitol-enriched medium seems to be correlated with subsequent DNA hypermethylation, which may be an adaptivereaction to abnormal environmental conditions related to high osmotic stress. This result has implications for vegetatively multiplied potato plants, because the methylation status could be inherited in the tuber progeny, and changed phenotypes may subsequently be expressed in following generations. Overall it is suggested that the methylation changes may not be induced by cryopreservation per sebut may be the results of the whole process of in vitro culture and regeneration phases.

19.6 THERMAL ANALYSIS

The state of water, meaning ice or glass formation, during cryopreservation is important for the success of the method. Whereas ice crystallisation within cells during cooling andwarming leads to membrane damage and cell death, the glass formation, which is a solidamorphous state of water in solutions, can alternatively preserve the cells.

DSC is a thermal analysis method that can be applied for measurement and determination of glass transitions and ice crystallisation during cryopreservation. It measures heat flow and water transitions in samples as a function of temperature and/or time.Transitions between liquid, amorphous glassy, and ice states can be detected by heat flow data through endo- and exothermic peaks or sigmoidal changes in the record diagram.The point of change from the liquid to the glass state is the glass transition temperature.Thermal profiles provide important information about the cryoprotective treatments required to obviate lethal ice formation and to stabilise glasses.

Using the DMSO droplet method in S. tuberosum 'Désirée' and S. demissum DSC analyses were conducted with freshly isolated shoot tips and following incubation in cryoprotectant solution (10% DMSO in MS medium with 3%. The explants were thermally characterised by an ice crystallisation peak during cooling, and melting peak of ice during rewarming. In addition the cryoprotected shoot tips of S. tuberosum 'Désirée' exhibited a glass transition at −117 °C. The cryoprotected accessions showed crystallisation and melting of icc, but thc hcat flow was much lower compared to freshly isolated shoot tips. This means that the cryoprotectant withdraws water from shoot tips. Also melting and crystallisation points were shifted to lower temperatures, which means that the cryoprotectant decreased the ice crystallisation temperature. It was concluded that the DMSO droplet method relies on ultra-rapid cooling and rewarming with marginal ice formation, which does not detain shoot tips from regeneration after rewarming. Survival and regeneration of shoot tips with marginal intracellular ice crystallisation was also reported by other authors. In contrast to the DMSO droplet method, the vitrification method uses PVS2 as cryoprotectant. Depending on incubation time, the cryoprotected potato shoot tips showed either crystallisation of ice together with glass transition or only glass transition at −108 °C without ice crystallisation at all. The longer the incubation time, the less ice crystallisation was found and as a result of 60 min incubation in PVS2, only glass transition was detected. Therefore, DSC analysis can be used to find optimum cryoprotectant incubation times, although it should additionally be considered that DSC instruments cannot manage the speed of cooling and rewarming used in cryopreservation experiments and there are differences between DSC analyses and actual cryopreservation experiments.

19.7 BIOCHEMICAL ANALYSIS AND PROTEOME RESEARCH

Biochemical characterization of explants can be assessed through determination of the plant protein expression. Proteomics approaches are used to analyse the complex functions of plants at different levels and are supported through the progress obtained in sensitive and rapid protein identification by mass spectrometry.

In potato, proteome studies have been carried out mainly on tuber tissue and in connection with investigation on the action of phytopathogens such as Pectobacterium atrosepticum and the aphid Macrosiphum euphorbiae. The application of protein expression studies for cryopreservation is a new research area. Used banana as a model to study protein expression associated with sucrose-mediated osmotic stress in the meristem. A protocol for two-dimensional gels was developed for small amounts of tissue. Polypeptide databases obtained by mass spectrometry were used for discriminative identification of proteins from closely related plant species. From those results, the conclusion was drawn that the maintenance of an osmoprotective intracellular sucrose concentration, the enhanced expression of particular genes of glycolysis and the conservation of the cell wall integrity are all essential to maintain homeostasis, and to acclimate to and to survive from dehydration influencing the explant prior to cryopreservation. Results were, however, not further evaluated and no identification of proteins was presented. First proteome analyses in connection with cryopreservation have been made for potato shoot tips. Analysed proteins of S. tuberosum 'Désirée' shoot tips and leaves after preculture of donor plants for 21 days on regular MS medium and MS complemented with 0.055, 0.11 and 0.22

M sorbitol. In their study, the regeneration rate after cryopreservation was increased from 50% to 80% with increased sorbitol concentra- tion in the medium. Differences in 2D-difference-gel-electrophoresis patterns were observed for samples precultured on various sorbitol-enriched media. In total, 63 and 15 differently expressed proteins were determined in the shoot tips and leaves, respectively. Of them 43 and 3, respectively, were identified by matrix-assisted laser desorption/ionisation mass spectrometry. The results indicated strong alteration of the primary metabolism and in carbon fixation. These results were additionally confirmed by parallel biochemical analyses. Increase in carbohydrates was measured in plants treated with up to 0.11 M sorbitol, whereas the concentration of 0.22 M sorbitol showed decreased levels of carbohydrates. In conclusion, better cryopreservation response could be obtained due to high internal sugar concentration and alterations in primary metabolism as a result of the cryoprotectant function of sugars. It is analysed analysed S. tuberosum 'Désirée' and S. demissum after two different preculture conditions followed by cryopreservation; 1 week culture at 22/ 8 °C day/night temperature in comparison to 1 week culture at constant temperature of 22 °C. The alternating temperature preculture of donor plants revealed higher survival and regeneration rates after cryopreservation. In protein expression, only minor changes were found between the preculture treatments in the two analysed species, meaning that 2.7% or 0.4% of the proteins were changed in S. demissum and S. tuberosum 'Désireé', respectively. Proteins were mostly down-regulated in their expression level after preculture with alternating temperature. The identified proteins belonged to the functional groups of metabolism, signal transduction, defence, transcription, energy and secondary metabolism.

19.8 CRYOTHERAPY

Besides long-term storage, cryopreservation can also be used for virus elimination, a method named cryotherapy. Cooling and freezing of explants down to LN temperatures facilitates viral elimination in larger meristems, because many viruses are unable to infect the meristem proper, and virus-free plants can be grown from excised apices of small size. Cryotherapy works as a micro-scalpel, where larger, more hydrated cells outside the meristem die because offreezing, and only small cells within and close to the meristem survive cryopreservation. Cryotherapy has been successfully applied to eliminate or reduce viruses in banana, grapevine, *Prunus*, raspberry, citrus and sweet potato. Also potato shoot tips were freed from viruses by cryotherapy. Applied encapsulation/ dehydration, encapsulation/ vitrification and DMSO droplet method to eliminate *Potato leafroll virus* (PLRV) and *Potato virus Y* (PVY) from virus-infected potato shoot tips. Each methodproduced similar rates of virus-free plants, between 83% and 86% for PLRV, and 91% and 95% for PVY. Frequencies of virus-free plantlets produced by cryotherapy were actually higher than those derived from meristem culture (56% for PLRV and62% for PVY) and from thermotherapy (50% for PLRV and 65% for PVY), and comparable to those by thermotherapy followed by meristem culture (90% for PLRVand 93% for PVY). Regrowth rates of cryopreserved shoot tips were 85–89% and, thus, higher than regrowth ability found by using meristem culture (55%) alone, and thermotherapy followed by meristem culture (50%), but comparable to the results obtained by thermotherapy (87%). In addition, the required working time for cryotherapy was much less (about 55 days) in comparison to thermotherapy or thermotherapy followed by shoot tip culture (87 days). In total, cryotherapy has shown its advantages being easier in handling and simpler in procedures for plant regeneration. It has also been verified that cryotherapy produced higher rates of regrowth and virus eradication and shorter time for production of virus-free plants was needed than in using other virus elimination methods. Since virus-free donor material is required for successful cryopreservation and conserva- tion of accessions, proposed the use of cryopreservation simultaneously for long-term storage of potato accessions and the production of virus-free plants.

Improvement in the development and successful application of potato shoot tip cryopreservation protocols have been made over the last 30 years. Many studies have been conducted on the influence of various factors affecting cryopreservation such as the physiological state of donor plants and shoot tips, as well as specific cryogenic factors, e.g., type of cryoprotectant(s) and speed of cooling and rewarming. So far, all known cryopreservation methods of plants already applied for shoot tip tissue, like two-step cooling, ultra-rapid cooling, DMSO droplet method (droplet freezing), encapsulation/dehydration, encapsulation/vitrification, vitrifica- tion and droplet vitrification, have been successfully employed for a wide range of species. It remains to be seen whether one particular technique can become the 'standard' for plant cryogenic storage (e.g. droplet vitrification) or whether, as it already exists, many protocols will remain in common use as long as they work effectively. Although many successful protocols have been established, none of the cryopreservation techniques can guarantee the conservation of all genotypes. A critical point in potato cryopreservation is the diverse response between different genotypes in terms of their regeneration capacities as a result of cryopreservation. Therefore, weakly reacting genotypes need to be maintained using tissue or field culture. In addition, certain genotypes may not even have tissue culture ability and need to be preserved as tubers. Finally, none of the conservation strategies (cryopreservation, tissue culture, field culture) can be completely safe, because materials may always be lost regardless of the conservation technique.

Studies on the genomic DNA stability conducted on morphological and molecularlevel have revealed no changes and thereby confirmed the successful conservation ofcultivar characteristics (true-to-type status) as a result of cryopreservation. However,no studies on the epigenetic level followed by cryopreservation have been publishedso far. In addition to long-term conservation, the cryopreservation methods have alsoeffectively been applied for virus elimination in potato (i.e. cryotherapy). Therefore,it can be seen as an alternative method to chemo- or thermotherapy with theadvantage that the cryotreated shoot tips can also be simultaneously stored for long-term conservation. Additional fundamental studies of freeze tolerance and influenceof LN on plant tissues need to be carried out to further understand the processes occurring after various steps in the cryopreservation protocols in order to develop more practical cryogenic methods.

RELATED TERMINOLOGY

Alpha and beta thalassemia: Inherited conditions that affect red blood cells. These conditions can affect any ethnicity, but are found most frequently in Asian, Mediterranean, and African populations, respectively. In their most severe forms, these conditions can cause still birth or death. In less severe forms, they can cause anemia, jaundice, and may require blood transfusions.

Alternatively the affected clothing can be flushed/soaked with water (e.g. using a safety shower) until the trapped LN2 is all vaporized Transport/shipment: For safety reasons, liquid nitrogen should not be transported by car or air. Vapour shippers can be transported by both cars and aero planes Ventilation When liquid nitrogen is used/held in enclosed spaces, the ventilation must be sufficient to prevent nitrogen gas accumulation (see 'Asphyxia').

Ampoule: Ampoules made of glass were used by many early cryopreservation studies. They were usually flame-sealed, or sealed with plastic caps. They have usually been replaced by plastic vials, or other freezing containers, in part because of the risk of explosion with attendant spread of glass shards. Mainly used now for slow cooling.

Andrology: The study of male fertility and infertility.

Asphyxia: Death from asphyxia has occurred in several laboratories around the world where liquid nitrogen is in routine use. Nitrogen gas is odourless and colourless. It forms 80% of the air we breathe; the remaining 20% is mainly the essential oxygen. Nitrogen gas is released from liquid nitrogen as it evaporates, and if ventilation is inadequate, it can displace the air, creating a hypoxic or anoxic environment where there is a high risk of asphyxia. The risk is highest in enclosed, poorly ventilated rooms. A person can become unconscious after just a few breaths of 100% nitrogen gas. A litre of liquid nitrogen produces ~1600 l of nitrogen gas (at room temperature and pressure). Liquid nitrogen storage vessels should therefore be stored in well-ventilated rooms. Ideally that room should have a regularly serviced oxygen meter/alarm. Travelling in a lift (elevator) together with a large nitrogen tank is not recommended.

Assisted insemination: While some people use the term artificial insemination, Cryobio prefers the term assisted insemination. This is a procedure where a prepared semen sample is placed into the uterus or cervix to attempt to establish a pregnancy.

Audio interview: This is an audio recording of the donor's response to various questions. The recording allows potential recipients to hear the donor's voice and his reactions and responses to the questions.

Biobank: A collection of biological samples like blood, tissue or DNA stored in freezers at low temperatures (ex: -80C) and/or data for use in research.

Box: Cardboard or plastic boxes are commonly used as storage devices to hold straws or vials within a tank. Not as widely used in assisted reproduction as goblets or canes. Mainly used in slow cooling of cell lines.

Burns: Tissue damage similar to heat-burns can be caused by exposure to extremely low temperatures, for example contact with liquid nitrogen at a temperature of "–196°C. Skin exposed to this temperature will blister in the same way as a burn. A full face visor, goggles or glasses should be worn when handling/dispensing liquid nitrogen because a splash to the eye could burn the cornea; appropriate personal protective equipment should be worn to protect the rest of the body. See also 'Frostbite' Cold tolerant containers: Most containers become brittle at cryogenic temperatures. Only use those specifically designed for cryopreservation. Containers such as Eppendorf tubes are not designed to tolerate low subzero temperatures and easily shatter/explode (see below)

Cane: Commonly used storage device. Usually a ~30 cm long aluminium strip with four to six paired projections designed to grip straws and goblets. A code (usually the patient ID) written on the cane facilitates retrieval of particular specimens in a tank. Alternative storage devices are boxes and goblets

Canister: Storage device for canes and/or goblets that fits inside a storage vessel. Liquid nitrogen tanks commonly have six to 10 canisters each one with a diameter that matches the opening (neck). The canisters are usually 'parked' inside the tank along the wall, but have an attached handle which allows them to be moved. Specimens (usually canes or goblets) can only be placed in a canister when it is moved to the position directly under the tank opening. The canister can then be raised out of the tank via the tank opening.

Capillary: Glass capillaries are brittle when used for cryopreservation but were used by Hochi et al. (2001) to show interactions between cooling and warming rates. They found for 7.2 mol/l ethylene glycol and 1.0 mol/l sucrose that 'capillaries of 2000, 1400, 1000, 630 and 440 μm diameters provided cooling rates of 2000, 3000, 5000, 8000 and 12 000 °C/min and warming rates of 5000, 8000, 17 000, 33 000 and 62 000°C/min, respectively.' They observed that a moderate warming rate (3000°C/

min) gave the highest post-thaw survival of bovine oocytes irrespective of the warming rate.

Carrier status: A person that has inherited a recessive allele for a genetic trait or mutation, but usually does not display that trait or show symptoms of the disease. Carriers can pass these genetic variants on to any resulting children.

Cell culture: the process where cells harvested from living tissue are grown in vitro (outside of their

Cell line: A permanently established cell line, which will proliferate indefinitely when maintained under controlled conditions. May be cryopreserved using specific protocols and cryoprotectant agents.

Client depositor: A male who freezes sperm for his own use in the future. A client depositor is not a donor.

Closed pulled straw: Describes a new way of loading embryos into an open pulled straw (OPS, see below) for rapid cooling which prevents liquid nitrogen from coming into direct contact with the cryoprotectant bead containing the oocyte/embryo. Cooling rate would be approximately the same as for OPS, i.e. >20 000°C/min .

Cryobank: A collection of (generally) reproductive tissues (i.e. semen, oocytes) stored at ultra-low temperatures achieved by using liquid nitrogen (i.e. -196 C).

Cryoloop, nylon loop, loop: A thin nylon loop. Oocytes and embryos can be suspended in a cryoprotectant film within the loop of nylon. Specialized vials have been developed to handle and store the loops. They achieve a very high cooling rate (Lane et al., 1999; Yeoman et al., 2001). Device initially developed for crystallography.

Cryopreservation: The act of freezing and thawing.

Cryopreservation: The process of cooling and storing tissue or cell samples at ultra-low temperatures (-196) in liquid nitrogen in such a way that viability is maintained (halts metabolic processes) and samples can be thawed for potential future uses. Preserved in a way that little or no water freezes within the cell, it can be thawed to avoid recrystallization and a cryoprotectant is administered at non-toxic levels, and not exceeding osmotic tolerance limits.

Cryoprotectant agent (CPA): a substance added to a biological sample to protect it from damage

Cryotop: A very fine strip onto which oocytes or embryos can be loaded. The strip is then immersed into liquid nitrogen or nitrogen vapour to give a very fast cooling rate (Hochi et al., 2003).

Cystic fibrosis: A hereditary condition in which the body produces an excess of thick mucus affecting the respiratory system and pancreas. Cystic fibrosis is the most common recessive genetic condition in Caucasian populations.

Cytomegalovirus: A virus in the same family of viruses as the herpes virus. Once exposed to CMV, the virus is always harbored in the body, either in an active or a dormant state. Symptoms are often mild and overlooked in healthy individuals. Donors are tested for active or contagious states of the virus while participating in the program. If detected, samples are not released for sale. If a donor has a positive CMV test, but it is in a dormant state, the samples are cleared to be released for sale. Physicians differ in opinions about the significance of CMV, therefore, recipients should consult their physician before proceeding with insemination.

Dewar: Originally a double-walled silvered glass vessel in which the space between the walls is evacuated, and used for storage of liquids at low temperature. Invented by the Scottish chemist Sir James Dewar. Now used more generally for any similar container in which liquid nitrogen can be held. Can refer to large storage tanks or small (e.g. 1 l) containers for short-term use, e.g. for bench-top work and to transport samples between rooms. Most commonly made from stainless steel. 'Thermos'-type domestic Dewars with glass walls should never be used, even for short-term work as they can explode because the caps have no pressure vent, and the walls are not sufficiently thick to cope with materials as cold as liquid nitrogen.

Directed donor or known donor: A donor who is known to a specific recipient. Some directed donors intend to be a parent of any resulting children, while some directed donors are donors only and do not intend to be a parent to any resulting children. Directed donors undergo screening and testing like open and anonymous donors, and their sperm is frozen and stored for use by a specific recipient.

Epididymal sperm: Spermatoza which is able to fertilize that is harvested directly from the epididymis, the tubular system attached to the testis. Spermatozoa are produced in the testicles and mature within the epididymis.

Explosions: Explosions easily happen if great care is not taken when using liquid nitrogen. The most common cause is entry of liquid nitrogen into a container (such as an unsuitable sample tube). This easily happens due to the reduction of pressure inside the vessel when the air inside is cooled by placing in liquid nitrogen. When the vessel is rewarmed, the nitrogen boils and the resulting gas pressure causes the container to explode. This is particularly common with containers not designed for use with liquid nitrogen. Flasks must never be of the 'thermos' type (silvered glass walls with a vacuum) as these can explode, filling whole laboratories with glass shards; use instead stainless steel flasks or polystyrene boxes with appropriate venting to prevent build-up of internal pressure.

Flexipet: A pipette tip which can be used as the container in which oocytes or embryos are cryopreserved. Rapid cooling only.

Forceps: Paired metal blades for holding specimens deep within the vapour phase or retrieving specimens from liquid nitrogen. Long handles or blades, or specialized tools such as the 'cryoclaw' are less likely to result in frost bite. Also used in icc-nucleation/seeding. May damage straws if not equilibrated to their temperature or if used with too much pressure.

Freezing machine(s)/ biological Freezer: For controlled rate cooling and warming of specimens. **Different types include:** (i) platform cooling devices and equivalent where specimens are gradually being lowered into nitrogen vapour (vapour cooling); (ii) gradient freezers which gradually move the specimen along an increasingly cold surface; (iii) chamber freezers in which controlled amounts of nitrogen vapour are injected into the specimen chamber; (iv) uncontrolled or passive cooling devices which are placed for example in a −80°C freezer or on dry ice; (v) chamber freezers in which the outside of the chamber is at −196°C and the specimen's temperature is regulated by controlled warming of the chamber wall. Machines differ in: size, portability, durability, flexibility (range of options), repeatability and controllability (e.g. over rates/times, number of programs etc.), purchase price, maintenance costs, number and size of specimens, nitrogen consumption, lowest controllable temperature, seeding procedure, ease of loading/unloading, ease of use, permanent record output and dependence on a reliable power supply and decibel output.

from the freezing process in cryopreservation. There are many types of cry protectants, and it is important to identify most effective type, concentrations/ quantities for each species/ cell type.

Frostbite: Skin exposed to liquid nitrogen or any other surface cold enough to cause the cells in the skin to freeze will hurt, or if the injury is bad enough it will blister. Injury is most commonly caused by cold metal surfaces, or ice, due to their high thermal conductivity. See also 'Burns'.

Gamete: male (spermatozoa) or female germ cell (oocyte) which, during sexual reproduction, fuses

Germplasm: The living genetic resources (genes from seeds or tissues) for an organism, which can be harvested and maintained for breeding or research purposes. Examples: Spermatozoa, Oocytes, Embryos.

Gloves: Used to protect the hands/fingers from frost bite/burns. The best sorts extend up the forearms, e.g. leather welder's gloves, or special gloves (e.g. Sigma 'cryogloves'). Gloves are not designed to allow the user to insert their hands into liquid nitrogen. The user should use long-handled forceps or tongs to hold specimens within the vapour phase or retrieve specimens from within the liquid nitrogen.

Goggles, visors: Used to protect the eyes/face in case of explosions or liquid nitrogen splashes.

Gonad: an organ that produces gametes (i.e. testis or ovary)

Grid: Very thin 5 mm diameter copper or gold mesh upon which oocytes and embryos can be placed; the fluid is then blotted away from below, and the grids rapidly cooled. They can achieve a very high cooling rate (Martino et al., 1996a,b). Initially developed for electron microscopy. Cooling rate; in nitrogen ~ 9000°C/min and up to 30000°C/min in nitrogen slush.

Heat sealer: Piece of equipment (e.g. autoclave bag sealer or plastic bag sealer) used to close and seal plastic straws Impulse sealer Piece of equipment used to close and seal CBS straws (Cryo Bio System by IMV Technologies, France).

Hemoglobin electrophoresis: A blood test used to check hemoglobin levels in the blood. Hemoglobin is a protein that carries oxygen in the blood. All Cryobio donors undergo hemoglobin electrophoresis. Abnormal results may suggest an inherited blood disorder or an increased risk of passion on an inherited blood disorder.

ICI: Intracervical Insemination. An assisted insemination procedure where the processed semen sample is deposited at the base of the cervix and the sperm cells swim through the cervix and into the uterus. Semen samples in an unwashed state (meaning they contain seminal fluid) can be used in an ICI because the cervix acts as a natural filter and removes the seminal fluid, allowing the sperm to travel to the uterus.

ICSI: Intracytoplasmic Sperm Injection. An assisted reproductive technology procedure like IVF with medicated ovulation and an egg retrieval. Then, instead of allowing fertilization to occur by combining sperm and eggs in a petri dish as in IVF, in ICSI a single sperm cell is injected into each egg to attempt fertilization. ICSI is often used for males with compromised motility, which may prevent the sperm cell from penetrating an egg on its own. Resulting embryos are cultured to a certain cell stage and then transferred to the uterus or frozen and stored.

In your own words: A one-page response to four essay questions that Cryobio requires of all donors. We photocopy the page to allow potential recipients to see the donor's handwriting and spelling.

Insulation: Material surrounding a storage tank to minimize heat transfer through the walls.

Inventory: Sometimes used to refer to the hardware placed within storage vessels to hold the **specimens Liquid nitrogen:** Nitrogen below its boiling point (−196°C). Often abbreviated LN2 or LN2

IUI: Intrauterine Insemination. An assisted insemination procedure where a processed semen sample is deposited directly into the uterus. Samples used for IUI cannot contain seminal fluid, because unwashed semen causes cramping in the uterus, so these samples must undergo a wash process to remove the fluid either prior to insemination or prior to freezing.

IVF: In Vitro Fertilization. An assisted reproductive technology procedure in which the female is medicated in order to produce multiple eggs in a single cycle. The eggs are retrieved and are fertilized by adding the male's sperm into the same petri dish with the eggs in the laboratory. The resulting embryos are cultured until they reach a certain cell stage and then the best embryo is transferred into the uterus. Any remaining embryos that are not transferred can be frozen and stored to use for future pregnancy attempts.

Karyotyping: A lab test looking for abnormalities in chromosomes that can lead to a genetic disposition to birth defects or other health or development issues. All Cryobio donors undergo karyotyping. Liquid nitrogen should not be transported in cars where nitrogen vapour can access the passenger compartment and cannot (legally) be shipped by air. Specialized vessels (see vapour shippers/dry shippers) which keep specimens at low subzero temperatures without giving off nitrogen gas are suitable for transport by car and by air.

Liquid nitrogen slush: When the liquid nitrogen is at a temperature lower than its boiling point it becomes more viscous. The most common strategy by which nitrogen slush is made is by applying a vacuum to liquid nitrogen. This increases the evaporation rate, which in turn lowers the temperature due to evaporative cooling.

Liquid nitrogen: The environment in which cryopreserved specimens are stored and shipped. Nitrogen is a liquid when its temperature is -196°C. Cryopreserved specimens are stored in liquid nitrogen in the laboratory. Nitrogen turns to vapor at a slightly elevated temperature, -185° C, than its liquid phase. Sperm is shipped in tanks filled with nitrogen at the vapor temperature since this phase facilitates shipping.

Loading: A term describing the process of transferring oocytes and embryos to the container in which they are to be cryopreserved (e.g. oocytes are loaded into straws). To reduce the risk of contamination during storage, the loading procedure should avoid all contact between the cryoprotectant solution (with the cells) and all surfaces that come in direct contact with liquid nitrogen or other specimens. Failure to avoid this contact could result in contamination—for example viruses have been transferred between samples carried in the liquid nitrogen.

Medical health history: A three-generation family health history of a donor. This typically includes the health history of the donor and his siblings, parents, aunts and uncles, and grandparents. The medical health history may contain information about other family members if it is available.

Minimum drop size: A vitrification method in which nanolitre-sized drops can be cooled at up to 130 000°C/min.

NAT: Nucleic Acid Test. A test to detect the genetic material of a virus within the body. The advantage of NAT over traditional antibody tests is that it provides results much sooner after infection than the traditional tests.

Nitrogen Level monitor: For tanks as a back-up to provide a warning about low nitrogen levels. Sophisticated versions can be linked to a phone dial-up facility. These devices are not failsafe and should be tested according to the manufacturer's recommendations, and routine physical checks of nitrogen levels made at appropriate intervals.

Nitrogen vapour: Usually referring to the cold nitrogen gas that accumulates above the surface of liquid nitrogen. Specimens can be held deep within the vapour phase as part of a cooling protocol. This gives a less abrupt cooling rate than plunging directly into liquid nitrogen (see vapour cooling) due to the higher temperatures and the lower thermal transfer between nitrogen vapour compared to transfer from LN2. This reduces the likelihood of physical damage (cracks and stress fractures) and may give as good or better results than plunging into liquid nitrogen. However, some cell types (e.g. early pig embryos) appear to benefit from the highest possible cooling rates. Specimens are also commonly held in the vapour phase as the first event in the thawing step, to reduce the risk of specimens exploding due to trapped nitrogen, and to reduce this likelihood of stress fractures (see stress fractures) (at very low temperatures (e.g. <–196°C) most material is brittle and is easily damaged by non-uniform sudden temperature changes. Cooling and warming is best performed deep within a stable layer of vapour where the temperature is <–130°C. To establish such a stable, cold, vapour layer the liquid nitrogen level must be well below the rim of the container ("15 cm) and left to stand for "10 min without being disturbed by air currents Open pulled straw (OPS): A type of container used to rapidly cool oocytes and embryos. Basically a plastic straw which has been narrowed and thinned by stretching, and then cut at its thinnest point to give two 'OPS' straws. These are usually loaded by capillary action. A straw which is pulled to narrow its diameter so that it will take up ~2 μl of cryoprotectant. When immersed deeply into the vapour phase or just let to touch the surface of liquid nitrogen the OPS gives a cooling rates of >20 000°C/min. Slower rates for OPS straws have, however, been measured (5300°C/min in nitrogen and 10 300°C/min in slush).

Nitrogen: Air contains almost 80% nitrogen gas. It is liquid below its boiling point of –196°C. Note the dangers associated with asphyxiation by nitrogen gas, and frost burns from the liquid normal biological context) – can be expanded and maintained.

Oocyte (immature and mature): the female germ cell found in the follicles of the ovarian cortex

Opening size: The diameter of the opening in a storage tank. A larger opening allows bigger canisters/inventory to be used, but also compromises the insulation rating, reducing the static holding time.

Oxygen meter: Used in liquid nitrogen storage facilities to provide warning of lowered oxygen levels in the air. Note that oxygen meters may need regular servicing and routine testing to remain functional.

Passive cooling device: A container which, when placed at a specific temperature (e.g. –80°C), cools at a known rate, e.g. 1°C/min. A non-controlled cooling method used for slow cooling exemplified by the isopropyl alcohol filled 'Mr frosty' (Nalgene) for cooling vials.

Safety valves: Storage vessels for nitrogen (not the specimen storage tanks) hold and dispense nitrogen under pressure. Safety valves are required to ensure that the pressure stays within safe limits. Pressurized vessels should be safety-tested (pressure-tested) at least every 10 years sample. Lyophilized samples may be stored at room temperatures.

Seal: Usually used in the context of straws. To prevent liquid nitrogen entering or the contents leaking out, straws can be sealed with heat (thermosoldering), PVA or PVP powder, identification plugs, ball bearings, or ultrasonically with an impulse sealer.

Semen: Fluid containing spermatozoa in suspension, secreted from the male gonads/ reproductive tract (also known as seminal fluid).

Seroconversion: The act of converting from negative to positive on a blood test. You can be infected with a virus but blood tests may still be negative until it seroconverts. The lag between infection and seroconversion was an issue with traditional viral testing, but it is not as much of an issue anymore.

Sickle cell anemia: The most common heritable blood disorder in the United States. It is a condition affecting the hemoglobin where the red blood cells take on a sickling shape instead of the characteristic round shape. People with sickle cell anemia can have anemia, frequent infections, or pain because of the condition. Sickle cell anemia is most prevalent in African, Hispanic, and Mediterranean ethnicities.

Sintered disk: A porous disk, for example made of brass, which can be applied to the end of nitrogen hoses to control the flow and reduce the likelihood of personal injury solidification of cells and the solution in which they are in without the formation of ice crystals.

Spills: A small spill of nitrogen is generally harmless unless it is to a particularly vulnerable part of the body (e.g. the eye). Larger spills can cause serious injury. Clothes that become soaked by liquid nitrogen should be removed or at least held away from the body.

Static holding time: The length of time taken for a liquid nitrogen storage vessel which is not used or re-filled to go from full to empty. Commonly measured in weeks. Old or damaged tanks may have a greatly reduced static holding time (can be as little as hours).

Static nitrogen consumption: The amount of nitrogen which boils off from a storage vessel which is not in use. Commonly 0.1–1 l a day. The rate of loss can be much higher in vessels which are opened frequently.

Straw: Traditionally supplied as 0.25 or 0.5 ml plastic straws. Now available in a range of materials and configurations. Materials include polyvinyl chloride (PVC), polyethylene terephthalate glycol (PETG) and high-security ionomeric resin (IR). Straws may or may not be sealed after inserting specimens (see seal). The likelihood of leakage and breakage in liquid nitrogen is lowest for soft straws (type 'CBS', Cryo Bio System by IMV Technologies, France) sealed with an 'impulse sealer', but we are not aware of any instance where these have been used for rapid cooling procedures. Straws which have been stretched and then cut at the narrowest point have a smaller diameter and thinner walls and are widely used for rapid cooling (see open pulled straws and closed pulled straws). Cooling rates of ~200°C/min (in vapour), >2000°C/min (plunging in nitrogen) and 4000 °C/min for plunging in nitrogen slush. Warming rate is ~2500 °C/min.

Tank: Usually refers to the container into which specimens are placed for long-term storage. Three main categories: liquid nitrogen tanks, nitrogen vapour tanks and ultralow temperature ("–130°C) electrical freezers. Can also be referred to as a vessel or a Dewar Vapour shipper: A specialized container to maintain specimens at low subzero temperatures during transport. They can be shipped in cars and aeroplanes. They are pre-cooled with liquid nitrogen, but the nitrogen is then emptied out fully and specimens inserted. Their well-insulated design allows a low subzero temperature to be maintained for days after the nitrogen has been emptied out.

Tay Sachs: A rare hereditary genetic disorder. There are different variations of the disease, with the most serious and common one leading to muscle weakness, blindness, paralysis, and ultimately death. Tay Sachs occurs in all ethnicities, but is more common in people of Ashkenazi Jewish, French Canadian, and Irish descent.

Vial: Container, usually of 1–5 ml capacity. Used mainly for slow cooling of cells or gonadal tissue. Only vials which are made to withstand cryopreservation should be used to reduce the risk of explosions. The size should be chosen to fit the canes, freezing machine and specimen. Vials differ in volume, the configuration of the base, the shape inside (U- or V-shape), internal or external thread, the writing area, the type of O ring, and whether the lids can be colour-coded. They are bulky but easy to load, and have a large surface on which to write ID details. Some manufacturers make vials with bar-coding printed on them. Vials have been used for some mouse rapid cooling protocols but fast cooling rates can only be achieved if specimens are dripped into vials precooled to a low subzero temperature. Vials take a long time to warm which means that they are not suitable for all cryoprotectant solutions. Their thermal inertia makes accidental thawing unlikely.

Vial: The container in which the semen sample is frozen and stored. Each Cryobio vial is labeled with the sperm bank name, the donor number, and the date the semen sample was collected.

Vitrification: a method of cryopreservation which involves ultra-rapid rates of cooling to produce.

Volume: In relation to liquid nitrogen storage vessels, the amount of liquid nitrogen that they will hold.

Waterproof pens: Water vapour condenses on cold surfaces, so it is very important that all information on surfaces which are going to become cold is written with pencil or indelible waterproof pens or using specialized labelling devices. As most waterproof ink is soluble in alcohol it is important that all information is copied or noted before the surface is wiped with alcohol (for example if the surfaces are wiped with alcohol to sterilize them before opening them).

Weight: Liquid nitrogen weighs approximately the same as water. Large storage vessels can be too heavy to move, in particular for small persons. Canisters, or racks, within the storage vessels, particularly those of large capacity, can be very heavy. Tall tanks exacerbate the problem. Use of appropriate lifting techniques, trolleys and other equipment must be considered.

REFERENCES

Abir R, Fisch B, Raz A, Nitke S and Ben-Rafael Z (1998) Preservation of fertility in women undergoing chemotherapy: current approach and future prospects. *J Assist Reprod Genet* 15, 469-477.

Academic Press, New York, pp. 189-218.

Acker JP (2007) The use of intracellular protectants in cell biopreservation. In Baust, JG and Baust, JM (eds) Advances in Biopreservation. Taylor and Francis, Boca Raton, pp. 299-320.

Acker JP and McGann LE (2003) Protective effect of intracellular ice during freezing? *Cryobiology* 46, 197-202.

Acker JP, Elliott JA and McGann LE (2001) Intercellular ice propagation: experimental evidence for ice growth through membrane pores. *Biophys* J 81, 1389-1397.

Acker JP, Larese A, Yang H, Petrenko A and McGann LE (1999) Intracellular ice formation is affected by cell interactions. *Cryobiology* 38, 363-371.

Adams SL, Hessian PA, Mladenov PV. Cryopreservation of sea urchin (Evechinus chloroticus) sperm. *CryoLetters*, 2004; 25(4): 287-

Adams SL, Hessian PA, Mladenov PV. The potential for cryopreserving larvae of the sea urchin, Evechinus chloroticus. *Cryobiology*. 2006; 52(1): 139-145.

Adams SL, Kleinhans FW, Mladenov PV and Hessian PA (2003) Membrane permeability characteristics and osmotic tolerance limits of sea urchin (Evechinus chloroticus) eggs. *Cryobiology* 47, 1-13.

Adams SL, Kleinhans FW, Mladenov PV, Hessian PA. Membrane permeability characteristics and osmotic tolerance limits of sea urchin (Evechinus chloroticus) eggs. Cryobiolog. 2003; 47: 1-13.

Adams SL, Smith JF, Roberts RD, Janke AR, Kaspar HF, Tervit HR, et al. Cryopreservation of sperm of the Pacific oyster (Crassostrea gigas): development of a practical method for commercial spat production. Aquaculture, 2004; 242: 271-282.

Agca Y (2000) Cryopreservation of oocyte and ovarian tissue. Ilar J 41, 207-220.

Agca Y, Liu J, Mullen S, Johnson-Ward J, Gould K, Chan A and Critser J (2004) Osmotic tolerance and membrane permeability characteristics of Rhesus (Macaca mulatta) spermatozoa. *Cryobiology* 49, 316-317.

Agca Y, Liu J, Rutledge JJ, Critser ES and Critser JK (2000) Effect of osmotic stress on the developmental competence of germinal vesicle and metaphase II stage bovine cumulus oocyte complexes and its relevance to cryopreservation. Mol Reprod Dev 55, 212-219.

Agca Y, Mullen S, Liu J, Johnson-Ward J, Gould K, Chan A and Critser J (2005) Osmotic tolerance and membrane permeability characteristics of rhesus monkey (Macaca mulatta) spermatozoa. *Cryobiology* 51, 1-14.

Aigner S, Van der Elst J, Siebzehnrubl E, Wildt L, Lang N and Van Steirteghem AC (1992) The influence of slow and ultra-rapid freezing on the organization of the meiotic spindle of the mouse oocyte. Hum Reprod 7, 857-864.

Alberts B, Johnson A, Lewis J, Raff M, Roberts K and Walter P (2002) Molecular *biology* of the cell. (4th), Garland Science, New York.

Alderdice DF. Osmotic and ionic regulation in teleost eggs and larvae. In: Fish Physiology, editor. W.S. Hoar, D.J. Randall. Academic Press, San Diego, 1988. p. 163-251.

Alesi R (2005) Infertility and its treatment--an emotional roller coaster. Aust Fam *Physician* 34, 135-138.

Al-Fageeh MB and Smales CM (2006) Control and regulation of the cellular responses to cold shock: the responses in yeast and mammalian systems. Biochem J 397, 247-259.

Al-Fageeh MB, Marchant RJ, Carden MJ and Smales CM (2006) The cold-shock response in cultured mammalian cells: harnessing the response for the improvement of recombinant protein production. Biotechnol Bioeng 93, 829-835.

Al-Hasani S, Diedrich K, van der Ven H, Reinecke A, Hartje M and Krebs D (1987) Cryopreservation of human oocytes. Hum Reprod 2, 695-700.

Aman RR and Parks JE (1994) Effects of cooling and rewarming on the meiotic spindle and chromosomes of in vitro-matured bovine oocytes. Biol Reprod 50, 103-110.

Anchordoguy TJ, Carpenter JF, Cecchini CA, Crowe JH and Crowe LM (1990) Effects of protein perturbants on phospholipid bilayers. Arch Biochem Biophys 283, 356-361.

Anchordoguy TJ, Cecchini CA, Crowe JH and Crowe LM (1991) Insights into the cryoprotective mechanism of dimethyl sulfoxide for phospholipid bilayers. Cryobiology 28, 467-473.

Anchordoguy, T.J., Rudolph, A.S., Carpenter, J.F. & Crowe, J.H. (1987). Modes of interaction of cryoprotectants with membrane phospholipids during freezing. Cryobiology 24:pp. 324-331.

Andersen AN, Gianaroli L, Felberbaum R, de Mouzon J and Nygren KG (2006) Assisted reproductive technology in Europe, 2002. Results generated from European registers by ESHRE. Hum Reprod 21, 1680-1697.

Arakawa, T., Carpenter, J.F., Kita, Y.A. & Crowe, J.H. (1990). The basis for toxicity of certain cryoprotectants - a hypothesis. Cryobiology 27:pp. 401-415.

Arav A, Zeron Y, Leslie SB, Behboodi E, Anderson GB and Crowe JH (1996) Phase transition temperature and chilling sensitivity of bovine oocytes. *Cryobiology* 33, 589-599.

Armitage WJ and Mazur P (1984) Osmotic tolerance of human granulocytes. Am *J Physiol* 247, C373-381.

Armitage WJ, Parmar N and Hunt CJ (1985) The effects of osmotic stress on human platelets. *J Cell Physiol* 123, 241-248.

Asada M and Fukui Y (2000) Effect on fertilization and development by re-culture after freezing and thawing of bovine oocytes matured in vitro. *Theriogenology* 54, 889-898.

Asada M, Ishibashi S, Ikumi S and Fukui Y (2002) Effect of polyvinyl alcohol (PVA) concentration during vitrification of in vitro matured bovine oocytes. *Theriogenology* 58, 1199-1208.

Asch R, Simerly C, Ord T, Ord VA and Schatten G (1995) The stages at which human fertilization arrests: microtubule and chromosome configurations in inseminated oocytes which failed to complete fertilization and development in humans. Hum Reprod 10, 1897-1906.

Ashwood-Smith MJ (1987) Mechanism of cryoprotectant action. In Bowler, K and Fuller, BJ (eds) Temperature and Animal Cells. The company of *biologists*, Ltd., Cambridge, pp. 395-406.

Azambuja RM, Kraemer DC and Westhusin ME (1998) Effect of low temperatures on in-vitro matured bovine oocytes. *Theriogenology* 49, 1155-1164.

Babiak I, Glogowsky, Brzuska JE, Szumiec J, Adamek J, Cryopreservation of sperm of common carp Cyprinus carpio. Aquaculture Res. 1995; 28: 567-571.

Baek, K-H. & Skinner, D.Z. (2003). Alteration of antioxidant enzyme gene expression during cold acclimation of near isogenic wheat lines. Plant Science 165:pp. 1221-1227.

Baker TG (1963) A Quantitative and Cytological Study of Germ Cells in Human Ovaries. Proc R Soc Lond B Biol Sci 158, 417-433.

Bank H (1973) Visualization of freezing damage. II. Structural alterations during warming. Cryobiology 10, 157-170.

Bankowski BJ, Lyerly AD, Faden RR and Wallach EE (2005) The social implications of embryo cryopreservation. Fertil Steril 84, 823-832.

Barbazuk WB, Korf I, Kadavi C, Heyen J, Tate S, Wun E, et al. (2000), The syntenic relationship of the zebrafish and human genomes. Genome Res. 2000; 10: 1351-1358.

Barrett J (2001) Thermal hysteresis proteins. Int J Biochem Cell Biol 33, 105-117.

Battaglia DE, Goodwin P, Klein NA and Soules MR (1996) Influence of maternal age on meiotic spindle assembly in oocytes from naturally cycling women. Hum Reprod 11, 2217-2222.

Baudot A and Odagescu V (2004) Thermal properties of ethylene glycol aqueous solutions. Cryobiology 48, 283-294.

Baust JG and Baust JM (eds) (2007) Advances in Biopreservation. Taylor and Francis Boca Raton

Baust JM, Van B and Baust JG (2000) Cell viability improves following inhibition of cryopreservation- induced apoptosis. In Vitro Cell Dev Biol Anim 36, 262-270.

Baust JM, Vogel MJ, Van Buskirk R and Baust JG (2001) A molecular basis of cryopreservation failure and its modulation to improve cell survival. Cell Transplant 10, 561-571.

Bautista JA, Dela Pena EC, Katagiri S, Takahashi Y and Kanagawa H (1998) In vitro viability of mouse oocytes vitrified in an ethylene glycol-based solution. Jpn J Vet Res 46, 13-18.

Baxter SJ and Lathe GH (1971) Biochemical effects of kidney of exposure to high concentrations of dimethyl sulphoxide. Biochem Pharmacol 20, 1079-1091.

Benagiano G and Gianaroli L (2004) The new Italian IVF legislation. Reprod Biomed Online 9, 117-125.

Benson EE (2004) Cryoconserving algal and plant diversity: historical perspectives and future challenges. In Fuller, BJ, Lane, N and Benson, EE (eds) Life in the Frozen State. CRC Press, Boca Raton, pp. 299-328.

Benson, E.E. & Bremner, D. (2004). Oxidative stress in the frozen plant: a free radical point of view. In: Life in the Frozen State, Fuller, B.J., Lane, N. & Benson, E.E. (Eds.), pp. 206– 241, CRC Press, ISBN 0-415-24700-4, Boca Raton, Florida.

Benson, E.E. & Whithers, L.A. (1987). Gas-chromatographic analysis of volatile hydrocarbon production by cryopreserved plant-tissue cultures - a nondestructive method for assessing stability. CryoLetters 8:pp. 35-46.

Benson, E.E. (1990). Free Radical Damage in Stored Plant Germplasm, International Board for Plant Genetic Resources, ISBN 92-9043-196-2, Rome, Italy.

Benson, E.E. (2008). Cryopreservation of phytodiversity: A critical appraisal of theory & practice. Critical Reviews in Plant Sciences 27:pp. 141-219.

Benson, E.E., Lynch, P.T. & Jones, J. (1992). The detection of lipid peroxidation products in cryoprotected and frozen rice cells: consequences for post-thaw survival. Plant Science 85:pp. 107-114.

Benson, E.E., Lynch, P.T. & Jones, J. (1995). The use of the iron chelating agent desferrioxamine in rice cell cryopreservation: a novel approach for improving recovery. Plant Science 110:pp. 249-258.

Benson, E.E., Reed, B.M., Brennan, R.M., Clacher, K.A. & Ross, D.A. (1996). Use of thermal analysis in the evaluation of cyropreservation protocols for Ribes nigrum L. germplasm. CryoLetters. 17:pp. 347–362.

Bernard A, Hunter JE, Fuller BJ, Imoedemhe D, Curtis P and Jackson A (1992)

Beyler SA, Meyer WR and Fritz MA (2000) Disposition of extra embryos. Fertil Steril 74, 213-215.

Bianchi V, Coticchio G, Distratis V, Di Giusto N, Flamigni C and Borini A (2007) Differential sucrose concentration during dehydration (0.2 mol/l) and rehydration (0.3 mol/l) increases the implantation rate of frozen human oocytes. Reprod Biomed Online 14, 64-71.

Bianchi V, Coticchio G, Fava L, Flamigni C and Borini A (2005) Meiotic spindle imaging in human oocytes frozen with a slow freezing procedure involving high sucrose concentration. Hum Reprod 20, 1078-1083.

Biggers JD (1998) Reflections on the culture of the preimplantation embryo. Int J Dev Biol 42, 879-884.

Birth of a healthy female after intracytoplasmic sperm injection of cryopreserved human oocytes. Fertil Steril 68, 724-726.

Blanco JM, Gee G, Wildt DE and Donoghue AM (2000) Species variation in osmotic, cryoprotectant, and cooling rate tolerance in poultry, eagle, and peregrine falcon spermatozoa. Biol Reprod 63, 1164-1171.

Blaxter JHS. Sperm storage and cross-fertilization of spring and autumn spawning herring. Nature. 1953; 172: 1189-1190.

Boiso I, Marti M, Santalo J, Ponsa M, Barri PN and Veiga A (2002) A confocal microscopy analysis of the spindle and chromosome configurations of human oocytes cryopreserved at the germinal vesicle and metaphase II stage. Hum Reprod 17, 1885-1891.

Bokor Z, Müller T, Bercsényi M, Horváth L, Urbányi B, Horváth A, Cryopreservation of sperm of two European percid species, the pikeperch (Sander lucioperca) and the Volga pikeperch (S. volgensis). Acta. Biol. Hung. 2007; 58(2): 199-20.

Borini A, Sciajno R, Bianchi V, Sereni E, Flamigni C and Coticchio G (2006).

Bos-Mikich A and Whittingham DG (1995) Analysis of the chromosome complement of frozen-thawed mouse oocytes after parthenogenetic activation. Mol Reprod Dev 42, 254-260.

Bos-Mikich A, Wood MJ, Candy CJ and Whittingham DG (1995) Cytogenetical analysis and developmental potential of vitrified mouse oocytes. Biol Reprod 53, 780-785.

Bouquet M, Selva J and Auroux M (1993) Cryopreservation of mouse oocytes: mutagenic effects in the embryo? Biol Reprod 49, 764-769.

Bouquet M, Selva J and Auroux M (1995) Effects of cooling and equilibration in DMSO, and cryopreservation of mouse oocytes, on the rates of in vitro fertilization, development, and chromosomal abnormalities. Mol Reprod Dev 40, 110-115.

Bowler K and Fuller BJ (eds) (1987) Temperature and Animal Cells. The company of biologists, Ltd. Cambridge Carpenter JF and Crowe JH (1988) The mechanism of cryoprotection of proteins by solutes. Cryobiology 25, 244-255.

Boyko, A. & Kovalchuk, I. (2008). Epigenetic control of plant stress response. Environmental and Molecular Mutagenesis 49:pp. 61-72.

Brownlie A, Donovan A, Pratt SJ, Paw BH, Oates AC, Brugnara C, et al. Positional cloning of the zebrafish sauternes gene: A model for congential sideroblastic anaeia. Nat .Genet, 1998; 20: 244-250.

Bunn, E., Turner, S., Panaia, M. & Dixon, K.W. (2007). The contribution of in vitro technology and cryogenic storage to conservation of indigenous plants, Australian Journal of Botany 55:pp. 345-355.

Calvi SL, Maisse G, Cryopreservation of carp (Cyprinus carpio) blastomeres. Aquat. Living Resour. 1999; 12: 71-74.

Calvi SL, Maisse G. Cryopreservation of rainbow trout (Oncorhynchus mykiss) blastomeres: Influence of embryo stage on post-thaw survival rate. Cryobiology. 1998; 36: 255-262.

Candy CJ, Wood MJ, Whittingham DG, Restoration of a normal reproductive lifespan after grafting of cryopreserved mouse ovaries. Hum. Reprod. 2000; 15(6): 1300-1304.

Cardona-Costa J, García-Ximénez F. Vitrification of zebrafish embryo blastomeres in microvolumes. Vitrification of zebrafish embryo blastomeres in microvolumes. CryoLetters. 2007; 28: 303-309.

Carroll J, Depypere H and Matthews CD (1990) Freeze-thaw-induced changes of the zona pellucida explains decreased rates of fertilization in frozen-thawed mouse oocytes. J Reprod Fertil 90, 547-553.

Castillo, N.R.F., Bassil, N.V., Wada, S. & Reed, B.M. (2010). Genetic stability of cryopreserved shoot tips of Rubus germplasm. In Vitro Cellular & Developmental Biology-Plant 46:pp. 246-256.

Caswell, K.L. & Kartha, K.K. (2009). Recovery of plants from pea and strawberry meristems cryopreserved for 28 years. CryoLetters 30:pp. 41-46.

Cattanach BM, Iddon CA, Charlton HM, Chiappa SA and Fink G (1977) Gonadotrophin-releasing hormone deficiency in a mutant mouse with hypogonadism. Nature 269, 338-340.

Centers for Disease Control and Prevention (1989) Postponed childbearing-United States, 1970-1987. Morb Mortal Wkly Rep 38, 810-816.

Centers for Disease Control and Prevention (1989) Postponed childbearing-United States, 1970-1987. Morb Mortal Wkly Rep 38, 810-816.

Chandrasekaran M, Pitt RE, Parks JE and Steponkus PL (1990) Stochastic behavior of intracellular ice formation in isoltated rye protoplasts and mature bovine oocytes. Cryobiology 27, 676.

Chang MC (1952) Fertilizability of rabbit ova and the effects of temperature in vitro on their subsequent fertilization and activation in vivo. Journal of Experimental Zoology 121, 351-370.

Chang MC (1953) Storage of unfertilized rabbit ova; subsequent fertilization and the probability of normal development. Nature 172, 353-354.

Channuntapipat, C., Sedgley, M. & Collins, G. (2003). Changes in methylation and structure of DNA from almond tissues during in vitro culture and cryopreservation. J. Amer. Soc. Hort. Sci. 128:pp. 890-897.

Chao NH, Chao WC, Liu KC, Liao IC. The properties of tilapia sperm and its cryopreservation. J. Fish. Biol. 1987; 30: 107-118.

Chapman D (1975) Phase transitions and fluidity characteristics of lipids and cell membranes. Q Rev Biophys 8, 185-235.

Chen C (1986) Pregnancy after human oocyte cryopreservation. Lancet 1, 884- 886.

Chen C (1988) Pregnancies after human oocyte cryopreservation. Ann N Y Acad Sci 541, 541-549.

Chen, X.-L., Li, J.-H., Xin, X., Zhang, Z.-E., Xin, P.-P. & Lu, X.-X. (2011). Cryopreservation of in vitro-grown apical meristems of Lilium by droplet-vitrification. South African Journal of Botany 77:pp. 397-403

Chian RC, Buckett WM and Tan SL (2004) In-vitro maturation of human oocytes. Reprod Biomed Online 8, 148-166.

Chian RC, Kuwayama M, Tan L, Tan J, Kato O and Nagai T (2004) High survival rate of bovine oocytes matured in vitro following vitrification. J Reprod Dev 50, 685-696.

Clinical outcome of oocyte cryopreservation after slow cooling with a protocol utilizing a high sucrose concentration. Hum Reprod 21, 512-517.

Cloud JG, Erdahl AL, Graham EF. Survival and continued normal development of fish embryos after incubation at reduced temperatures. Trans. Am. Fish. Soc, 1988; 117: 503-506.

Clouthier DE, Avarbock MR, Maika SD, Hammer RE, Brinster RL. Rat spermatogenesis in mouse testis. Nature. 1996; 381: 418-421.

Cobo A, Rubio C, Gerli S, Ruiz A, Pellicer A and Remohi J (2001) Use of fluorescence in situ hybridization to assess the chromosomal status of embryos obtained from cryopreserved oocytes. Fertil Steril 75, 354-360.

Cocks FH and Brower WE (1974) Phase diagram relationships in cryobiology. Cryobiology 11, 340-358.

Cooper DK, Gollackner B and Sachs DH (2002) Will the pig solve the transplantation backlog? Annu Rev Med 53, 133-147.

Coticchio G, Bonu M, Borini A and Flamigni C (2004) Oocyte cryopreservation: a biological perspective. Eur J Obstet Gynecol Reprod Biol 1, S2-7.

Cozzi E and White DJ (1995) The generation of transgenic pigs as potential organ donors for humans. Nat Med 1, 964-966.

Critser JK and Russell RJ (2000) Genome resource banking of laboratory animal models. Ilar J 41, 183-186.

Critser JK, Agca Y and Gunasena KT (1997) The Cryobiology of Mammalian Oocytes. In Karow, AM and Critser, JK (eds) Reproductive Tissue Banking, Scientific Principles. Academic Press, San Diego, pp. 329-357.

Critser JK, Agca Y and Gunasena KT (1997) The Cryobiology of Mammalian Oocytes. In Karow, AM and Critser, JK (eds) Reproductive Tissue Banking, Scientific Principles. Academic Press, San Diego, pp. 329-357.

Crowe JH, Carpenter JF and Crowe LM (1998) The role of vitrification in anhydrobiosis. Annu Rev Physiol 60, 73-103.

Crowe JH, Crowe LM, Carpenter JF, Rudolph AS, Wistrom CA, Spargo BJ and Anchordoguy TJ (1988) Interactions of sugars with membranes. Biochim Biophys Acta 947, 367-384.

Crowe JH, Crowe LM, Tablin F, Wolkers W and Oliver AE (2004) Stabilization of cells diring freeze-drying: the trehalose myth. In Fuller, BJ, Lane, N and Benson, EE (eds) Life in the Frozen State. CRC Press, Boca Raton, pp. 581-601.

Crowe, J.H., Crowe, L.M., Carpenter, J.F., Rudolph, A.S., Wistrom, C.A., Spargo, B.J. & Anchordoguy, T.J. (1988). Interactions of sugars with membranes. Biochimica et Biophysica Acta 947:pp. 367-84.

Cryoprotectant toxicity and cryoprotectant toxicity reduction: in search of molecular mechanisms. Cryobiology 27, 247-268.

Dai, F., Huang, Y., Zhou, M. & Zhang G. (2009). The influence of cold acclimation on antioxidant enzymes and antioxidants in sensitive and tolerant barley cultivars. Biologia Plantarum 153:pp. 257-262.

Damak S, Su H, Jay NP and Bullock DW (1996) Improved wool production in transgenic sheep expressing insulin-like growth factor 1. Biotechnology (N Y) 14, 185-188.

Day BN (2000) Reproductive biotechnologies: current status in porcine reproduction. Anim Reprod Sci 60-61, 161-172.

Day, J.G. & Stacey, G.N. (2007). Cryopreservation and Freeze-Drying Protocols. Humana Press Inc., ISBN 978-1-58829-377-0, Totowa, NJ.

Day, J.G., Harding, K.C., Nadarajan, J. & Benson, E.E. (2008). Cryopreservation, conservation of bioresources at ultra low temperatures. In: Molecular Biomethods Handbook. Walker J.M & Rapley R. (Eds.), pp. 917-947, Humana Press, ISBN 978-1- 60327-374-9, Totowa, NJ.

De Loecker R and Penninckx F (1987) Osmotic effects of rapid dilution of cryoprotectants II. Effects on human erythrocyte hemolysis. Cryo-Letters 8, 140- 145.

De Santis L, Cino I, Rabellotti E, Papaleo E, Calzi F, Fusi FM, Brigante C and Ferrari A (2007) Oocyte cryopreservation: clinical outcome of slow-cooling protocols differing in sucrose concentration. Reprod Biomed Online 14, 57-63.

Decruse, S.W., Seeni, S. & Pushpangadan, P. (1999). Cryopreservation of alginate coated shoot tips of in vitro grown Holostemma annulare (Roxb.) K. Schum, an endangered medicinal plant: Influence of preculture and DMSO treatment on survival and regeneration. CryoLetters 20:pp. 243-250.

Development of alternative plant vitrification solutions in droplet-vitrification procedures. CryoLetters 30:pp. 320-34.

Diedrich K, Al-Hasani S, Van der Ven H and Krebs D (1987) Successful in vitro fertilizationof frozen-thawed rabbit and human oocytes. Proceedings of the 5th world congress on IVF and ET 652-670.

Diller KR (1979) Intracellular freezing of glycerolized red cells. Cryobiology 16, 125-131.

Diller KR (1997) Pioneers in cryobiology: Nikolay Aleksandrovich Maximov (1890-1952). Cryo-Letters 18, 81-92.

Dinnyes A, Dai Y, Jiang S and Yang X (2000) High developmental rates of vitrified bovine oocytes following parthenogenetic activation, in vitro fertilization, and somatic cell nuclear transfer. Biol Reprod 63, 513-518.

Dinnyés A, Urbánvi B, Baranvai B, Maqvary I. Chilling sensitivity of carp (Cyprinus carpio) embryos at different developmental stages in the presence or absence of cryoprotectants: work in progress. Theriogenology. 1998; 50(1): 1-13.

Doebbler GF and Rinfret AP (1962) The influence of protective compounds and cooling and warming conditions on hemolysis of erythrocytes by freezing and thawing. Biochim Biophys Acta 58, 449-458.

Domar AD (1997) Stress and infertility in women. In Leiblum, SR (eds) Infertility: Psychological issues and counseling strategies. John Wiley and Sons, New York, pp. 67-82.

Domar AD, Broome A, Zuttermeister PC, Seibel M and Friedman R (1992) The prevalence and predictability of depression in infertile women. Fertil Steril 58, 1158-1163.

Dowdle, J., Ishikawa, T., Gatzek, S., Rolinski, S. & Smirnoff, N. (2007). Two genes in Arabidopsis thaliana encoding GDP l galactose phosphorylase are required for ascorbate biosynthesis and seedling viability. The Plant Journal 52:pp. 673-689.

Dowgert MF and Steponkus PL (1983) Effect of Cold Acclimation on Intracellular Ice Formation in Isolated Protoplasts. Plant Physiol 72, 978-988.

Dowgert, M.F. & Steponkus, P.L. (1984). Behavior of the plasma membrane of isolated protoplasts during a freeze-thaw cycle. Plant Physiol 75:pp. 1139-1151.

Drobnis EZ, Crowe LM, Berger T. Anchordoruy TJ, Overstree tJW, Crow JH. Cold shock damage is due to lipid phase transitions in cell membranes: a demonstration using sperm as a model. J. Exp. Zoo . 1993; 265: 432-437.

Drokin SI. Phospholipid distribution and fatty acid composition of phosphatidylcholine and phosphatidylethanolamine in sperm of some freshwater and marine species of fish. Aquat. Living Resour. 1993; 6: 49-56.

Dunkel-Schetter C and Lobel M (1991) Psychological reactions to infertility. In Stanton, AL and Dunkel-Schetter, C (eds) Infertility: Perspectives from stress and coping research. Plenum Press, New York, pp. 29-57.

DunnRS, McLachlan J. Cryopreservation of echinoderm sperm. Can. J. Zool. 1973; 51: 666-669.

Durlinger AL, Kramer P, Karels B, de Jong FH, Uilenbroek JT, Grootegoed JA and Themmen AP (1999) Control of primordial follicle recruitment by anti- Mullerian hormone in the mouse ovary. Endocrinology 140, 5789-5796.

Edmunds TVan Patten SMPollock JHanson EBernasconi RHiggins EManavalan PZiomek CMeade HMcPherson JM, et al. (1998) Transgenically produced human antithrombin: structural and unctional comparison to human plasmaderived antithrombin. Blood 91, 4561-4571.

El-Toukhy T, Khalaf Y and Braude P (2006) IVF results: optimize not maximizc. Am J Obstet Gynecol 194, 322-331.

Emmelot, P. & Vanhoeven, R.P. (1975). Phospholipid unsaturation and plasma membrane organization. Chemistry and Physics of Lipids 14:pp. 236-246.

Engelmann, F. (2004). Plant cryopreservation: Progress and prospects. In Vitro Cellular & Developmental Biology-Plant 40:pp. 427-433.

Environ Mol Mutagen 28, 167-175.

Eroglu A, Toth TL and Toner M (1998) Alterations of the cytoskeleton and polyploidy induced by cryopreservation of metaphase II mouse oocytes. Fertil Steril 69, 944-957.

Evans MJ and Kaufman MH (1981) Establishment in culture of pluripotential cells from mouse embryos. Nature 292, 154-156.

Fabre, J. & Dereuddre, J. (1990). Encapsulation-dehydration: A new approach to cryopreservation of Solanum shoot-tips. CryoLetters 11:pp. 423-426.

Fahy GM (1980) Analysis of “solution effects” injury: rabbit renal cortex frozen in the presence of dimethyl sulfoxide. Cryobiology 17, 371-388.

Fahy GM (1986) The relevance of cryoprotectant “toxicity” to cryobiology. Cryobiology 23, 1-13.

Fahy GM and Karow AM, Jr. (1977) Ultrastructure-function correlative studies for cardiac cryopreservation. V. Absence of a correlation between electrolyte toxicity and cryoinjury in the slowly frozen, cryoprotected rat heart. Cryobiology 14, 418-427.

Fahy GM, Lilley TH, Linsdell H, Douglas MS and Meryman HT (1990)

Fahy GM, MacFarlane DR, Angell CA and Meryman HT (1984) Vitrification as an approach to cryopreservation. Cryobiology 21, 407-426.

Fajer AB (1983) Changes in glucose metabolism in premeiotic and meiotic ovaries of the hamster. J Reprod Fertil 69, 101-104.

Fang, J.Y., Wetten, A. & Johnston, J. (2008). Headspace volatile markers for sensitivity of cocoa (Theobroma cacao L.) somatic embryos to cryopreservation. Plant Cell Reports 27: pp. 453-461.

Feichtinger W, Benko I and Kemeter P (1988) Freezing human oocytes using rapid techniques. Ann N Y Acad Sci 101-110.

Fertilization and embryonic development of human oocytes after cooling. Hum Reprod 7, 1447-1450.

Fiéni F, Beckers JP, Buggin M, Bruyas JF, Perrin J, Daubié M and Tainturier D (1995) Evaluation of cryopreservation techniques for goat embryos.

Fleck, R.A., Benson, E.E., Bremner, D.H. & Day, J.G. (2000). Studies of free radical-mediated cryoinjury in the unicellular green alga Euglena gracilis using a non-destructive hydroxyl radical assay: a novel approach for developing protistan cryopreservation strategies. Free Radic Res 32:pp. 157-70.

Flemming W (1879) Beiträge zur Kenntnisse der Zelle und ihrer Lebenserscheinungen. Archiv fuer Mikroskopische Anatomie 18, 151-259.

Foote RH. (1975) Semen quality from the bull to the freezer: An assessment. Theriogenology. 1975; 3: 219-235.

Forabosco A, Sforza C, De Pol A, Vizzotto L, Marzona L and Ferrario VF (1991) Morphometric study of the human neonatal ovary. Anat Rec 231, 201-208.

Foubert, I., Dewettinck, K., Van de Walle, D., Dijkstra, A.J. & Quinn, P.J. (2007). Physical properties: structural and physical characteristics, In: The Lipid Handbook with CD- ROM, Gunstone, F.D., Harwood J.L. & Dijkstra A.J. (Eds.), pp. 471-534, CRC Press, Taylor & Francis Group, ISBN 0-8493-9688-3, LLC, Boca Raton, FL.

Foyer, C.H. & Noctor, G. (2009). Redox regulation in photosynthetic organisms: signaling, acclimation, and practical implications. Antioxidants & Redox Signaling 11:pp. 861- 905.

French AJ, Greenstein JL, Loveland BE and Mountford PS (1998) Current and future prospects for xenotransplantation. Reprod Fertil Dev 10, 683-696.

Fry RC, Simpson TL and Squires TJ (1998) Ultrasonically guided transvaginal oocyte recovery from calves treated with or without GnRH. Theriogenology 49, 1077-1082.

Fujihira T, Nagai H and Fukui Y (2005) Relationship between equilibration times and the presence of cumulus cells, and effect of taxol treatment for vitrification of in vitro matured porcine oocytes. Cryobiology 51, 339-343.

Fujita J (1999) Cold shock response in mammalian cells. J Mol Microbiol Biotechnol 1, 243-255.

Fuku E, Kojima T, Shioya Y, Marcus GJ and Downey BR (1992) In vitro fertilization and development of frozen-thawed bovine oocytes. Cryobiology 29, 485-492.

Fuku EJ, Liu J and Downey BR (1995) In vitro viability and ultrastructural changes in bovine oocytes treated with a vitrification solution. Mol Reprod Dev 40, 177-185.

Fuller B and Paynter S (2004) Fundamentals of cryobiology in reproductive medicine. Reprod Biomed Online 9, 680-691.

Fuller BJ (2004) Cryoprotectants: the essential antifreezes to protect life in the frozen state. Cryo-letters 25, 375-388.

Fuller BJ, Lane N and Benson EE (eds) (2004) Life in the Frozen State. CRC Press Boca Raton

Furt, F., Simon-Plas, F. & Sebastien, M. (2011). Lipids of the plat plasma membrane, In: The Plant Plasma Membrane. Murphy, A.S., Peer, W. & Schulz, B. (Eds.), pp. 3-30, Springer-Verlag, ISBN 978-3-642-13430-2, Berlin Heidelberg.

Gakhova EN, Krasts IV, Naidenko T, Saveleva BI. Embryonic development of the sea urchin after low- temperature preservation. Ontogenez. 1988; 19: 175-180.

Gamez-Pastrana, R., Martinez-Ocampo, Y., Beristain, C.I. & Gonzalez-Arnao, M.T. (2004). An improved cryopreservation protocol for pineapple apices using encapsulation-vitrification. CryoLetters 25:pp. 405-414.

Gao D, Mazur P and Critser JK (1997) Fundamental Cryobiology of Mammalian Spermatozoa. In Karow, AM and Critser, JK (eds) Reproductive Tissue Banking, Scientific Principles. Academic Press, San Diego, pp. 263-328.

Gao DY, Chang Q, Liu C, Farris K, Harvey K, McGann LE, English D, Jansen J and Critser JK (1998) Fundamental cryobiology of human hematopoietic progenitor cells. I: Osmotic characteristics and volume distribution. Cryobiology 36, 40-48.

Gao DY, Liu J, Liu C, McGann LE, Watson PF, Kleinhans FW, Mazur P, Critser ES and Critser JK (1995) Prevention of osmotic injury to human spermatozoa during addition and removal of glycerol. Hum Reprod 10, 1109-1122.

Gearheart WW, Smith C and Teepker G (1989) Multiple ovulation and embryo manipulation in the improvement of beef cattle: relative theoretical rates of genetic change. J Anim Sci 67, 2863-2871.

Gelfand VI and Bershadsky AD (1991) Microtubule dynamics: mechanism, regulation, and function. In (eds) Annu Rev Cell Biol. Vol 7. pp. 93-116.

George MA, Johnson MH and Howlett SK (1994) Assessment of the developmental potential of frozen-thawed mouse oocytes. Hum Reprod 9, 130- 136.

Ghetler Y, Yavin S, Shalgi R and Arav A (2005) The effect of chilling on membrane lipid phase transition in human oocytes and zygotes. Hum Reprod 20, 3385-3389.

Gonzalez-Arnao, M.T., Panta, A., Roca, W.M., Escobar, R.H. & Engelmann, F. (2008). Development and large scale application of cryopreservation techniques for shoot and somatic embryo cultures of tropical crops. Plant Cell Tissue and Organ Culture 92:pp. 1-13.

Gook DA, Osborn SM and Johnston WI (1993) Cryopreservation of mouse and human oocytes using 1,2-propanediol and the configuration of the meiotic spindle. Hum Reprod 8, 1101-1109.

Gook DA, Schiewe MC, Osborn SM, Asch RH, Jansen RP and Johnston WI (1995) Intracytoplasmic sperm injection and embryo development of human oocytes cryopreserved using 1,2-propanediol. Hum Reprod 10, 2637-2641.

Gordon I (2003) Preservation of embryos and oocytes. In Laboratory production of Cattle Embryos, Second Edition. CAB International, pp. 277-302

Gordon-Kamm, W.J. & Steponkus, P.L. (1984). Lamellar-to-hexagonal II phase transitions in the plasma membrane of isolated protoplasts after freeze-induced dehydration. Proc Natl Acad Sci USA 81:pp. 6373-6377.

Gosden R and Nagano M (2002) Preservation of fertility in nature and ART. Reproduction 123, 3-11.

Gosden RG (2001) Trade-offs in cancer and reproduction. Hum Reprod Update 7, 360-362.

Gosden RG, Baird DT, Wade JC, Webb R. Restoration of fertility to oophorectomised sheep by ovarian autografts stored at -196 o C. Hum. Reprod. 1994; 9: 597-603.

Greenlee RT, Hill-Harmon MB, Murray T and Thun M (2001) Cancer statistics, 2001. CA Cancer J Clin 51, 15-36.

Grunwald, C. & Saunders, P.F. (1978). Function of sterols [and discussion]. Philosophical Transactions of the Royal Society of London B Biological Sciences 284:pp. 541-558.

Grunwald, C. (1975). Plant sterols. Annual Review of Plant Physiology and Plant Molecular Biology 26:pp. 209-236.

Guan M, Rawson DM, Zhang T. Cryopreservation of zebrafish (Danio rerio) oocytes using improved controlled slow cooling protocols. Cryobiology. 2008; 56: 204-208.

Guenther JF, Seki S, Kleinhans FW, Edashige K, Roberts DM and Mazur P (2006) Extra- and intra-cellular ice formation in Stage I and II Xenopus laevis oocytes. Cryobiology 52, 401-416.

Gupta MK, Uhm SJ and Lee HT (2007) Cryopreservation of immature and in vitro matured porcine oocytes by solid surface vitrification. Theriogenology 67, 238-248.

Gupta, S. & Reed, B.M. (2006). Cryopreservation of shoot tips of blackberry and raspberry by encapsulation-dehydration and vitrification. CryoLetters 27:pp. 29.42.

Guy, C.L., Niemi, K.J. & Brambl, R. (1985). Altered gene expression during cold acclimation of spinach. Proc Natl Acad Sci USA 82: pp. 3673-3677.

Gwo JC, Chen CW, Cheng HY. Semen cryopreservation of small abalone (Haliotis diversicolor supertexa). Theriogenology. 2002; 58:1563-1578.

Gwo JC. Cryopreservation of aquatic invertebrate seman: A review. Aquaculture Res. 2000; 31: 259-271.

Haga Y. On the subzero temperature preservation of fertilized eggs of rainbow trout. Bull. Jpn. Soc. Sci. Fish. 1982; 48: 1569-1572.

Hagedorn M, Hsu EW, Pilatus U, Wildt DE, Rall WF, Blackband SJ. Magnetic resonance microscopy and spectroscopy reveal kinetics of cryoprotectant permeation in a multicompartmental biological system. Proc. Natl. Acad .Sci., USA 1996; 93(15): 7454-7459.

Hagedorn M, Kleinhans FW, Freitas R, Liu J, Hsu EW, Wildt DE, et al. Water distribution and permeability of zebrafish embryos, (Brachydanio rerio). J. Exp. Zool. 1997; 278: 356-371.

Hagedorn M, Pan R, Cox EF, Hollingsworth L, Krupp D, Lewis TD, Coral larvae conservation: physiology and reproduction. Cryobiology. 1997; 52: 33-47.

Halliwell, B. & Gutteridge, J.M.C. (2007). Free Radicals in Biology and Medicine (Fourth edition), Oxford University Press Inc., ISBN 978-0-19-856868-1, Oxford.

Halmagyi, A., Fischer-Klüver, G., Mix-Wagner, G. & Schumacher, H.M. (2004). Cryopreservation of Chrysanthemum morifolium (Dendranthema grandiflora Ramat.) using different approaches. Plant Cell Rep 22:pp. 371-375.

Hamano S, Koikeda A, Kuwayama M and Nagai T (1992) Full-term development of in vitro-matured, vitrified and fertilized bovine oocytes. Theriogenology 38, 1085-1090.

Hamaratoglu F, roglu A, Toner M, Sadler KC. Cryopreservation of starfish oocytes. Cryobiolog. 2005; 50: 38-47.

Hamilton, K.N., Turner, S.R. & Ashmore, S.E. (2009). Cryopreservation, In: Plant Germplasm Conservation in Australia: Strategies and Guidelines for Developing, Managing and Utilising Ex Situ Collections, Offord, C.A & Meagher, P.F. (Eds.), pp. 129-128, Australian Network for Plant Conservation Inc., ISBN 978-0-9752191-1-9, Canberra.

Hao YHYong HYMurphy CNWax DSamuel MRieke ALai LLiu ZDurtschi DCWelbern VR, et al. (2006) Production of endothelial nitric oxide synthase (eNOS) over-expressing piglets. Transgenic Res 15, 739-750.

Hao, Y.J., You, C.X. & Deng, X.X. (2002a). Analysis of ploidy and the patterns of amplified fragment length polymorphism and methylation sensitive amplified polymorphism in strawberry plants recovered from cryopreservation. CryoLetters 23:pp. 37-46.

Hao, Y.J., You, C.X. & Deng, X.X. (2002b). Effects of cryopreservation on developmental competency, cytological and molecular stability of citrus callus. CryoLetters 23:pp. 27-35.

Harding, K. (2004). Genetic integrity of cryopreserved plant cells: a review. CryoLetters 25:pp. 3-22.

Harding, K.N., Johnston, J.W. & Benson, E.E. (2009). Exploring the physiological basis of cryopreservation success and failure in clonally propagated in vitro crop plant germplasm. Agricultural and Food Science 18:pp. 103-116.

Harris CL, Toner M, Hubel A, Cravalho EG, Yarmush ML and Tompkins RG (1991) Cryopreservation of isolated hepatocytes: intracellular ice formation under various chemical and physical conditions. Cryobiology 28, 436-444.

Hartmann, M.A. (1998). Plant sterols and the membrane environment. Trends in Plant Science 3:pp. 170-175.

Harvey B, Ashwood-Smith MJ. Cryopretectant penetration and supercooling in the eggs of samonid fishes. Cryobiology. 1982; 19: 29-40.

Harvey B. Cooling of embryonic cells, isolated blastoderms and intact embryos of the zebra fish Brachydanio rerio to -196°C. Cryobiology. 1983; 20: 440-447.

Hassold T and Hunt P (2001) To err (meiotically) is human: the genesis of human aneuploidy. Nat Rev Genet 2, 280-291.

Hassold T, Abruzzo M, Adkins K, Griffin D, Merrill M, Millie E, Saker D, Shen J and Zaragoza M (1996) Human aneuploidy: incidence, origin, and etiology.

Hayden JH, Bowser SS and Rieder CL (1990) Kinetochores capture astral microtubules during chromosome attachment to the mitotic spindle: direct visualization in live newt lung cells. J Cell Biol 111, 1039-1045.

Hazel JR (1984) Effects of temperature on the structure and metabolism of cell membranes in fish. Am J Physiol 246, R460-470.

Helliot, B., Madur, D., Dirlewanger, E. & De Boucaud, M.T. (2002). Evaluation of genetic stability in cryopreserved Prunus. In Vitro Cellular & Developmental Biology-Plant 38:pp. 493-500.

Hew CL and Yang DS (1992) Protein interaction with ice. Eur J Biochem 203, 33- 42.

Hewitson L, Simerly C and Schatten G (1997) Inheritance defects of the sperm centrosome in humans and its possible role in male infertility. Int J Androl 20, 35-43.

Hipkiss, A.R., Preston, J.E., Himswoth, D., Worthington, V.C. & Abbot, N.J. (1997). Protective effects of carnosine against malondialdehyde-induced toxicity towards cultured rat brain endothelial cells. Neuroscience Letters 238:pp. 135-138.

Hirai, D. & Sakai, A. (1999). Cryopreservation of in vitro-grown meristems of potato (Solanum tuberosum L.) by encapsulation-vitrification. Potato Research 42:pp. 153-160.

Hirai, D., Shirai, K., Shirai, S. & Sakai, A. (1998). Cryopreservation of in vitro grown meristems of strawberry (Fragaria x ananassa Duch.) by encapsulation-vitrification. Euphytica 101:pp. 109-115.

Hirshfield AN (1991) Development of follicles in the mammalian ovary. Int Rev Cytol 124, 43-101.

Ho RK, Kimmel CB, Commitment of cell fate in the early zebrafish embryo. Science. 1993; 261: 109-111. Hong Y, Winkler C, Schartl M. Production of medakafish chimeras from a stable embryonic stem cell line. Proc. Natl. Acad. Sci. USA 1998; 95: 3679-3684.

Hodzic, A., Rappolt, M., Amenitsch, H., Laggner, P. & Pabst, G. (2008). Differential modulation of membrane structure and fluctuations by plant sterols and cholesterol. Biophysical Journal 94:pp. 3935-3944.

Hoffman DI, Zellman GL, Fair CC, Mayer JF, Zeitz JG, Gibbons WE and Turner TG, Jr. (2003) Cryopreserved embryos in the United States and their availability for research. Fertil Steril 79, 1063-1069.

Hornicek FJ, Woll JE and Kasprisin D (eds) (2002) Standards for Tissue Banking. American Association of Tissue Banks McLean, VA Hotamisligil S, Toner M and Powers RD (1996) Changes in membrane integrity, cytoskeletal structure, and developmental potential of murine oocytes after vitrification in ethylene glycol. Biol Reprod 55, 161-168.

Huang EJ, Manova K, Packer AI, Sanchez S, Bachvarova RF and Besmer P (1993) The murine steel panda mutation affects kit ligand expression and growth of early ovarian follicles. Dev Biol 157, 100-109.

Hunter J, Bernard A, Fuller B, McGrath J and Shaw RW (1992) Plasma membrane water permeabilities of human oocytes: the temperature dependence of water movement in individual cells. J Cell Physiol 150, 175-179.

Hunter JE, Fuller BJ, Bernard A and Shaw RW (1995) The effect of cooling and hypertonic exposure on murine oocyte function, fertilization, and development. Cryobiology 32, 318-326.

Hyams JS and Lloyd CW (1994) Microtubules. Wiley-Liss, New York.

Improved development by Taxol pretreatment after vitrification of in vitro matured porcine oocytes. Reproduction 131, 795-804.

Infertility due to growth arrest of ovarian follicles in Sl/Slt mice. Dev Biol 126, 71-79.

Inouye M and Phadtare S (2004) Cold shock response and adaptation at nearfreezing temperature in microorganisms. Sci STKE 2004, pe26.

Interactions of cooling rate, warming rate and protective additive on the survival of frozen mammalian cells. In Wolstenholme, GEW and O'Connor, M (eds) The Frozen Cell. J and A Churchill, London, pp. 69-58.

Irimia D and Karlsson JO (2005) Kinetics of intracellular ice formation in onedimensional arrays of interacting biological cells. Biophys J 88, 647-660.

Isachenko V, Isachenko E, Michelmann HW, Alabart JL, Vazquez I, Bezugly N and Nawroth F (2001) Lipolysis and ultrastructural changes of intracellular lipid vesicles after cooling of bovine and porcine GV-oocytes. Anat Histol Embryol 30, 333-338.

Isachenko V, Montag M, Isachenko E, Nawroth F, Dessole S and van der Ven H (2004) Developmental rate and ultrastructure of vitrified human pronuclear oocytes after step-wise versus direct rehydration. Hum Reprod 19, 660-665.

Isayeva, A, Zhang T, Rawson DM, Studies on chilling sensitivity of zebrafish (Danio rerio) oocytes. Cryobiology. 2004; 49(2): 114-122.

Jamieson ME, Coutts JR and Connor JM (1994) The chromosome constitution of human preimplantation embryos fertilized in vitro. Hum Reprod 9, 709-715.

Johnson MH (1989) The effect on fertilization of exposure of mouse oocytes to dimethyl sulfoxide: an optimal protocol. J In Vitro Fert Embryo Transf 6, 168-175.

Johnson MH and Pickering SJ (1987) The effect of dimethylsulphoxide on the microtubular system of the mouse oocyte. Development 100, 313-324.

Johnson MH and Pickering SJ (1987) The effect of dimethylsulphoxide on the microtubular system of the mouse oocyte. Development 100, 313-324.

Johnson MH, Pickering SJ and George MA (1988) The influence of cooling on the properties of the zona pellucida of the mouse oocyte. Hum Reprod 3, 383- 387.

Johnston, J.W., Benson, E.E. & Harding, K. (2009). Cryopreservation induces temporal DNA methylation epigenetic changes and differential transcriptional activity in Ribes germplasm. Plant Physiology and Biochemistry 47:pp. 123-131.

Jokipii, S., Ryynänen, L., Kallioc, P.T., Aronen, T. & Häggmana, H. (2004). A cryopreservation method maintaining the genetic fidelity of a model forest tree, Plant Science. 166:pp. 799-806.

Joly C, Bchini O, Boulekbache H, Testart J and Maro B (1992) Effects of 1,2- propanediol on the cytoskeletal organization of the mouse oocyte. Hum Reprod 7, 374-378.

Kaczmarczyk, A., Houben, A., Keller, E.R.J. & Mette, M.F. (2010). Influence of cryopreservation on the cytosine methylation state of potato genomic DNA. CryoLetters 31:pp. 380-391.

Kaczmarczyk, A., Rokka, V.-M. & Keller, E.R.J. (2011a). Potato shoot tip cryopreservation, a review. Potato Research 54:pp. 45-79.

Kaczmarczyk, A., Turner, S.R., Bunn, E., Mancera, R.L. & Dixon, K.W. (2011b). Cryopreservation of threatened native Australian species - what have we learned and where to from here? In Vitro Cellular & Developmental Biology-Plant 47:pp. 17-25.

Kahn RA (1978) Biochemical changes in frozen platelets. In Greenwalt, TJ and Jamieson, GA (eds) The blood platelet in transfusion therapy. Alan R. Liss, New York, pp. 167-180.

Kaity, A., Ashmore, S.E., Drew, R.A. & Dulloo, M.E. (2008). Assessment of genetic and epigenetic changes following cryopreservation in papaya. Plant Cell Reports 27:pp. 1529-1539.

Kang X, Li G, Mu S, Guo M, Ge S. (2009), Acrosome reaction of Chinese mitten-handed crab Eriocheir sinensis (Crustacea: Decapoda) spermatozoa: Promoted by long-term cryopreservation. Aquaculture. 2009; 295: 195-199.

Karlsson JO (2001) A theoretical model of intracellular devitrification. Cryobiology 42, 154-169.

Karlsson JO and Toner M (1996) Long-term storage of tissues by cryopreservation: critical issues. Biomaterials 17, 243-256.

Karlsson JO, Cravalho EG and Toner M (1994) A model of diffusion-limited ice growth inside biological cells during freezing. Journal of Applied Physiology 75, 4442-4450.

Karlsson JO, Cravalho EG, Borel Rinkes IH, Tompkins RG, Yarmush ML and Toner M (1993) Nucleation and growth of ice crystals inside cultured hepatocytes during freezing in the presence of dimethyl sulfoxide. Biophys J 65, 2524-2536.

Karow AM (1997) Implications of Tissue Banking for Human Reproductive Medicine. In Karow, AM and Critser, JK (eds) Reproductive Tissue Banking, Scientific Principles. Academic Press, San Diego, pp. 441-464.

Karow AM and Critser JK (eds) (1997) Reproductive Tissue Banking. Academic Press San Diego. 472p.

Karow AM, Jr. (1969) Cryoprotectants--a new class of drugs. J Pharm Pharmacol 21, 209-223.

Karsenti E and Vernos I (2001) The mitotic spindle: a self-made machine. Science 294, 543-547.

Kazem R, Thompson LA, Srikantharajah A, Laing MA, Hamilton MP and Templeton A (1995) Cryopreservation of human oocytes and fertilization by two techniques: in-vitro fertilization and intracytoplasmic sperm injection. Hum Reprod 10, 2650-2654.

Kedem O and Katchalsky A (1958) Thermodynamic analysis of the permeability of biological membranes to non-electrolytes. Biochimica et Biophysica Acta 27, 229-246.

Keller, E.R.J., Senula, A., Leunufna, S. & Grube, M. (2006). Slow growth storage and cryopreservation - tools to facilitate germplasm maintenance of vegetatively propagated crops in living plant collections. International Journal of Refrigeration-Revue Internationale Du Froid 29:pp. 411-417.

Kelly RB (1990) Microtubules, membrane traffic, and cell organization. Cell 61, 5-7.

Kent, B., Garvey, C.J., Lenne, T., Porcar, L, Garamus, V.M. & Bryant, G. (2010). Measurement of glucose exclusion from the fully hydrated DOPE inverse hexagonal phase. Soft Matter 6:pp. 1197-1202.

Kim NH, Cho SK, Choi SH, Kim EY, Park SP and Lim JH (2000) The distribution and requirements of microtubules and microfilaments in bovine oocytes during in vitro maturation. Zygote 8, 25-32.

Kim NH, Chung HM, Cha KY and Chung KS (1998) Microtubule and microfilament organization in maturing human oocytes. Hum Reprod 13, 2217-2222.

Kim NH, Moon SJ, Prather RS and Day BN (1996) Cytoskeletal alteration in aged porcine oocytes and parthenogenesis. Mol Reprod Dev 43, 513-518.

Kim SHKu SYSung KCKang MJKim SAKim HSOh SKJee BCSuh CSChoi YM, et al. (2006) Simplified EM grid vitrification is a convenient and efficient method for mouse mature oocyte cryopreservation. Yonsei Med J 47, 399-404.

Kim, H.H., Lee, Y.G., Shin, D.J., Ko, H.C., Gwag, J.G., Cho, E.G. & Engelmann, F. (2009).

Kleinhans FW (1998) Membrane permeability modeling: Kedem-Katchalsky vs. a two-parameter formalism. Cryobiology 37, 271-289.

Knight CA, DeVries AL and Oolman LD (1984) Fish antifreeze protein and the freezing and recrystallization of ice. Nature 308, 295-296.

Kola I, Kirby C, Shaw J, Davey A and Trounson A (1988) Vitrification of mouse oocytes results in aneuploid zygotes and malformed fetuses. Teratology 38, 467- 474.

Koller BH, Hagemann LJ, Doetschman T, Hagaman JR, Huang S, Williams PJ, First NL, Maeda N and Smithies O (1989) Germ-line transmission of a planned alteration made in a hypoxanthine phosphoribosyltransferase gene by homologous recombination in embryonic stem cells. Proc Natl Acad Sci U S A 86, 8927-8931.

Komender J (2004) Cell and tissue preservation and storage for transplantation. Present and future. Ann Transplant 9, 88-90.

Kono T, Kwon OY and Nakahara T (1991) Development of vitrified mouse oocytes after in vitro fertilization. Cryobiology 28, 50-54.

Kopeika E, Kopeika J, Zhang T. Cryopreservation of fish sperm. Methods Mol. Biol. 2007; 368: 203-17.

Kopeika J, Zhang T, Rawson DM, Elgar GS. Effect of cryopreservation on mitochondrial DNA of zebrafish (Danio rerio) blastomere cells. Mutat. Res. 2005; 570: 49-61.

Korber C, Scheiwe MW, Boutron P and Rau G (1982) The influence of hydroxyethyl starch on ice formation in aqueous solutions. Cryobiology l9, 478- 492.

Koseoglu M, Eroglu A, Toner M, Sadler KC. Starfish oocytes form intracellular ice at unusually high temperatures. Cryobiology. 2001; 43: 248-259.

Koshimoto C, Gamliel E and Mazur P (2000) Effect of osmolality and oxygen tension on the survival of mouse sperm frozen to various temperatures in various concentrations of glycerol and raffinose. Cryobiology 41, 204-231.

Kranner, I., Birtic, S., Anderson, K.M. & Pritchard, H.W. (2006). Glutathione half-cell reduction potential: A universal stress marker and modulator of programmed cell death? Free Radical Biology and Medicine 40:pp. 2155-2165.

Kreck, C.A., Mandumpal, J.B. & Mancera, R.L. (2011). Prediction of the glass transition in aqueous solutions of simple amides by molecular dynamics simulations. Chemical Physics Letters 501:pp. 273-277.

Kristiansen E and Zachariassen KE (2005) The mechanism by which fish antifreeze proteins cause thermal hysteresis. Cryobiology 51, 262-280.

Kruip TA, Pieterse MC, van Beneden TH, Vos PL, Wurth YA and Taverne MA (1991) A new method for bovine embryo production: a potential alternative to superovulation. Vet Rec 128, 208-210.

Kuleshova L, Gianaroli L, Magli C, Ferraretti A and Trounson A (1999) Birth following vitrification of a small number of human oocytes: case report. Hum Reprod 14, 3077-3079.

Kumar TR, Wang Y, Lu N and Matzuk MM (1997) Follicle stimulating hormone is required for ovarian follicle maturation but not male fertility. Nat Genet 15, 201-204.

Kuroda H, Terada N, Nakayama H, Matsumoto K and Kitamura Y (1988)

Kusuda S, Teranishi T, Koide N, Nagai T, Arai K, Yamaha E. Pluripotency of cryopreserved blastomeres of the goldfish. J. Exp. Zool, 2004; 301A: 131-138.

Kusuda S, Teranishi T, Koide N. Cryopreservation of chum salmon blastomeres by the straw method. Cryobiology. 2002; 45: 60-67.

Kuwayama M, Vajta G, Kato O and Leibo SP (2005) Highly efficient vitrification method for cryopreservation of human oocytes. Reprod Biomed Online 11, 300-308.

Lahnsteiner F. Semen cryopreservation in the Salmonidae and in the northern pike; in special issue: cryopreservation of gametes in aquatic species. Aquacult. Res. 2000; 31: 245-258.

Lai LKang JXLi RWang JWitt WTYong HYHao YWax DMMurphy CNRieke A, et al. (2006) Generation of cloned transgenic pigs rich in omega-3 fatty acids. Nat Biotechnol 24, 435-436.

Lai LKolber-Simonds DPark KWCheong HTGreenstein JLIm GSSamuel MBonk ARieke ADay BN, et al. (2002) Production of alpha-1,3-galactosyltransferase knockout pigs by nuclear transfer cloning. Science 295, 1089-1092.

Landfald B and Strom AR (1986) Choline-glycine betaine pathway confers a high level of osmotic tolerance in Escherichia coli. J Bacteriol 165, 849-855.

Lane M and Gardner DK (2001) Vitrification of mouse oocytes using a nylon loop. Mol Reprod Dev 58, 342-347.

Le Gal F and Massip A (1999) Cryopreservation of cattle oocytes: effects of meiotic stage, cycloheximide treatment, and vitrification procedure. Cryobiology 38, 290-300.

Leibo SP (2007) Origins of techniques in human and animal embryology. Theriogenology 67, 2-5.

Leibo SP, Mazur P and Jackowski SC (1974) Factors affecting survival of mouse embryos during freezing and thawing. Exp Cell Res 89, 79-88.

Leibo SP, McGrath JJ and Cravalho EG (1978) Microscopic observation of intracellular ice formation in unfertilized mouse ova as a function of cooling rate. Cryobiology 15, 257-271.

Lenne, T., Bryant, G., Garvey, C.J., Kelderling, U. & Koster, K.L. (2006). Location of sugars in multilamellar membranes at low hydration. Physica B-Condensed Matter 385-86:pp. 862-864.

Lenne, T., Bryant, G., Holcomb, R. & Koster, K.L. (2007). How much solute is needed to inhibit the fluid to gel membrane phase transition at low hydration? Biochimica et Biophysica Acta-Biomembranes 1768:pp. 1019-1022.

Lenne, T., Garvey, C.J., Koster, K.L. & Bryant, G. (2009). Effects of sugars on lipid bilayers during dehydration - SAXS/WAXS measurements and quantitative model. Journal of Physical Chemistry B, 113:pp. 2486-2491.

Leunufna, S. & Keller, E.R.J. (2005). Cryopreservation of yams using vitrification modified by including droplet method: effects of cold acclimation and sucrose. CryoLetters 26:pp. 93-102.

Levi Setti PE, Albani E, Novara PV, Cesana A and Morreale G (2006) Cryopreservation of supernumerary oocytes in IVF/ICSI cycles. Hum Reprod 21, 370-375.

Lim JM, Ko JJ, Hwang WS, Chung HM and Niwa K (1999) Development of in vitro matured bovine oocytes after cryopreservation with different cryoprotectants. Theriogenology 51, 1303-1310.

Lin C, Tsai S. The effect of chilling and cryoprotectants on hard coral (Echinopora spp.) oocytes during short- term low temperature preservation. Theriogenology. 2012; 77: 1257-1261.

Lin C, Zhang T, Kuo FW, Tsai S. Gorgonian coral (Junceella juncea and Junceella fragilis) oocyte chilling sensitivity in the context of adenosine triphosphate response. Cryoletters. 2011; 32: 141-148.

Lin C, Zhang T, Rawson DM. Cryopreservation of zebrafish (Danio rerio) blastomeres by controlled slow cooling. CryoLetter. 2009; 30(2): 132-141.

Lin S, Long W, Chen J, Hopkins N. Production of germ-line chimeras in zebrafish by cell transplants from genetically pigmented to albino embryos. Proc. Natl. Acad. Sci. USA 1992; 89: 4519-4523.

Lin TT, Chao NH, Tung HT. Factors affecting survival of cryopreserved oyster (Crassostrea gigas) embryos. Cryobiology. 1999; 39(2): 192-196.

Lindsay M, Gil GC, Cadiz A, Velander WH, Zhang C and Van Cott KE (2004) Purification of recombinant DNA-derived factor IX produced in transgenic pig milk and fractionation of active and inactive subpopulations. J Chromatogr A 1026, 149-157.

Liu K, Chou T, Lin H. Cryosurvival of goldfish embryos after subzero freezing. Aquat. Living Resour. 1993; 6: 145-153.

Liu RH, Sun QY, Li YH, Jiao LH and Wang WH (2003) Effects of cooling on meiotic spindle structure and chromosome alignment within in vitro matured porcine oocytes. Mol Reprod Dev 65, 212-218.

Liu XH, Zhang T, Rawson DM. Differential scanning calorimetry studies of intraembryonic freezing and cryoprotectant penetration in zebrafish (Danio rerio) embryos. J. Exp. Zool. 2001; 290: 299-310.

Lovelock JE (1953a) The haemolysis of human red blood cells by freezing and thawing. Biochim Biophys Acta 10, 414-426.

Lovelock JE (1953b) The mechanism of the protective action of glycerol against haemolysis by freezing and thawing. Biochim Biophys Acta 11, 28-36.

Luyet BJ and Gehenio PM (1940) Life and death at low temperatures. Biodynamica, Normandy, Missouri 341p.

Lynch, D.V. & Steponkus, P.L. (1987). Plasma membrane lipid alterations associated with cold acclimation of winter rye seedlings (Secale cereale L. cv Puma). Plant Physiol 83:pp. 761-767.

Lynch, P.T., Siddika, A., Johnston, J.W., Trigwell, S.M., Mehra, A., Benelli, C., Lambardi, M. & Benson, E.E. (2011). Effects of osmotic pretreatments on oxidative stress, antioxidant profiles and cryopreservation of olive somatic embryos. Plant Science 181:pp. 47-56.

Maddock BC. A technique to prolong the incubation period of brown trout ova. Progressive Fish Cult. 1974; 36: 219-222.

Magistrini M and Szollosi D (1980) Effects of cold and of isopropyl-Nphenylcarbamate on the second meiotic spindle of mouse oocytes. Eur J Cell Biol 22, 699-707.

Magistrini M and Szöllösi D (1980) Effects of cold and of isopropyl-Nphenylcarbamate on the second meiotic spindle of mouse oocytes. Eur J Cell Biol 22, 699-707.

Magyary I, Urbányi B, Horváth L. Cryopreservation of common carp (Cyprinus carpio L.) sperm II Optimal conditions for fertilization. J. Appl. Ichthyol. 1996; 12: 117-119.

Mallon, R., Bunn, E., Turner, S.R. & Gonzalez, M.L. (2008). Cryopreservation of Centaurea ultreiae (Compositae) a critically endangered species from Galica (Spain). CryoLetters 29:pp. 363-370.

Mandal, B.B. & Dixit-Sharma, S. (2007). Cryopreservation of in vitro shoot tips of Dioscorea deltoidea Wall., an endangered medicinal plant: effect of cryogenic procedure and storage duration. CryoLetters 28:pp. 461-470.

Mandelbaum J, Junca AM, Tibi C, Plachot M, Alnot MO, Rim H, Salat-Baroux J and Cohen J (1988) Cryopreservation of immature and mature hamster and human oocytes. Ann N Y Acad Sci 541, 550-561.

Mandumpal, J.B., Kreck, C.A. & Mancera, R.L. (2011). A molecular mechanism of solvent cryoprotection in aqueous DMSO solutions. Physical Chemistry Chemical Physics 13:pp. 3839-3842.

Mapletoft RJ, Hasler JF. Assisted reproductive technologies in cattle: a review. Rev. Sci. Tech. 2005; 24(1): 393-403.

Marin, M.L., Gogorcena, Y., Ortiz, J. & Duranvila, N. (1993). Recovery of whole plants of sweet orange from somatic embryos subjected to freezing-thawing treatments. Plant Cell Tissue and Organ Culture 34:pp. 27-33.

Marina F and Marina S (2003) Comments on oocyte cryopreservation. Reproductive Biomedicine Online6, 401-2.

Marnett, L.J. (1999). Chemistry and biology of DNA damage by malondialdehyde. IARC scientific publications 150:pp. 1-17.

Marsan, M.P., Bellet-Amalric, E., Muller, I., Zaccai, G. & Milon, A. (1998). Plant sterols: a neutron diffraction study of sitosterol and stigmasterol in soybean phosphatidylcholine membranes. Biophysical Chemistry 75:pp. 45-55.

Martin, M.C. & González-Benito, M. E. (2006) Sequence comparison in somaclonal variant of cryopreserved Dendranthema grandiflora shoot apices. Cryobiology 53:pp. 424.

Martino A, Songsasen N and Leibo SP (1996) Development into blastocysts of bovine oocytes cryopreserved by ultra-rapid cooling. Biol Reprod 54, 1059-1069.

Mazur P (1963) Kinetics of water loss from cells at subzero temperatures and the likelihood of intracellular freezing. Journal of General Physiology 47, 47-69.

Mazur P (1965) The role of cell membranes in the freezing of yeast and other single cells. Ann N Y Acad Sci 125, 658-76.

Mazur P (1970) Cryobiology: the freezing of biological systems. Science 168, 939-949.

Mazur P (1977) The role of intracellular freezing in the death of cells cooled at supraoptimal rates. Cryobiology 14, 251-272.

Mazur P (1984) Freezing of living cells: mechanisms and implications. Am J Physiol 247, C125-142.

Mazur P (2004) Principles of Cryobiology. In Fuller, BJ, Lane, N and Benson, EE (eds) Life in the Frozen State. CRC Press, Boca Raton, pp. 3-65.

Mazur P and Cole KW (1985) Influence of cell concentration on the contribution of unfrozen fraction and salt concentration to the survival of slowly frozen human erythrocytes. Cryobiology 22, 509-536.

Mazur P and Cole KW (1989) Roles of unfrozen fraction, salt concentration, and changes in cell volume in the survival of frozen human erythrocytes. Cryobiology 26, 1-29.

Mazur P and Rigopoulos N (1983) Contributions of unfrozen fraction and of salt concentration to the survival of slowly frozen human erythrocytes: influence of warming rate. Cryobiology 20, 274-289.

Mazur P and Schmidt JJ (1968) Interactions of cooling velocity, temperature, and warming velocity on the survival of frozen and thawed yeast. Cryobiology 5, 1-17.

Mazur P and Schneider U (1986) Osmotic responses of preimplantation mouse and bovine embryos and their cryobiological implications. Cell Biophys 8, 259- 285.

Mazur P, Leibo SP, Farrant J, Chu EHY, Hanna Jr MG and Smith LH (1970)

Mazur P, Pinn IL, Seki S, Kleinhans FW and Edashige K (2005a) Effects of hold time after extracellular ice formation on intracellular freezing of mouse oocytes.Cryobiology 51, 235-239.

Mazur P, Rall WF and Rigopoulos N (1981) Relative contributions of the fraction of unfrozen water and of salt concentration to the survival of slowly frozen human erythrocytes. Biophys J 36, 653-675.

Mazur P, Schneider U, Mahowald AP. Characteristics and kinetics of subzero chilling injury in Drosophila embryos. Cryobiology. 1992; 29: 39-68.

Mazur P, Seki S, Pinn IL, Kleinhans FW and Edashige K (2005b) Extra- and intracellular ice formation in mouse oocytes. Cryobiology 51, 29-53.

Mazur P. Freezing of living cells: mechanisms and implications. Am. J. Physiol. 1984; 247: C125-C142.

McEvoy TG, Coull GD, Broadbent PJ, Hutchinson JS and Speake BK (2000) Fatty acid composition of lipids in immature cattle, pig and sheep oocytes with intact zona pellucida. J Reprod Fertil 118, 163-170.

McNitt JI, Patton NM, Lukefahr SD and Cheeke PR (1996) Rabbit Production. (Seventh), Interstate Publishers, Inc., Danville 477p.

McWilliams RB, Gibbons WE and Leibo SP (1995) Osmotic and physiological responses of mouse zygotes and human oocytes to mono- and disaccharides. Hum Reprod 10, 1163-1171.

Men H, Agca Y, Mullen SF, Critser ES and Critser JK (2004) Osmotic stress on the cellular actin filament organization of in vitro produced porcine embryos.

Men H, Agca Y, Mullen SF, Critser ES and Critser JK (2005) Osmotic tolerance of in vitro produced porcine blastocysts assessed by their morphological integrity and cellular actin filament organization. Cryobiology 51, 119-129.

Men H, Monson RL and Rutledge JJ (2002) Effect of meiotic stages and maturation protocols on bovine oocyte's resistance to cryopreservation. Theriogenology 57, 1095-1103.

Men HS, Chen JC, Ji WZ, Shang EY, Yang SC and Zou RJ (1997) Cryopreservation of Kunming mouse oocytes using slow cooling, ultrarapid cooling and vitrification protocols. Theriogenology 47, 1423-1431.

Menning BE (1980) The emotional needs of infertile couples. Fertil Steril 34, 313- 319.

Meryman HT (1970) The exceeding of a minimum tolerable cell volume in hypertonic suspensions as a cause of freezing injury. In Wolstenholme, GEW and O'Connor, M (eds) The Frozen Cell. J and A Churchill, London, pp. 51-64.

Meryman HT (1971) Osmotic stress as a mechanism of freezing injury. Cryobiology 8, 489-500.

Meryman HT (eds) (1966) Cryobiology. Academic Press, New York 477p.

Meryman HT, Williams RJ and Douglas MS (1977) Freezing injury from "solution effects" and its prevention by natural or artificial cryoprotection. Cryobiology 14, 287-302.

Meryman, H.T. (1974). Freezing injury and its prevention in living cells Annual Review of Biophysics and Bioengineering 3:pp. 341-363.

Messinger SM and Albertini DF (1991) Centrosome and microtubule dynamics during meiotic progression in the mouse oocyte. J Cell Sci 100, 289-298.

Mitchison T and Kirschner M (1984) Dynamic instability of microtubule growth. Nature 312, 237-242.

Mix-Wagner, G., Schumacher, H.M. & Cross, R.J. (2003). Recovery of potato apices after several years of storage in liquid nitrogen. CryoLetters 24:pp. 33-41.

Møller, I.M., Jensen, P.E. & Hansson, A. (2007). Oxidative modifications to cellular components in plants. Annu. Rev. Plant Biol. 58:pp. 459-481.

Morris GJ and Watson PF (1984) Cold-Shock injury - a comprehensive bibliography. Cryo-Letters 5, 352-372.

Morton NE, Jacobs PA, Hassold T and Wu D (1988) Maternal age in trisomy. Ann Hum Genet 52, 227-235.

Muldrew K and McGann LE (1990) Mechanisms of intracellular ice formation. Biophys J 57, 525-532.

Muldrew K and McGann LE (1994) The osmotic rupture hypothesis of intracellular freezing injury. Biophys J 66, 532-541.

Muldrew K, Acker JP, Elliott JAW and McGann LE (2004) The water to ice transition: implications for living cells. In Fuller, BJ, Lane, N and Benson, EE (eds) Life in the Frozen State. CRC Press, Boca Raton, pp. 67-108.

Mullen SF, Agca Y, Broermann DC, Jenkins CL, Johnson CA and Critser JK (2004) The effect of osmotic stress on the metaphase II spindle of human oocytes, and the relevance to cryopreservation. Hum Reprod 19, 1148-1154.

Mycock, D.J., Wesley-Smith, J. & Berjak, P. (1995). Cryopreservation of somatic embryos of four species with and without cryoprotectant pre-treatment. Annals of Botany 75: pp. 331-336.

Myers SP, Pitt RE, Lynch DV and Steponkus PL (1989) Characterization of intracellular ice formation in Drosophila melanogaster embryos. Cryobiology 26, 472-484.

Na, H.-Y. & Kondo, K. (1996). Cryopreservation of tissue-cultured shoot primordia from shoot apices of cultured protocorms in Vanda pumila following ABA preculture and desiccation. Plant Science 118:pp. 195-201.

Nagashima H, Cameron RDA, Kuwayama M, Young M, Beebe L, Blackshaw AW and Nottle MB (1999) Survival of porcine delipated oocytes and embryos after cryopreservation by freezing or vitrification. Journal of Reproduction and Development 45, 167-176.

Nagashima H, Kashiwazaki N, Ashman RJ, Grupen CG, Seamark RF and Nottle MB (1994) Removal of cytoplasmic lipid enhances the tolerance of porcine embryos to chilling. Biol Reprod 51, 618-622.

Nakagata N (1993) Production of normal young following transfer of mouse embryos obtained by in vitro fertilization between cryopreserved gametes. J Reprod Fertil 99, 77-80.

Nash T (1966) Chemical constitution and physical properties of compounds able to protect living cells against damage due to freezing and thawing. In Meryman, HT (eds) Cryobiology. Academic Press, New York, pp. 179-211.

Nasmyth K (2001) Disseminating the genome: joining, resolving, and separating sister chromatids during mitosis and meiosis. Annu Rev Genet 35, 673-745.

National center for chronic disease prevention and health promotion: 2002. Assisted Reproductive Technology Success Rates. (2004) Centers for Disease Control and Prevention 522 p.

Navara CS, First NL and Schatten G (1996) Phenotypic variations among paternal centrosomes expressed within the zygote as disparate microtubule lengths and sperm aster organization: correlations between centrosome activity and developmental success. Proc Natl Acad Sci U S A 93, 5384-5388.

Nes, W.R. (1974). Role of sterols in membranes. Lipids 9:pp. 596-612.

Newton H (1998) The cryopreservation of ovarian tissue as a strategy for preserving the fertility of cancer patients. Hum Reprod Update 4, 237-247.

Nicholas FW (1996) Genetic improvements through reproductive technology. Animal Reproduction Science 42, 205-214.

Nilsson EE, Cloud JG. Rainbow trout chimeras produced by injection of blastomeres into recipient blastulae. Proc. Natl .Acad .Sci. USA 1992; 89: 9425-9428.

Nishizawa, S., Sakai, A., Amano, Y. & Matsuzawa, T. (1993). Cryopreservation of Asparagus officinalis L. embryogenic suspension cells and subsequent plant regeneration by vitrification. Plant Science 91:pp. 67-73.

Nurse P (2000) A long twentieth century of the cell cycle and beyond. Cell 100, 71-78.

Obert, B., Benson, E.E., Millam, S., Pretová, A. & Bremner, D.H. (2005). Moderation of morphogenetic and oxidative stress responses in flax in vitro cultures by hydroxynonenal and desferrioxamine. Journal of Plant Physiology 162:pp. 537-547.

Odintsova N, Kiselev K, Sanina N, Kostetsky E. Cryopreservation of primary cell cultures of marine invertebrates. CryoLetters. 2001; 22: 299-310.

Oktay K (2001) Ovarian tissue cryopreservation and transplantation: preliminary findings and implications for cancer patients. Hum Reprod Update 7, 526-534.

Olive PJW, Wang WB. Cryopreservation of Nereis virens (Polychaeta, Annelida) lavae: the mechanism of cryopreservation a different metazoan. Cryobiology. 1997; 34: 284-294.

O'Neil L, Paynter SJ and Fuller BJ (1997) Vitrification of mature mouse oocytes: improved results following addition of polyethylene glycol to a dimethyl sulfoxide solution. Cryobiology 34, 295-301.

O'Neil L, Paynter SJ, Fuller BJ, Shaw RW and DeVries AL (1998) Vitrification of mature mouse oocytes in a 6 M DMSO solution supplemented with antifreeze glycoproteins: the effect of temperature. Cryobiology 37, 59-66.

O'Neill GT, Rolfe LR and Kaufman MH (1991) Developmental potential and chromosome constitution of strontium-induced mouse parthenogenones. Mol Reprod Dev 30, 214-219.

Otoi T, Tachikawa S, Kondo S, Takagi M and Suzuki T (1994) Developmental competence of bovine oocytes frozen at different cooling rates. Cryobiology 31, 344-348.

Otoi T, Yamamoto K, Koyama N and Suzuki T (1995) In vitro fertilization and development of immature and mature bovine oocytes cryopreserved by ethylene glycol with sucrose. Cryobiology 32, 455-460.

Otoi T, Yamamoto K, Koyama N, Tachikawa S and Suzuki T (1998) Cryopreservation of mature bovine oocytes by vitrification in straws. Cryobiology 37, 77-85.

Otoi T, Yamamoto K, Koyama N, Tachikawa S, Murakami M, Kikkawa Y and Suzuki T (1997) Cryopreservation of mature bovine oocytes following centrifugation treatment. Cryobiology 34, 36-41.

Pace MM and Graham EF (1974) Components in egg yolk which protect bovine spermatozoa during freezing. Journal of Animal Science 39, 1444-1449.

Packer, J.E., Slater, T. & Willson, R. (1979). Direct observation of a free radical interaction between vitamin E and vitamin C. Nature 278:pp. 737-738.

Palasz AT, Mapletoft RJ. Cryopreservation of mammalian embryos and oocytes: Recent advances. Biotechno. Adv. 1996; 14: 127-149.

Paniagua-Chavez CG, Tiersch TR. Laboratory studies of cryopreservation of sperm and trochophore larvae of the eastern oyster. Cryobiology. 2001; 43(3): 211-223.

Panis, B. & Lambardi, M. (2005). Status of cryopreservation technologies in plants (crops and forest trees). Proceedings of the International Workshop on 'The Role of Biotechnology', pp. 43-54, Villa Gualino, Turin, Italy, 5.-7. March 2005.

Panis, B., Piette, B. & Swennen, R. (2005). Droplet vitrification of apical meristems: a cryopreservation protocol applicable to all Musaceae. Plant Science 168:pp. 45 55.

Panis, B., Strosse, H., Van den Hende, S. & Swennen, R. (2002). Sucrose preculture to simplify cryopreservation of banana meristem cultures. CryoLetters 23:pp. 375-384.

Papis K, Shimizu M and Izaike Y (2000) Factors affecting the survivability of bovine oocytes vitrified in droplets. Theriogenology 54, 651-658.

Park KE, Kwon IK, Han MS and Niwa K (2005) Effects of partial removal of cytoplasmic lipid on survival of vitrified germinal vesicle stage pig oocytes. JReprod Dev 51, 151-160.

Parks JE (1997) Hypothermia and Mammalian Gametes. In Karow, AM and Critser, JK (eds) Reproductive Tissue Banking, Scientific Principles. Academic Press, San Diego, pp. 229-261.

Parks JE and Ruffing NA (1992) Factors affecting low temperature survival of mammalian oocytes. Theriogenology 37, 5973.

Partelli, F.L., Vieira, H.D., Pais, I.P., Quartin, V.L., Campos, P.S., Fortunato, A., Eichler, P., Viana, A.P., Ribeiro, A. & Ramalho, J.C. (2009). Chloroplast membrane lipids from Coffea sp. under low positive temperatures. Proceedings of 22nd International Conference on Coffee Science, pp. 891-898, ASIC 2008, ISBN 2900212219, Campinas, SP, Brazil, September, 2008.

Paulet, F., Engelmann, F. & Glaszmann, J.C. (1993). Cryopreservation of apices of in vitro plantlets of sugarcane (Saccharum sp hybrids) using encapsulation dehydration. Plant Cell Reports 12:pp. 525-529.

Paunescu, A. (2009). Biotechnology for endangered plant conservation: a critical overview Romanian Biotechnological Letters 14:pp. 4095-4103.

Paynter S, Cooper A, Thomas N and Fuller B (1997) Cryopreservation of Multicellular Embryos and Reproductive Tissue. In Karow, AM and Critser, JK (eds) Reproductive Tissue Banking, Scientific Principles. Academic Press, San Diego, pp. 359-397.

Pearl M, Arav A. Chilling sensitivity in zebrafish (Brachiodanio rerio) oocytes is related to lipid phase transition. CryoLetters. 2000; 21: 171-178.

Pedro PB, Zhu SE, Makino N, Sakurai T, Edashige K and Kasai M (1997) Effects of hypotonic stress on the survival of mouse oocytes and embryos at various stages. Cryobiology 35, 150-158.

Pegg DE (1987) Mechanisms of freezing damage. In Bowler, K and Fuller, BJ (eds) Temperature and Animal Cells. The company of biologists, Ltd., Cambridge, pp. 363-378.

Pegg DE (1989) Viability assays for preserved cells, tissues, and organs. Cryobiology 26, 212-231.

Pegg DE and Diaper MP (1988) On the mechanism of injury to slowly frozen erythrocytes. Biophys J 54, 471-488.

Pegg DE and Diaper MP (1991) The effect of initial tonicity on freeze/thaw injury to human red cells suspended in solutions of sodium chloride. Cryobiology 28, 18-35.

Penrose L (1933) The relative effects of paternal and maternal age on mongolism. Journal of Genetics 27, 219-224.

Phadtare S and Inouye M (2004) Genome-wide transcriptional analysis of the cold shock response in wild-type and cold-sensitive, quadruple-csp-deletion strains of Escherichia coli. J Bacteriol 186, 7007-7014.

Phadtare S, Alsina J and Inouye M (1999) Cold-shock response and cold-shock proteins. Curr Opin Microbiol 2, 175-180.

Pickering SJ and Johnson MH (1987) The influence of cooling on the organization of the meiotic spindle of the mouse oocyte. Hum Reprod 2, 207-216.

Pickering SJ, Braude PR and Johnson MH (1991) Cryoprotection of human oocytes: inappropriate exposure to DMSO reduces fertilization rates. Hum Reprod 6, 142-143.

Pickering SJ, Braude PR, Johnson MH, Cant A and Currie J (1990) Transient cooling to room temperature can cause irreversible disruption of the meiotic spindle in the human oocyte. Fertil Steril 54, 102-108.

Pincus G (1936) The eggs of mammals. Macmillan, New York Porcu E, Fabbri R, Seracchioli R, Ciotti PM, Magrini O and Flamigni C (1997)

Plachinta M, Zhang T, Rawson DM. Studies on cryoprotectant toxicity to zebrafish (Danio rerio) oocytes. CryoLetters. 2004; 25(6): 415-424.

Polge C, Smith AU, Parkes AS. Revival of spermatozoa after vitrification and dehydration at low temperatures. Nature. 1949; 164: 666.

Posada MN, Kolp L and Garcia JE (2001) Fertility options for female cancer patients: facts and fiction. Fertil Steril 75, 647-653.

Prather RS (2006) Cloned transgenic heart-healthy pork? Transgenic Res 15, 405-407.

Prather RS, Hawley RJ, Carter DB, Lai L and Greenstein JL (2003) Transgenic swine for biomedicine and agriculture. Theriogenology 59, 115-123.

Presicce GA, Jiang S, Simkin M, Zhang L, Looney CR, Godke RA and Yang X (1997) Age and hormonal dependence of acquisition of oocyte competence for embryogenesis in prepubertal calves. Biol Reprod 56, 386-392.

Pukazhenthi B, Noiles E, Pelican K, Donoghue A, Wildt D and Howard J (2000) Osmotic effects on feline spermatozoa from normospermic versus teratospermic donors. Cryobiology 40, 139-150.

Quinn PJ (1985) A lipid-phase separation model of low-temperature damage to biological membranes. Cryobiology 22, 128-146.

Quinn PJ, Chow PY and White IG (1980) Evidence that phospholipid protects ram spermatozoa from cold shock at a plasma membrane site. J Reprod Fertil 60, 403-407.

Rajkovic A, Pangas SA and Matzuk MM (2006) Follicular Development: Mouse, Sheep, and Human Models. In Neill, JD (eds) Follicular Development: Mouse, Sheep, and Human Models Vol 1. Elsevier, New York, pp. 383-423.

Rall WF and Fahy GM (1985) Ice-free cryopreservation of mouse embryos at - 196 degrees C by vitrification. Nature 313, 573-575.

Rall WF, Mazur P and McGrath JJ (1983) Depression of the ice-nucleation temperature of rapidly cooled mouse embryos by glycerol and dimethyl sulfoxide. Biophys J 41, 1-12.

Raymond JA, Wilson P and DeVries AL (1989) Inhibition of growth of nonbasal planes in ice by fish antifreezes. Proc Natl Acad Sci U S A 86, 881-885.

Rayos AA, Takahashi Y, Hishinuma M and Kanagawa H (1994) Quick freezing of unfertilized mouse oocytes using ethylene glycol with sucrose or trehalose. J Reprod Fertil 100, 123-129.

Reed, B.M. (2008). Plant Cryopreservation: A Practical Guide, Springer, ISBN 978-0-387-72275-7, New York.

References Cited

Reinhoud, P.J., Versteege, I., Kars, I., van Iren, F. & Kijne, J.W. (2000). Physiological and molecular changes in tobacco suspension cells during development of tolerance to cryopreservation by vitrification, In: Cryopreservation of Tropical Plant Germplasm. Current Research Progress and Application, Engelmann, F. & Takagi, H. (Eds.), pp. 57- 66, International Plant Genetic Resources Institute, ISBN 92-9043-428-7, Rome.

Reproduction, Fertility, and Development 12, 177.

Reproduction, Nutrition, Development 35, 367-373.

Rho GJ, Kim S, Yoo JG, Balasubramanian S, Lee HJ and Choe SY (2002) Microtubulin configuration and mitochondrial distribution after ultra-rapid cooling of bovine oocytes. Mol Reprod Dev 63, 464-470.

Rieder CL and Cole RW (2002) Cold-shock and the Mammalian cell cycle. Cell Cycle 1, 169-175.

Roach, T., Ivanova, M., Beckett, R.P., Minibayeva, F.V., Green, I., Pritchard, H.W. & Kranner, I. (2008). An oxidative burst of superoxide in embryonic axes of recalcitrant sweet chestnut seeds as induced by excision and desiccation. Physiologia Plantarum 133:pp. 131-139.

Rojas C, Palomo MJ, Albarracin JL and Mogas T (2004) Vitrification of immature and in vitro matured pig oocytes: study of distribution of chromosomes, microtubules, and actin microfilaments. Cryobiology 49, 211-220.

Rudolph AS and Crowe JH (1985) Membrane stabilization during freezing: the role of two natural cryoprotectants, trehalose and proline. Cryobiology 22, 367- 377.

Ruffing NA, Steponkus PL, Pitt RE and Parks JE (1993) Osmometric behavior, hydraulic conductivity, and incidence of intracellular ice formation in bovine oocytes at different developmental stages. Cryobiology 30, 562-580.

Rutledge JJ, Monson RL, Northey DL and Leibfried-Rutledge ML (1999) Seasonality of cattle embryo production in a temperate region. Theriogenology 51, 330 (Abstract).

Sakai, A. & Engelmann, F. (2007). Vitrification, encapsulation-vitrification and droplet-vitrification: a review. CryoLetters 28:pp. 151-172.

Sakai, A. (1965). Survival of plant tissue at super-low temperatures III. Relation between effective prefreezing temperatures and the degree of frost hardiness. Plant Physiol. 40:pp. 882-887.

Sakai, A., Hirai, D. & Niino, T. (2008). Development of PVS-based vitrification and encapsulation-vitrification protocols, In: Plant Cryopreservation: A Practical Guide, Reed, B. (Ed.), pp. 33-57, Springer, ISBN 978-0-378-72275-7, New York.

Sakai, A., Kobayashi, S. & Oiyama, I. (1990). Cryopreservation of nucellar cells of navel orange (Citrus sinensis Osb. var. brasiliensis Tanaka) by vitrification. Plant Cell Reports 9:pp. 30-33.

Sapundzhiev E. Conservation of ancient breed small ruminants as frozen embryos. Bulg. J. Vet. Med. 2008; 11: 251-255.

Sathananthan AH, Kirby C, Trounson A, Philipatos D and Shaw J (1992) The effects of cooling mouse oocytes. J Assist Reprod Genet 9, 139-148.

Sathananthan AH, Ng SC, Trounson AO, Bongso A, Ratnam SS, Ho J, Mok H and Lee MN (1988) The effects of ultrarapid freezing on meiotic and mitotic spindles of mouse oocytes and embryos. Gamete Res 21, 385-401.

Sathananthan AH, Trounson A, Freemann L and Brady T (1988) The effects of cooling human oocytes. Hum Reprod 3, 968-977.

Saunders KM and Parks JE (1999) Effects of cryopreservation procedures on the cytology and fertilization rate of in vitro-matured bovine oocytes. Biol Reprod 61, 178-187.

Saxton WM, Stemple DL, Leslie RJ, Salmon ED, Zavortink M and McIntosh JR (1984) Tubulin dynamics in cultured mammalian cells. J Cell Biol 99, 2175-2186.

Schalkoff ME, Oskowitz SP and Powers RD (1989) Ultrastructural observations of human and mouse oocytes treated with cryopreservatives. Biol Reprod 40, 379-393.

Schatten G (1994) The centrosome and its mode of inheritance: the reduction of the centrosome during gametogenesis and its restoration during fertilization. Dev Biol 165, 299-335.

Schatten G and Schatten H (1987) Cytoskeletal alterations and nuclear architectural changes during mammalian fertilization. Curr Top Dev Biol 23, 23-54.

Schellander K, Peli J, Schmoll F and Brem G (1994) Effects of different cryoprotectants and carbohydrates on freezing of matured and unmatured bovine oocytes. Theriogenology 42, 909-915.

Schnieke AE, Kind AJ, Ritchie WA, Mycock K, Scott AR, Ritchie M, Wilmut I, Colman A and Campbell KH (1997) Human factor IX transgenic sheep produced by transfer of nuclei from transfected fetal fibroblasts. Science 278, 2130-2133.

Schnorr JA, Muasher SJ and Jones HW, Jr. (2000) Evaluation of the clinical efficacy of embryo cryopreservation. Mol Cell Endocrinol 169, 85-89.

Schoots AFM, Stikkelbroeck JJM, Bekhuis JF, Denuce JM. Hatching in teleost fishes: fine structure changes in the egg envelope during enzymatic breakdown in vivo and in vitro. J. Ultrastruct. Res. 1982; 80: 185- 196.

Scott AP, Baynes SM. A review of the biology, handing and storage of salmonid spermatozoa. J. Fish Biol. 1980; 17: 707-739.

Scottez, C., Chevreau, E., Godard, N., Arnaud, Y., Duron, M. & Dereuddre, J. (1992). Cryopreservation of cold acclimated shoot tips of pear in vitro cultures after encapsulation-dehydration. Cryobiology 29:pp. 691-700.

Sen A, Brain AP, Quinn PJ and Williams WP (1982) Formation of inverted lipid micelles in aqueous dispersions of mixed sn-3-galactosyldiacylglycerols induced by heat and ethylene glycol. Biochim Biophys Acta 686, 215-224.

Sen-Rong, H. & Ming-Hua, Y. (2009). High-efficiency vitrification protocols for cryopreservation of in vitro grown shoot tips of rare and endangered plant Emmenopterys henryi Oliv. Plant Cell Tissue and Organ Culture 99:pp. 217-226.

Senula, A., Keller, E.R.J., Sanduijav, T. & Yohannes, T. (2007). Cryopreservation of cold-acclimated mint (Mentha spp.) shoot tips using a simple vitrification protocol. CryoLetters 28:pp. 1-12.

Sharp J and Mobraaten LE (1997) To save or not to save: the role of repositories in a period of rapidly expanding development of genetically engineered strains of mice. In LM, H (eds) Transgenic Animals: Generation and Use. Harwood Academic Publishers, Switzerland, pp. 525-532.

Shaw JM, Oranratnachai A and Trounson AO (2000) Fundamental cryobiology of mammalian oocytes and ovarian tissue. Theriogenology 53, 59-72.

Shaw PW, Bernard AG, Fuller BJ, Hunter JH and Shaw RW (1992) Vitrification of mouse oocytes using short cryoprotectant exposure: effects of varying exposure times on survival. Mol Reprod Dev 33, 210-214.

Shaw PW, Fuller BJ, Bernard A and Shaw RW (1991) Vitrification of mouse oocytes: improved rates of survival, fertilization, and development to blastocysts. Mol Reprod Dev 29, 373-378.

Sherman JK and Lin TP (1959) Temperature shock of cold-storage of unfertilized mouse eggs. Fertil Steril 10, 384-396.

Shi LY, Jin HF, Kim JG, Mohana Kumar B, Balasubramanian S, Choe SY and Rho GJ (2006) Ultra-structural changes and developmental potential of porcine oocytes following vitrification. Anim Reprod Sci In Press Shi WQ, Zhu SE, Zhang D, Wang WH, Tang GL, Hou YP and Tian SJ (2006)

Shimada K and Asahina E (1975) Visualization of intracellular ice crystals formed in very rapidly frozen cells at -27 degree C. Cryobiology 12, 209-218.

Sjostrom M (1975) Ice crystal growth in skeletal muscle fibres. J Microsc 105, 67- 80.

Skyba, M., Urbanova, M., Kapchina-Toteva, V., Kosuth, J., Harding, K. & Cellarova, E. (2010). Physiological, biochemical and molecular characteristics of cryopreserved Hypericum perforatum L. shoot tips. CryoLetters 31:pp. 249-260.

Somero GN and Yancey PH (1997) Osmolytes and cell-volume regulation: physiological and evolutionary principles. In Hoffman, JF and Jamieson, JD (eds) Handbook of Physiology. Section 14: Cell Physiology. Oxford University Press, New York, pp. 441-484.

Somfai T, Dinnyes A, Sage D, Marosan M, Carnwath JW, Ozawa M, Kikuchi K and Niemann H (2006) Development to the blastocyst stage of parthenogenetically activated in vitro matured porcine oocytes after solid surface vitrification (SSV). Theriogenology 66, 415-422.

Songsasen N, Yu I, Murton S, Paccamonti DL, Eilts BE, Godke RA and Leibo SP (2002) Osmotic sensitivity of canine spermatozoa. Cryobiology 44, 79-90.

Songsasen N, Yu IJ, Ratterree MS, VandeVoort CA and Leibo SP (2002) Effect of chilling on the organization of tubulin and chromosomes in rhesus monkey oocytes. Fertil Steril 77, 818-825.

Sonmezer M and Oktay K (2004) Fertility preservation in female cancer patients: a comprehensive approach. In Tulandi, T and Gosden, RG (eds) Preservation of Fertility. Taylor and Francis, London, pp. 177-190.

Soodak H and Iberall A (1978) Osmosis, diffusion, convection. Am J Physiol 235, R3-17.

Stachecki JJ and Cohen J (2004) An overview of oocyte cryopreservation. Reprod Biomed Online 9, 152-163.

Stachecki JJ, Cohen J and Willadsen S (1998) Detrimental effects of sodium during mouse oocyte cryopreservation. Biol Reprod 59, 395-400.

Stachecki JJ, Cohen J and Willadsen SM (1998) Cryopreservation of unfertilized mouse oocytes: the effect of replacing sodium with choline in the freezing medium. Cryobiology 37, 346-354.

Stachecki JJ, Cohen J, Schimmel T and Willadsen SM (2002) Fetal development of mouse oocytes and zygotes cryopreserved in a nonconventional freezing medium. Cryobiology 44, 5-13.

Stachecki JJ, Munne S and Cohen J (2004) Spindle organization after cryopreservation of mouse, human, and bovine oocytes. Reprod Biomed Online 8, 664-672.

Steponkus PL, Dowgert MF and Gordon-Kamm WJ (1983) Destabilization of the plasma membrane of isolated plant protoplasts during a freeze-thaw cycle: the influence of cold acclimation. Cryobiology 20, 448-465.

Steponkus, P.L. (1984). Role of the plasma membrane in freezing injury and cold acclimation. Annual Review of Plant Physiology 35:pp. 543-584.

Sterzik K, Rosenbusch B, Grab D, Wahl A, Beier HM and Lauritzen C (1992) Numerical chromosome anomalies after fertilization of freeze-thawed mouse oocytes. Arch Gynecol Obstet 251, 133-138.

Storey KB and Storey JM (1988) Freeze tolerance in animals. Physiol Rev 68, 27-84.

Storey KB and Storey JM (1992) Natural freeze tolerance in ectothermic vertebrates. Annu Rev Physiol 54, 619-637.

Storey KB, Baust JG and Wolanczyk JP (1992) Biochemical modification of plasma ice nucleating activity in a freeze-tolerant frog. Cryobiology 29, 374-384.

Stoss J, Donaldson EM. Studies on cryopreservation of eggs from rainbow trout (Salmo gairdneri) and coho salmon (Oncorhynchus Kisutch). Aquaculture. 1983; 31: 51-65.

Streb, P., Shang, W., Feierabend, J. & Bligny, R. (1998). Divergent strategies of photoprotection in high-mountain plants. Planta 207:pp. 313-324.

Strussmann CA, Nakatsugawa H, Takashima F, Hasobe M, Suzuki T, Takai R. Cryopreservation of isolated fish blastomeres: effects of cell stage, cryoprotectant concentration, and cooling rate on post thawing survival. Cryobiology. 1999; 39: 252-261.

Summers K (1974) ATP-induced sliding of microtubules in bull sperm flagella. J Cell Biol 60, 321-324.

Sun QY and Schatten H (2007) Centrosome inheritance after fertilization and nuclear transfer in mammals. Adv Exp Med Biol 591, 58-71.

Suquet M, Dreanno C, Fauvel C, Cosson J, Billard R. Cryopreservation of sperm in marine fish. Aquaculture Res. 2000; 31(3): 231-243.

Surrey ES and Quinn PJ (1990) Successful ultrarapid freezing of unfertilized oocytes. J In Vitro Fert Embryo Transf 7, 262-266.

Suzuki T, Komada H, Takai R, Arii R, Kozima TT, (Relation between toxicity of cryoprotectant DMSO and concentration in several fish embryos. Fish Sci. 1995; 61: 193-197.

Szollosi D (1972) Changes of some cell organells during oogenesis in mammals. In Biggers, JD and Schuetz, AW (eds) oogenesis. University Park Press, Baltimore, pp. 47-64.

Szöllösi D (1972) Changes of some organelles during oogenesis in mammals. In Biggers, JD and Schultz, AW (eds) Changes of some organelles during oogenesis in mammals University Park Press, Baltimore, pp. 47-64.

Takagi, H., Thinh, N.T., Islam, O.M., Senboku, T. & Sakai, A. (1997). Cryopreservation of in vitro-grown shoot tips of taro (Colocasia esculenta (L) Schott) by vitrification 1. Investigation of basic conditions of the vitrification procedure. Plant Cell Reports 16:pp. 594-599.

Takeuchi Y, Yoshizaki G, Takeuchi T. Production of germ-line chimeras in rainbow trout by blastomere transplantation. Mol. Reprod .Dev. 2001; 59: 380-389.

Tanaka, D., Niino, T., Isuzugawa, K., Hikage, T. & Uemura, M. (2004). Cryopreservation of shoot apices of in-vitro grown gentian plants: Comparison of vitrification and encapsulation-vitrification protocols. CryoLetters 25:pp. 167-176.

Tervit HR, Adams SL, Roberts RD, McGowan LT, Pugh PA, Smith JF, et al. Successful cryopreservation of Pacific oyster (Crassostrea gigas) oocytes. Cryobiology. 2005; 51(2): 142-151.

Terwilliger TC and Solomon AK (1981) Osmotic water permeability of human red cells. J Gen Physiol 77, 549-570.

Thomas KR and Capecchi MR (1987) Site-directed mutagenesis by gene targeting in mouse embryo-derived stem cells. Cell 51, 503-512.

Thomashow, M.F. (1999). Plant cold acclimation: Freezing tolerance genes and regulatory mechanisms. Annual Review of Plant Physiology and Plant Molecular Biology 50:pp. 571-599.

Tiersch TR, Yang H, Jenkins JA, Dong Q. Sperm cryopreservation in fish and shellfish. Soc. Reprod. Fertil. Suppl. 2007; 65: 493-508.

Todorow SJ, Siebzehnrubl ER, Koch R, Wildt L and Lang N (1989) Comparative results on survival of human and animal eggs using different cryoprotectants and freeze-thawing regimens. I. Mouse and hamster. Hum Reprod 4, 805-811.

Toner JP, Brzyski RG, Oehninger S, Veeck LL, Simonetti S and Muasher SJ (1991) Combined impact of the number of pre-ovulatory oocytes and cryopreservation on IVF outcome. Hum Reprod 6, 284-289.

Toner M, Cravalho EG and Huggins CE (1990) Thermodynamics and kinetics of intracellular ice formation during freezing of biological cells. Journal of Applied Physiology 69, 1582-1593.

Toner M, Cravalho EG, Karel M and Armant DR (1991) Cryomicroscopic analysis of intracellular ice formation during freezing of mouse oocytes without cryoadditives. Cryobiology 28, 55-71.

Toner M, Cravalho EG, Stachecki J, Fitzgerald T, Tompkins RG, Yarmush ML and Armant DR (1993) Nonequilibrium freezing of one-cell mouse embryos. Membrane integrity and developmental potential. Biophys J 64, 1908-1921.

Toshimori K, Tsuzumi F. The morphology and the function of the oocyte chorion in the teleost, Plecoglossus altivelis. J. Electron Microscopy. 1976; 25: 210.

Touchell, D.H., Dixon, K.W. & Tan, B. (1992). Cryopreservation of shoot-tips of Grevillea scapigera (Proteaceae) - a rare and endangered plant from Western-Australia. Australian Journal of Botany 40:pp. 305-310.

Touchell, D.H., Turner, S.R., Bunn, E. & Dixon, K.W. 2002. Cryostorage of somatic tissues of endangered Australian species, In: Cryopreservation of Plant Germplasm. Towill, L.E. & Bajaj, Y.P.S. (Eds.), pp. 357-372, Springer Verlag, ISBN 7055-8630, Berlin Heidelberg, Germany.

Trounson A and Mohr L (1983) Human pregnancy following cryopreservation, thawing and transfer of an eight-cell embryo. Nature 305, 707-709.

Tsai S, Lin C. Effects of cryoprotectant on the embryos of banded coral shrimp (Stenopus hispidus), preliminary studies to establish freezing protocols. CryoLetters. 2009; 30(5): 373-381.

Tsai S, Rawson DM, Zhang T. Development of cryopreservation protocols for early stage zebrafish (Danio rerio) ovarian follicles using controlled slow cooling. Theriogenology. 2009; 71: 1226-1233.

Tsai S, Rawson DM, Zhang T. Development of in vitro culture method for early stage zebrafish (Danio rerio) ovarian follicles for use in cryopreservation studies. Theriogenology. 2010; 74: 290-303.

Tsai S, Rawson DM, Zhang T. Studies on cryoprotectant toxicity to early stage zebrafish (Danio rerio) ovarian follicle. CryoLetters. 2008; 29(6): 477- 483.

Tsai S, Spikings E, Huang IC, Lin C. Study on the mitochondrial activity and membrane potential after exposing later stage oocytes of two gorgonian corals (Junceella juncea and Junceella fragilis) to cryoprotectants. CryoLetters, 2011; 32(1): 1-12.

Tsai S, Spikings E, Kuo FW, Lin C. Use of an adenosine triphosphate assay, and simultaneous staining with fluorescein diacetate and propidium iodide, to evaluate the effects of cryoprotectants on hard coral (Echinopora sp.) oocytes. Theriogenology. 2010; 73: 605-611.

Tsai S, Spikings E, Lin C. Effects of the controlled slow cooling procedure on freezing parameters and ultrastructural morphology of Taiwan shoveljaw carp (Varicorhinus barbatulus) sperm. Aquat. Living Resour. 2010; 23: 119-124.

Tsvetkova LI, Cosson J, Linhart O, Billard R. Motility and fertilizing capacity of fresh and frozen-thawed spermatozoa in sturgeons Acipenser baeri and A. ruthenus. J. Appl. Ichthyol. 1996; 12: 107-112.

Tucker M, Wright G, Morton P, Shanguo L, Massey J and Kort H (1996) Preliminary experience with human oocyte cryopreservation using 1,2- propanediol and sucrose. Hum Reprod 11, 1513-1515.

Turk JR and Laughlin MH (2004) Physical activity and atherosclerosis: which animal model? Can J Appl Physiol 29, 657-683.

Turner, S. 2001. Cryopreservation of somatic germplasm of selected Australian monocodyledonous taxa (Haemodoraceae), PhD Thesis, Curtin University of Technology, Perth, Western Australia.

Turner, S., Senaratna, T., Touchell, D., Bunn, E., Dixon, K. & Tan, B. (2001). Stereochemical arrangement of hydroxyl groups in sugar and polyalcohol molecules as an important factor in effective cryopreservation. Plant Science 160:pp. 489-497.

Uchendu, E.E., Leonard, S.W., Traber, M.G. & Reed, B.M. (2010). Vitamins C and E improve regrowth and reduce lipid peroxidation of blackberry shoot tips following cryopreservation. Plant Cell Reports 29:pp. 25-35.

Uemura, M. & Steponkus, P.L. (1994). A contrast of the plasma membrane lipid composition of oat and rye leaves in relation to freezing tolerance. Plant Physiology 104:pp. 479- 496.

Uemura, M., Joseph, R.A. & Steponkus, P.L. (1995). Cold-acclimation of Arabidopsis thaliana - effect on plasma membrane lipid composition and freeze-induced lesions. Plant Physiology 109:pp. 15-30.

Urbanova, M., Kosuth, J. & Cellarova, E. (2006). Genetic and biochemical analysis of Hypericum perforatum L. plants regenerated after cryopreservation. Plant Cell Reports 25:pp. 140-147.

Vajta G and Nagy ZP (2006) Are programmable freezers still needed in the embryo laboratory? Review on vitrification. Reprod Biomed Online 12, 779-796.

Vajta G, Holm P, Kuwayama M, Booth PJ, Jacobsen H, Greve T and Callesen H (1998) Open Pulled Straw (OPS) vitrification: a new way to reduce cryoinjuries of bovine ova and embryos. Mol Reprod Dev 51, 53-58.

Valdez MDJr, Miyamoto A, Hana T, Edashige K, Kasai M. Sensitivity of chilling of medaka (Orysias latipes) embryos at various developmental stage. Theriogenology. 2005; 64: 112-122.

Valojerdi MR and Salehnia M (2005) Developmental potential and ultrastructural injuries of metaphase II (MII) mouse oocytes after slow freezing or vitrification. J Assist Reprod Genet 22, 119-127.

Van Blerkom J (1996) Sperm centrosome dysfunction: a possible new class of male factor infertility in the human. Mol Hum Reprod 2, 349-354.

Van Blerkom J and Davis PW (1994) Cytogenetic, cellular, and developmental consequences of cryopreservation of immature and mature mouse and human oocytes. Microsc Res Tech 27, 165-193.

Van der Elst J, Amerijckx Y and Van Steirteghem A (1998) Ultra-rapid freezing of mouse oocytes lowers the cell number in the inner cell mass of 5 day old in-vitro cultured blastocysts. Hum Reprod 13, 1595-1599.

Van der Elst J, Nerinckx S and Van Steirteghem AC (1993) Association of ultrarapid freezing of mouse oocytes with increased polyploidy at the pronucleate stage, reduced cell numbers in blastocysts and impaired fetal development. J Reprod Fertil 99, 25-32.

Van der Elst J, Van den Abbeel E, Nerinckx S and Van Steirteghem A (1992) Parthenogenetic activation pattern and microtubular organization of the mouse oocyte after exposure to 1,2-propanediol. Cryobiology 29, 549-562.

Van der Straten KM, Leung LK, Rossini R, Johnston SD. Cryopreservation of spermatozoa of black marlin, Makaira indica (Teleostei: Istiophoridae). CryoLetters. 2006; 27(4): 203-209.

van Meer, G., Voelker, D.R. & Feigenson, G.W. (2008). Membrane lipids: where they are and how they behave. Nature Reviews Molecular Cell Biology 9:pp. 112-124.

van Uem JF, Siebzehnrubl ER, Schuh B, Koch R, Trotnow S and Lang N (1987) Birth after cryopreservation of unfertilized oocytes. Lancet 1, 752-753.

Vazquez Illanes MD and Storey KB (1993) 6-Phosphofructo-2-kinase and control of cryoprotectant synthesis in freeze tolerant frogs. Biochim Biophys Acta 1158, 29-32.

Vidal, N., Sanchez, C., Jorquera, L., Ballester, A. & Vieitez, A.M. (2005). Cryopreservation of chestnut by vitrification of in vitro-grown shoot tips. In Vitro Cellular & Developmental Biology-Plant 41:pp. 63-68.

Vincent C and Johnson MH (1992) Cooling, cryoprotectants, and the cytoskeleton of the mammalian oocyte. Oxf Rev Reprod Biol 14, 73-100.

Vincent C, Pickering SJ and Johnson MH (1990) The hardening effect of dimethylsulphoxide on the mouse zona pellucida requires the presence of an oocyte and is associated with a reduction in the number of cortical granules present. J Reprod Fertil 89, 253-259.

Vincent C, Pickering SJ, Johnson MH and Quick SJ (1990) Dimethylsulphoxide affects the organisation of microfilaments in the mouse oocyte. Mol Reprod Dev 26, 227-235.

Volarcik K, Sheean L, Goldfarb J, Woods L, Abdul-Karim FW and Hunt P (1998) The meiotic competence of in-vitro matured human oocytes is influenced by donor age: evidence that folliculogenesis is compromised in the reproductively aged ovary. Hum Reprod 13, 154-160.

Volk, G.M. & Walters, C. (2006). Plant vitrification solution 2 lowers water content and alters freezing behavior in shoot tips during cryoprotection. Cryobiology 52:pp. 48-61.

Wakamatsu Y, Pristyaznhyuk I, Ju B, Niwa K, Ladygina T, Kinoshit M, et al. Fertile and diploid nuclear transplants derived from embryonic cells of a small laboratory fish, medaka (Oryzias Latipes). Proc. Natl. Acad. Sci. USA 2001; 98: 1071-1076.

Wallace WH, Anderson RA and Irvine DS (2005) Fertility preservation for young patients with cancer: who is at risk and what can be offered? Lancet Oncol 6, 209-218.

Walters E and Critser JK History of Sperm Cryopreservation In Pacey, A and Tomlinson, MJ (eds) Practical Guide for Sperm Banking. Cambridge University Press, Cambridge, In Press.

Walters E, Men H, Agca Y, Mullen S, Critser E and Critser J (2004) Osmotic tolerance of mouse spermatozoa from various genetic backgrounds. Cryobiology 49, 344.

Walters EM, Men H, Agca Y, Mullen SF, Critser ES and Critser JK (2005) Osmotic tolerance of mouse spermatozoa from various genetic backgrounds: Acrosome integrity, membrane integrity, and maintenance of motility. Cryobiology 50, 193-205.

Wang WH, Meng L, Hackett RJ, Odenbourg R and Keefe DL (2001) Limited recovery of meiotic spindles in living human oocytes after cooling-rewarming observed using polarized light microscopy. Hum Reprod 16, 2374-2378.

Wang, Z.C. & Deng, X.X. (2004). Cryopreservation of shoot-tips of citrus using vitrification: effect of reduced form of glutathione. CryoLetters 25:pp. 43-50.

Ward KA and Brown BW (1998) The production of transgenic domestic livestock: successes, failures and the need for nuclear transfer. Reprod Fertil Dev 10, 659-665.

Wassarman PM and Albertini DF (1994) The Mammalian Ovum. In Knobil, E and Neill, J (eds) The Physiology of Reproduction. Raven Press, New York, pp. 79-122.

Watson PF (1981) The effects of cold shock on sperm cell membranes. In Morris, GJ and Clarke, A (eds) Effects of low temperatures on biological membranes.

Watson PF and Morris GJ (1987) Cold shock injury in animal cells. In Bowler, K and Fuller, BJ (eds) Temperature and animal cells. The Company of Biologist Limited, Cambridge, pp. 311-340.

Wheeler MB and Walters EM (2001) Transgenic technology and applications in swine. Theriogenology 56, 1345-1369.

Whittingham DG (1971) Survival of Mouse Embryos after Freezing and Thawing. Nature 233, 125-126.

Whittingham DG (1974) Embryo banks in the future of developmental genetics. Genetics 78, 395-402.

Whittingham DG (1977) Fertilization in vitro and development to term of unfertilized mouse oocytes previously stored at --196 degrees C. J Reprod Fertil 49, 89-94.

Whittingham DG and Wood MJ (1983) Reproductive Physiology. In Foster, HL, Sall, JD and Fox, JG (eds) The mouse in biomedical research. Vol III Normative Biology, Immunology, and Husbandry. Academic Press, New York, pp. 138-164.

Whittingham DG, Leibo SP and Mazur P (1972) Survival of mouse embryos frozen to -196 degrees and -269 degrees C. Science 178, 411-414.

Williams RJ and Shaw SK (1980) The relationship between cell injury and osmotic volume reduction: II. Red cell lysis correlates with cell volume rather than intracellular salt concentration. Cryobiology 17, 530-539.

Williams VR, Mattice WL and Williams HB (1978) Basic Physical Chemistry for the Life Sciences. (Third), W. H. Freeman and Co., San Francisco 553p.

Wilmut I (1972) The effect of cooling rate, warming rate, cryoprotective agent and stage of development on survival of mouse embryos during freezing and thawing. Life Sci II 11, 1071-1079.

Wilson PW (1994) A model for thermal hysteresis utilizing the anisotropic interfacial energy of ice crystals. Cryobiology 31, 406-412.

Wise, R.R. & Naylor, A.W. (1987). Chilling-enhanced photooxidation: evidence for the role of singlet oxygen and superoxide in the breakdown of pigments and endogenous antioxidants. Plant Physiology 83:pp. 278.

Wise, R.R. (1995). Chilling-enhanced photooxidation: the production, action and study of reactive oxygen species produced during chilling in the light. Photosynthesis Research 45:pp. 79-97.

Wittmann T, Hyman A and Desai A (2001) The spindle: a dynamic assembly of microtubules and motors. Nat Cell Biol 3, E28-34.

Wolanczyk JP, Storey KB and Baust JG (1990) Ice nucleating activity in the blood of the freeze-tolerant frog, Rana sylvatica. Cryobiology 27, 328-335.

Wolfe, J. & Bryant, G. (1999). Freezing, drying, and/or vitrification of membrane-solute-water systems. Cryobiology 39:pp. 103-129.

Woods EJ, Benson JD, Agca Y and Critser JK (2004) Fundamental cryobiology of reproductive cells and tissues. Cryobiology 48, 146-156.

Wu, Y.J., Zhao, Y.H., Engelmann, R. & Zhou, M.D. (2001). Cryopreservation of kiwi shoot tips. CryoLetters 22:pp. 277-284.

Xie G and Timasheff SN (1997) The thermodynamic mechanism of protein stabilization by trehalose. Biophys Chem 64, 25-43.

Yamaha E, Mizuno T, Hasebe Y, Yamazaki F. Chimeric fish produced by exchanging upper parts of blastoderms in goldfish blastulae. Fish. Sci. 1997; 63: 514-519.

Yoon, J.-W., Kim, H.-H., Ko, H.-C., Hwang, H.-S., Hong, E.-S., Cho, E.-G. & Engelmann, F. (2006). Cryopreservation of cultivated and wild potato varieties by droplet vitrification: effect of subculture of mother-plants and of preculture of shoot tips. CryoLetters 27:pp. 211-222.

Young, I. & McEneny, J. (2001). Lipoprotein oxidation and atherosclerosis. Biochemical Society Transactions 29:pp. 358-362.

Yuge M, Otoi T, Nii M, Murakami M, Karja NW, Rajaei F, Agung B, Wongsrikeao P, Murakami M and Suzuki T (2003) Effects of cooling ovaries before oocyte aspiration on meiotic competence of porcine oocytes and of exposing in vitro matured oocytes to ambient temperature on in vitro fertilization and development of the oocytes. Cryobiology 47, 102-108.

Zade-Oppen AM (1968) Posthypertonic hemolysis in sodium chloride systems. Acta Physiol Scand 73, 341-364.

Zampolla T, Rawson DM, Zhang T. Development of new viability assessment methods for zebrafish (Danio rerio) oocytes. Cryobiology. 2006; 58:16.

Zarghami, R., Pirseyedi, S.M., Hasrak, Sh. & Pakdaman, B.S. (2008). Evaluation of genetic stability in cryopreserved Solanum tuberosum. African Journal of Biotechnology 7:pp. 2798-2802.

Zenzes MT, Bielecki R, Casper RF and Leibo SP (2001) Effects of chilling to 0 degrees C on the morphology of meiotic spindles in human metaphase II oocytes. Fertil Steril 75, 769-777.

Zhang T, Isayeva A, Adams SL, Rawson DM. Studies on membrane permeability of zebrafish (Danio rerio) oocytes in the presence of different cryoprotectants. Cryobiology. 2005; 50: 285-293.

Zhang T, Liu XH, Rawson DM. Effects of methanol and developmental arrest on chilling injury in zebrafish (Danio rerio) embryos. Theriogenology. 2003; 59(7): 1545-1556.

Zhang T, Rawson DM, Morris GJ. Cryopreservation of pre-hatch embryos of zebrafish (Brachydanio rerio). Aquat. Living Resour. 1993; 6: 145-153.

Zhang T, Rawson DM. Feasibility studies on vitrification of zebrafish (Brachydanio rerio) embryos. Cryobiology. 1996; 33: 1-13.

Zhang T, Rawson DM. Studies on chilling sensitivity of zebrafish (Brachydanio rerio) embryos. Cryobiology. 1995; 32: 239-246.

Zhang TT, Rawson DM. Permeability of dechorionated 1-cell and 6-somite stage zebrafish (Brachydanio rerio) embryos to water and methanol. Cryobiology. 1998; 37: 13-21.

Zhao, S. & Blumwald, E. (1998). Changes in oxidation reduction state and antioxidant enzymes in the roots of jack pine seedlings during cold acclimation. Physiologia Plantarum 104: pp.134-142.

Zhu, G.Y., Geuns, J.M.C., Dussert, S., Swennen, R. & Panis, B. (2006). Change in sugar, sterol and fatty acid composition in banana meristems caused by sucrose-induced acclimation and its effects on cryopreservation. Physiologia Plantarum 128:pp. 80-94.

Zieger MA, Woods EJ, Lakey JR, Liu J and Critser JK (1999) Osmotic tolerance limits of canine pancreatic islets. Cell Transplant 8, 277-284.

Zuelke KA (1998) Transgenic modification of cows milk for value-added processing. Reprod Fertil Dev 10, 671-676.

INDEX

N

O

P

www.ingramcontent.com/pod-product-compliance
Lightning Source LLC
LaVergne TN
LVHW062215200926
848484LV00006B/1403